T0176849

PHYSICS OF FLOW IN POROUS MEDIA

An invaluable reference for graduate students and academic researchers, this book introduces the basic terminology, methods and theory of the physics of flow in porous media. Geometric concepts, such as percolation and fractals, are explained, and simple simulations are created, providing readers with both the knowledge and the analytical tools to deal with real experiments. It covers the basic hydrodynamics of porous media and how complexity emerges from it, as well as establishing key connections between hydrodynamics and statistical physics. Covering current concepts and their uses, this book is of interest to applied physicists and computational/theoretical Earth scientists and engineers seeking a rigorous theoretical treatment of this topic. *Physics of Flow in Porous Media* fills a gap in the literature by providing a physics-based approach to a field that is mostly dominated by engineering approaches.

JENS FEDER was Professor of Physics at the University of Oslo and author of the classic text, *Fractals* (1988). He had broad research interests including condensed matter physics, fluid dynamics and complex systems, and published many accomplished papers on these topics among others. He also successfully mentored a large number of developing researchers throughout his career. Feder was a member of the Norwegian Academy of Science and Letters.

EIRIK GRUDE FLEKKØY is Professor of Physics at the University of Oslo and Professor of Chemistry at the Norwegian University of Science and Technology. He has published a number of articles and written several books in statistical physics and hydrodynamics. He currently co-leads the Centre of Excellence Porous Media Laboratory (PoreLad) and is a member of the Royal Norwegian Society for Science and Letters.

ALEX HANSEN is a Professor of Physics at the Norwegian University of Science and Technology. Since 2017 he has been the Director of the Centre of Excellence Porous Media Laboratory (PoreLab). He has an outstanding and extensive publication record in physics of porous media, complex matter physics and computational physics. Hansen is a member of the Norwegian Academy of Science and Letters and the Royal Norwegian Society for Science and Letters. He has an honorary doctorate from the University of Rennes, and he is honorary faculty at the Indian Institute of Technology at Guwahati.

PHYSICS OF FLOW IN POROUS MEDIA

JENS FEDER
University of Oslo

EIRIK GRUDE FLEKKØY
University of Oslo

ALEX HANSEN
Norwegian University of Science and Technology

CAMBRIDGE
UNIVERSITY PRESS

University Printing House, Cambridge CB2 8BS, United Kingdom

One Liberty Plaza, 20th Floor, New York, NY 10006, USA

477 Williamstown Road, Port Melbourne, VIC 3207, Australia

314–321, 3rd Floor, Plot 3, Splendor Forum, Jasola District Centre, New Delhi – 110025, India

103 Penang Road, #05–06/07, Visioncrest Commercial, Singapore 238467

Cambridge University Press is part of the University of Cambridge.

It furthers the University's mission by disseminating knowledge in the pursuit of education, learning, and research at the highest international levels of excellence.

www.cambridge.org
Information on this title: www.cambridge.org/9781108839112
DOI: 10.1017/9781108989114

© Cambridge University Press 2022

First published 2022

Printed in the United Kingdom by TJ Books Limited, Padstow Cornwall

A catalogue record for this publication is available from the British Library.

Library of Congress Cataloging-in-Publication data
Names: Feder, Jens, author. | Flekkøy, Eirik Grude, author. | Hansen, Alex, 1955– author.
Title: Physics of flow in porous media / Jens Feder, University of Oslo,
Eirik Grude Flekkøy, University of Oslo, Alex Hansen,
Norwegian University of Science and Technology.
Description: Cambridge, United Kingdom ; New York, NY :
Cambridge University Press, 2022. | Includes bibliographical references and index.
Identifiers: LCCN 2021056297 (print) | LCCN 2021056298 (ebook) |
ISBN 9781108839112 (hardback) | ISBN 9781108989114 (epub)
Subjects: LCSH: Porous materials–Fluid dynamics. |
BISAC: SCIENCE / Earth Sciences / Hydrology
Classification: LCC QC173.4.P67 F43 2021 (print) | LCC QC173.4.P67 (ebook) |
DDC 620.1/16–dc23/eng20220207
LC record available at https://lccn.loc.gov/2021056297
LC ebook record available at https://lccn.loc.gov/2021056298

ISBN 978-1-108-83911-2 Hardback

Contents

Note: asterisks(*) denote sections belonging to the advanced course (please see the preface for more details).

Colour plates can be found between pages 84 and 85.

Preface

This book started out a very long time ago as notes for a course on porous media given by Jens Feder at the physics department at the University of Oslo. The oldest version of these notes dates back to 1984. Jens was an excellent teacher, and these notes developed over the years to a rather comprehensive volume, but one that reflected the time in which they were written. The eighties were the heyday of fractals, and the enthusiasm for finding ever new fractal structures was great. Jens passed away in 2019. By that time, Eirik Grude Flekkøy had been teaching the course on porous media for some years, and the notes had been taken over by him. The notes were continuously updated to reflect the shifting focus in porous media research, and the idea to turn the notes into a textbook was born. Alex Hansen at the Department of Physics at the Norwegian University of Science and Technology in Trondheim was invited to join the team to develop the notes into a balanced text on the physics of flow in porous media. The text grew into a volume that no longer would fit into a one-semester course. Rather, it now covers two courses, one intro-ductory, and one advanced – the chapters and sections belonging to the advanced course being marked by an asterisk. We have for the asterisk-marked material not hesitated to take the reader to the research frontier of today where we have found this appropriate. Perhaps these parts of the book will age faster than the rest?

The notion of what constitutes "fundamental physics" has shifted greatly over the last century. At the beginning of the century, the search for the fundamental would systematically lead to ever larger, or ever smaller, length and time scales. However, during the 1980s and 1990s, the field of complexity emerged with its paradigm that fundamental laws of nature may also be about structure arising in many-particle systems. Statistical physics was the tool needed to reveal the cooperative behavior of such systems, which is independent of microscopic details and therefore general. Jens Feder was a leader in this field and a co-funder of the Cooperative Phenomena Group at the University of Oslo. Jens held the view that there is no clear boundary between basic and applied science. Clearly, without basic science, there would be no science to apply. But in many cases, fundamental questions will arise exactly from applied problems, in which case basic science will arise from applied science, as well as vice versa.

This book has developed from this cultural background, as well as from our teaching in PoreLab, the center of excellence that we founded in 2017. It reflects the fact that while the subject matters of our field change, the classical tools of statistical physics and hydrodynamics do not. Likewise, many of the more novel algorithmic tools that rely on a bottom-up type of modeling have become so established and applicable that we consider them necessary parts in the education of our graduate students. Therefore, we derive the Navier–Stokes equations of the 1840s, as well as the connection to lattice gases and the lattice Boltzmann models of the 1980s. Fractals are introduced as a classical tool, and particle models for diffusion are introduced along with the corresponding continuum equations.

Throughout, we aim to give the students a combination of analytical and numerical skills. Their background is assumed to include fundamental calculus and thermodynamics/statistical physics. But, since hydrodynamics is often disregarded as a basic topic in physics departments, no pior knowledge of this topic is assumed. The excersises at the end of each chapter are both analytical and numerical and range from problems that fill in minor steps lacking in the text to more comprehensive simulation problems. In the last chapter, there is a series of lattice Boltzmann-based simulation problems, each of which the students would spend around two weeks to solve. We have used these as end-of-term projects replacing normal exams. Along with a shorter mid-term problem on percolation, these projects have been the basis of student evaluation.

The physics of porous media deals with model systems, either experimentally, computationally or theoretically. These model systems are stripped of any irrelevant contents. As Einstein put it, *Everything should be as simple as possible, but not simpler.* This is the approach we have taken in this book.

Research on the fundamentals of porous media has recently made tremedous progress based on the developement of imaging techniques and the increase in computational resources. It is possible to follow the motion of immisicble fluids at the pore level and to have this motion reproduced in computer simulations. However, since these techniques by themselves do not provide the concepts or theory needed to understand the phenomena at hand, simple models and analytical approaches are now needed more than ever.

There are so many to thank: Øyvind Aker, Jan Øystein Haavig Bakke, Dick Bedeaux, Carl Fredrik Berg, HyeJeong Cheon, Magnus Aa. Gjennestad, Bjørn Hafskjold, Hursanay Fyhn, Signe Kjelstrup, Henning A. Knudsen, Federico Lanza, Knut Jørgen Måløy, Håkon Pedersen, Thomas Ramstad, Alberto Rosso, Stéphane Roux, Subhadeep Roy, Isha Savani, Jean Schmittbuhl, Santanu Sinha, Erik Skjetne,

Per Arne Slotte, Laurent Talon, Glen Tørå, Pål Eric Øren and many more. Their input through discussions and collaborations has been invaluable.

We also thank students Halvor Melkild and Shayla Viet for a great job with preparing the manuscript and figures for publication.

But, the two of us who have brought this book to completion direct a special thanks to Liv Feder, Jens' widow, for allowing us to build on Jens' work. She has also made many of the figures scattered around in the text.

1

Introduction

The flow of fluids in porous media is of great technological importance. Clearly the flow of groundwater or of oil, gas and water in reservoirs has a very significant economical aspect.

The information about flows in porous media is spread over many fields of science and technology. For instance, in biochemistry and microbiology the separation of macromolecules on packed columns of "chromatographic" materials is a standard technique. The separation of pieces of DNA by electrophoresis in gel layers is a common analytical and preparative method in biotechnology. The understanding of how these methods work in scientific terms is, however, rather limited.

In this book we will try to learn about various physical aspects of flows in porous media. The properties of porous media is a part of condensed matter physics. Indeed, much of the current effort in solid state physics and in statistical physics concentrates on the properties of disordered systems, fluctuations, *phase-transitions* and *percolation*. Associated problems include the precipitation and clustering phenomena that may occur both in bulk fluids and in fluids in porous media. Other areas of relevant active scientific inquiry include fuel cell technologies, polymer and colloidal chemistry and physics, interfacial phenomena, and the adsorbtion of molecules on interfaces and solids. Also, most biological systems involve flow through porous media in some way. This is exemplified by the blood capillaries in a human hand or the flow of air through the nose of a reindeer.

As a guiding principle we will try to make the connection between the microscopic and the macroscopic aspects of porous media.

Also, we will proceed from the simple to the more complex, starting with basic geometric concepts and basic hydrodynamics. Then we proceed to use this pore-scale description to derive a description that pertains to larger scales, like Darcy's law, before we introduce fluids that mix or create interfaces with surface tension between them.

We will study the flow of fluids through capillaries, also with droplets or other phases present. Capillary pressure, wetting and interfacial tension are important for the multiphase flow in porous media. Also because of the large specific surface area of porous media, the surface adsorbtion of molecules is of great interest and will be discussed in some detail.

Later we will discuss the percolation problem. In particular we will discuss the statistical, structural and transport properties of percolating systems. This will include a discussion of critical behavior and the scaling properties of various physical quantities.

The simultaneous flow of several fluids in porous media will be discussed in detail, and we will demonstrate both immiscible and miscible flow and describe the types of pattern they give rise to. Toward the end we turn to numerical simulations of multiphase flow. These simulations too are based on the *bottom-up* apporach, that is, the idea that macroscopic processes and their description should be based on the underlying processes at the micro level. The lattice gas and lattice Boltzmann models that we introduce to simulate the fluid flow equations are based on a particle picture of the fluids, and it is shown how their behavior links to the hydrodynamic behavior at the larger scale.

2
Geometry of Porous Media

In the discussion of porous media, we immediately get involved with the question of how to describe the geometry of the medium. Any soil or sand consists of mineral grains of various sizes that are packed together. For a complete geometric description, one would need the shape, position and orientation of all the grains. For the fluid flow problem in porous media, we are mostly interested in the void space between particles. The individual particles may be loose (unconsolidated) or cemented together in geological or laboratory processes that form the porous medium.

The simplest geometrical measure of a porous medium is the *porosity* ϕ,

$$\phi = \frac{\text{Volume of pore space}}{\text{Total sample volume}}.$$ (2.1)

In oil production, porosities in the range 0.05 to 0.4 are of practical interest. The measurement of the porosity of samples of reservoir rocks is, of course, of prime interest since the recoverable hydrocarbons are found in the pore space.

The porosity ϕ is the most important geometric characterization of a porous medium. As will be discussed later, porous media are also characterized by their specific surface S and by their permeability, relative permeability, dispersion and other geometric characteristics.

As a start, we consider different packings of spheres. Such packings may be ordered or disordered, overlapping or nonoverlapping. Since they are easy to characterize and also to generate on a computer, they serve as useful models for more complex porous geometries.

2.1 Three-Dimensional Packing of Spheres

The packing of spheres has a long history. Kepler (1941, 1966) discussed in 1611 the packing of spheres in his essay *"De Nive Sexangula,"* that is, *On the Six-Cornered Snowflake*. Kepler's discussion of packings is interesting.

3

Kepler discussed *periodic* packings of spheres and noted that the *simple cubic lattice* (SC) is less dense than what is called the *face-centered cubic lattice* (FCC). The filling factors for these packings are

$$c = (1 - \phi) = \frac{\pi}{6} = 0.5235\ldots, \quad \text{SC}, \tag{2.2}$$

$$c = (1 - \phi) = \frac{\pi}{3\sqrt{2}} = 0.7404\ldots, \quad \text{FCC}. \tag{2.3}$$

Kepler further concluded that the packing on a triangular base leads to the *same* close packing based on the square lattice, that is, the face-centered cubic lattice. Kepler discussed many other geometrical features of packings but concluded that he could not explain why snowflakes are six-cornered. In fact, ice crystallizes in many forms, depending on pressure and temperature. Ice forms hydrogen-bonded crystals that are not closely packed. However, the common form of ice has a hexagonal base. The macroscopic form of snow crystals depends on humidity and temperature in the atmosphere, and many varieties have been found. (See the discussion by Mason (Kepler, 1966)).

Kepler explained clearly regular lattices, the foundation of crystallography. However, he did not discuss other close packings, notably the hexagonal close-packed lattice, and packings with stacking faults, which have the same density. Mason (Kepler, 1966, pp. 47–56) states that Thomas (1599), an Englishman, really preceded Kepler and was the first to postulate that closest packing was achieved when 1 ball was surrounded by 12 others. Harriot also understood the difference between the hexagonal and cubic close packings.

The close-packing of spheres is a well known model of solids. Among the 100 or so elements in the periodic table, approximately 55 form solids that maximize the packing of the atoms in that the number of spheres representing atoms are arranged in such a way that the volume fraction of the spheres is the largest of any possible arrangement of spheres.

Both the *face-centered cubic* (FCC) and the *hexagonal close-packed* (HCP) lattices have the maximum density, with a volume fraction c that is larger than the *simple cubic* (SC) lattice. The densities and the porosities are given by

$$c = (1 - \phi) = \frac{\pi}{3\sqrt{2}} = 0.7404\ldots, \quad \text{FCC}. \tag{2.4}$$

The close-packed structures may be generated by the procedure described in Figure 2.1. However, there is a family of structures that may be generated by choosing arbitrary sequences, such as –ABACBCBACBCAB–, that all have the same porosity. The FCC structure has been attributed to Kepler (Stewart, 1991; Zallen, 1983). The mathematician Wu-Yi Hsiang (1993) has published a paper that proves that the close-packed lattices represent the maximal packing density. However, the

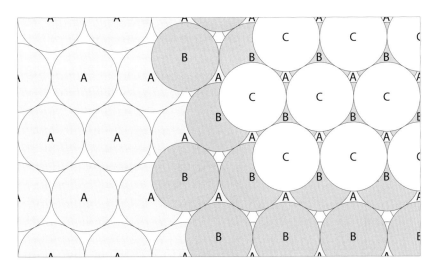

Figure 2.1 Close packing of spheres is obtained when a triangular layer of spheres (A) is overlaid by a next triangular layer (B) positioned with centers at the symmetry centers between the spheres in the previous layer. The next layer may now be positioned directly over (A) or in the position (C) as indicated in the figure. The sequence –ABCABCABC– corresponds to face-centered cubic close packing. The sequence –ABABAB– is the hexagonal close-packed structure.

proof is controversial. Thomas C. Hales (1994) has criticized Hsiang's work (1995) and has recently published a proof of his own on the internet (Hales, 1998).

2.1.1 Random Close Packing of Spheres

Random close packing (RCP) has a long history. Many experiments by Bernal (Bernal, 1964, 1965; Bernal and Mason, 1960) and also by Scott and Kilgour (1969) conclude that

$$c = (1 - \phi) = 0.6366 \pm 0.0005. \tag{2.5}$$

Thus, the porosity is $\phi = 0.3634$. A sphere must make contact with at least three others to be stable (in a gravitational field).

A simple way to characterize random packings of spheres and fluids is by the radial distribution function $G(r) = n(r)dr/4\pi r^2$, where $n(r)dr$ is the number of spheres that have a distance r from a sphere at the origin, in the range $[r, r + dr]$. The radial distribution function may be interpreted as the conditional probability that with a sphere at the origin, one finds another sphere in a shell of width dr at a distance r. For liquids, $G(r)$ has been measured by X-ray and neutron-scattering experiments, and Bernal measured $G(r)$ for a random packing of steel balls by measuring the location of a very large number of spheres in a random packing.

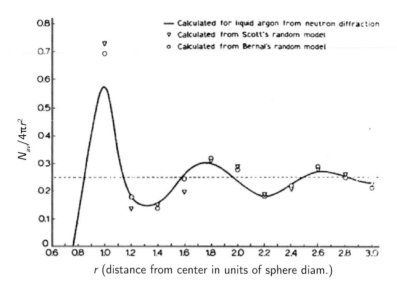

Figure 2.2 The radial distribution of random close packings of equal spheres. (N_{av} is the average number of spheres in intervals of 0.2 of sphere diameter.) (After Bernal (1965).)

His results are shown in Figure 2.2. The RCP model captures much of the structure in the correlation function of argon.

2.1.2 Random Loose Packing of Spheres

If spheres are just poured into a container, the density will be somewhat less (see the review of granular media by Jaeger and Nagel (1992)). Scott (1960) noted different ways of packing balls into containers: "There is a range of random packings lying between two well defined limits. The limits are called here 'dense random packing' and 'loose random packing'." Scott measured the packing density of spheres in cylinders with dimpled walls as a function of filling height h. The packing density was extrapolated to $1/h = 0$, and the resulting values plotted as shown in Figure 2.3. For loose packings, the balls fill the vessel essentially by rolling down a slope of random-packed balls. The dense packings are obtained by gentle shaking and tapping. "In both types of packings, the balls are always rigid in the sense that uniform pressure over the top will not alter the packing."

As the distance between the particles is increased, there are fewer "paths" that can transmit force. These paths and the distribution of forces can be observed experimentally, see Figure 2.4 (Liu et al., 1995). The force distribution was measured in a separate experiment where 3.5 mm glass spheres in a container 90 mm in diameter and 75 mm high were pressed down with a 310 N force on the top. A layer of

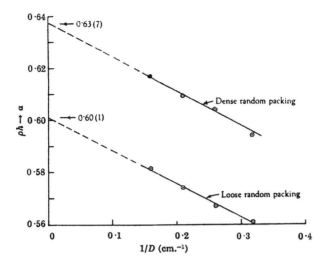

Figure 2.3 The packing density in cylinder of "infinite" length obtained from extrapolation, plotted against reciprocal of the cylinder diameter D. (After Scott (1960).)

carbon paper on the container bottom left marks that were used to measure the vertical component of the force, f, between individual spheres and the bottom. The measured probability density, $P(f)$, for this force had a simple exponential distribution:

$$P(f) = C \exp(-f/f_0), \tag{2.6}$$

with $C = 736$ and $f_0 = 0.98$ N. Thus, most of the load is carried by only a few spheres. The large fluctuations in force make granular packs highly inhomogeneous, as illustrated in the force paths shown in Figure 2.4.

The loosest random packing that is still mechanically stable under a given force F is called *random loose packing* (RLP). In sand piles, the force F is due only to the weight of the particles and thus to the acceleration of gravity g. Experiments by Onoda and Liniger (1990) show that in the limit $g \to 0$, one finds $c_{\mathrm{RLP}} = 0.555 \pm 0.005$. A random packing that has a density larger than the RLP density must *expand* when the packing is deformed. In recent computer simulations, packing densities of spheres in the range 0.5447–0.6053 were found for various algorithms for generating the packings (Jullien et al., 1992a). Thus, the lowest density was below the RLP limit. We summarize the properties of different packings in Table 2.1.

Packings of sand or particles has many interesting features and many practical applications. Recently, much interest has centered on *granular materials* or the "granular state" (Jaeger and Nagel, 1992). Granular materials are peculiar in that

Geometry of Porous Media

Table 2.1. *Common packings of spheres in two and three dimensions.*

Two dimensions	Symbol	c	ϕ
Square lattice		$\pi/4 \simeq 0.7853$	0.2147
Triangular lattice		$\pi/(2\sqrt{3}) \simeq 0.9068$	0.0931
Random sequential adsorption	RSA	0.547	0.453
Random close packed	RCP	0.772	0.228

Three dimensions	Symbol	c	ϕ
Simple cubic lattice	SC	$\pi/6 \simeq 0.5235$	0.4765
Body-centered cubic lattice	BCC	$\pi\sqrt{3}/8 \simeq 0.6802$	0.3198
Face-centered cubic lattice	FCC	$\pi/(3\sqrt{2}) \simeq 0.7404$	0.2596
Hexagonal close-packed lattice	HCP	$\pi/(3\sqrt{2}) \simeq 0.7404$	0.2596
Random close packed	RCP	0.6366	0.3634
Random loose packed	RLP	0.555	0.445
Upper limit (tetrahedron)		$\sqrt{2}(3\arccos(1/3) - \pi) \simeq 0.7796$	0.2204

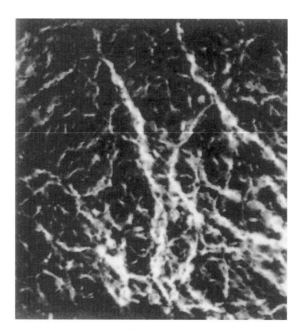

Figure 2.4 Forces between 3 mm Pyrex spheres visualized by birefringence. The spheres are in a container filled with a water–glycerol mixture that has the same refractive index as Pyrex. The system is placed between crossed circular polarizers; therefore no light is transmitted unless the spheres become birefringent due to stress. The bright regions therefore show the force paths that carry the load placed on a piston on top of the bead pack. The box had sides 70×70 mm and was 40 mm thick. (From Liu et al. (1995).)

they behave as solids when at rest but more or less like fluids when they move. The top surface of a granular material is stable as long as the slope is less then the *angle of repose* θ_r. When the slope is increased beyond some maximum angle, which is characteristic of the material, a relatively thin surface layer flows almost like a liquid, leaving the remainder of the heap at rest. The flow consists of avalanches that move particles down the slope. The avalanche dynamics may have scaling properties and provide examples of *self-organized critical behavior* (SOC) (Frette et al., 1996). A remarkable feature of granular media is that the pressure at the bottom of a container filled with a granular material is independent of the height h to which the container is filled. In an hourglass, there is an approximately linear relation between the filling height and the time it takes to empty the top container. If the two bulbs of an hourglass are not evacuated, then air must flow from the bottom container to the top container as the sand flows down. The resulting two-phase flow through the small orifice is unstable, and the sand flows periodically – in effect the hourglass ticks (Wu et al., 1993).

For powders consisting of particles of different sizes, one finds that the big particles move to the top by shaking (Jullien et al., 1992b).

The packing of sand, silt and clay is only the first step in the formation of sedimentary rocks. Sedimentation takes place in water, which has to be expelled in order for the compaction of particles to proceed. Further processes involving the dissolution and regrowth of various minerals contribute to geological compaction. Biological material that sediments together with sand, silt and clays may be buried deeply enough for the temperature and pressure to increase sufficiently to form *source rocks* in which the biological material may be transformed into oil and gas. Oil in source rocks may seep out by *primary migration*, which is a poorly understood process. The oil then moves in water-filled sands to the trap, if it exists, or to the surface, by the process of *secondary migration*, a process that is relatively well understood. In Chapter 11, we will discuss different examples of such migration processes in some detail.

2.2 Poisson Porous Media

2.2.1 Poisson Probabilities

A simple way to generate a (mathematical) model porous medium is to place spheres of radius r at *random* positions in space. This process of placing points, that is, the centers of spheres, with uniform probability in space revert is a useful model for many processes.

Consider a volume V, where N points are placed randomly and independently, as shown in Figure 2.5. The density of points representing centers of hyperspheres is

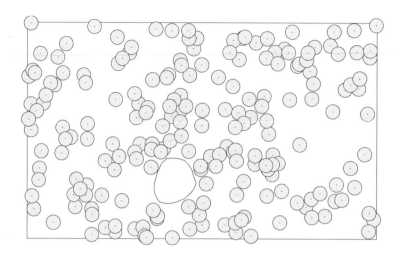

Figure 2.5 The Poisson process of placing at random N points in a volume V so that none are found in the volume v. In this illustration, there are $N = 200$ discs placed in a 'volume' $V = 477.5$ in units of the disc area, and $n = N/V = 0.4189$ in units of inverse disc area. Thus, the porosity is $\phi = \exp(-n\pi r^2) = \exp(-0.4189) = 0.658$.

$$n = \frac{N}{V}. \tag{2.7}$$

The probability that a sphere center lands in a volume v is v/V. Therefore, the probability for *not* placing a sphere in a given volume v is $(1 - v/V)$. The probability, P_0, for having *no* sphere centered in a volume v after having placed N spheres independently into the container of volume V is given by $[(V - v)/V]^N = [1 - (nv)/N]^N$. As N and V are increased, holding n and v fixed, we find

$$P_0 = \lim_{N \to \infty} \left(1 - \frac{nv}{N}\right)^N = \exp(-nv). \tag{2.8}$$

The probability that there are *no* points in a volume of radius ℓ in d-dimensional space is

$$P_0(\ell) = \exp\left(-c_d n \ell^d\right), \tag{2.9}$$

where the geometrical factor c_d is 2, π and $4\pi/3$ for one, two and three dimensions respectively. As a function of ℓ, Eq. (2.9) simply gives the fraction of distances between neighbors (that is, the distance between their centers) that are greater than ℓ. Thus, the fraction of distances between points $x < \ell$ is

$$P(x < \ell) = 1 - \exp\left(-c_d n \ell^d\right). \tag{2.10}$$

This is also the *probability* to find a distance $x < \ell$. By differentiating this expression we find the probability distribution of distances (probability density) between particles,

$$p(\ell) = dc_d n \ell^{d-1} \exp\left(-c_d n \ell^d\right). \tag{2.11}$$

This result can be clarified by another derivation. To simplify the notation let us discuss the three-dimensional case. Let there be $n = N/V$ particles per unit volume. The probability $p(r)dr$ that the nearest neighbor to a particle occurs at a distance in the range $[r, r + dr)$ must be the product of the probability that there are no particles within a sphere of radius r and that there is one particle in the spherical shell between r and $r + dr$. Therefore the function $p(r)$ must satisfy the relation

$$p(r) = \left(1 - \int_0^r p(r')dr'\right) 4\pi r^2 n. \tag{2.12}$$

Taking the derivative of this equation with respect to r, we find

$$\frac{\mathrm{d}}{\mathrm{d}r}\left(\frac{p(r)}{4\pi r^2 n}\right) = -4\pi r^2 n \frac{p(r)}{4\pi r^2 n}. \tag{2.13}$$

By introducing the variable $x = p(r)/4\pi r^2 n$, this may be integrated as

$$\int_1^{p(r)/4\pi r^2 n} \frac{\mathrm{d}x}{x} = -\int_0^r \mathrm{d}r\, 4\pi r^2 n, \tag{2.14}$$

and we find

$$p(r) = 4\pi r^2 n \exp\left(-\frac{4}{3}\pi r^3 n\right). \tag{2.15}$$

We find that the average distance to the nearest neighbor is

$$\langle r \rangle = \int_0^\infty \mathrm{d}r\, r p(r) = \Gamma\left(\frac{4}{3}\right)\left(\frac{3}{4\pi n}\right)^{1/3}, \tag{2.16}$$

where Γ is the Gamma function. The probability density for the distance to the nearest sphere in three dimensions is shown in Figure 2.7.

2.2.2 Porosity of Overlapping (Hyper-)Spheres

As is usual in statistics, simpler expressions may be found by phrasing the questions carefully. Following Weissberg (1963) and Strieder and Aris (1973), we note that the porosity ϕ is the probability that a random point is *not* contained by any sphere in the medium. Thus the porosity equals the probability that a randomly selected point

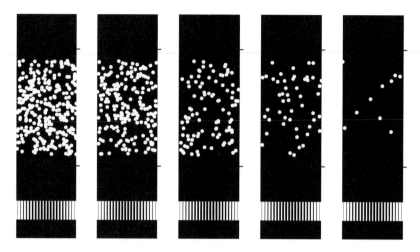

Figure 2.6 Quasi-two-dimensional models of porous media used in permeability measurements. Fluid can flow in the black regions. The grid at the entrance is used to distribute the flow evenly over the channel cross section.

in the porous medium has no sphere *center* within a distance a equal to the sphere radius. Therefore ϕ is given by Eq. (2.8) with $v = 4\pi a^3/3$ in three dimensions,

$$\phi = \exp\left(-n\frac{4\pi a^3}{3}\right), \quad d = 3. \tag{2.17}$$

This argument holds in any dimension for random packings of penetrable (hyper-)spheres. The correct result in two dimensions is

$$\phi = \exp\left(-n\pi a^2\right), \quad d = 2. \tag{2.18}$$

We have checked that this formula works quite well even for rather small models, as shown in Figure 2.6.

2.2.3 Void–Void Correlation Function

The geometry of a porous medium is completely specified by the *characteristic function X*:

$$X(\mathbf{r}) = \begin{cases} 1 & \text{if } \mathbf{r} \in \text{pore space,} \\ 0 & \text{if } \mathbf{r} \in \text{solid matrix.} \end{cases} \tag{2.19}$$

The characteristic function cannot be determined in practice, but a partial description can be obtained by considering various averages of X (Debye and Bueche, 1949; Strieder and Aris, 1973; Weissberg, 1963). The characteristic function of the

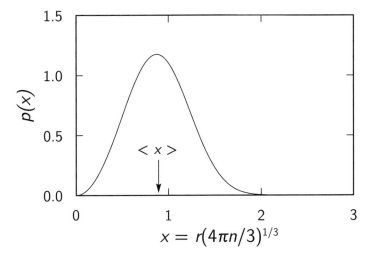

$$x = r(4\pi n/3)^{1/3}$$

Figure 2.7 The probability density for the reduced distance $x = r(4\pi n/3)^{1/3}$ to the nearest sphere in a three-dimensional Poisson porous medium with a number density n of spheres.

pore space clearly contains too much information to be of any practical interest except for well defined small models or periodic structures. The simplest average property is the *porosity* ϕ, that is, the void fraction, given by

$$\phi = \langle X(\mathbf{r}) \rangle = \frac{1}{V} \int_V d^3\mathbf{r}\, X(\mathbf{r}). \tag{2.20}$$

Information about the *structure* of the porous medium is contained in the average of $X(\mathbf{x})X(\mathbf{x} + \mathbf{r})$ with respect to $\mathbf{x}$ for *fixed* separation $\mathbf{r}$. The resulting average is the *void–void correlation function* $G(\mathbf{r})$ defined by

$$G(\mathbf{r}) = \frac{1}{V} \int_V d^3\mathbf{x}\, X(\mathbf{x})X(\mathbf{x} + \mathbf{r}). \tag{2.21}$$

The correlation function $G(\mathbf{r})$ is the probability that a line segment, having a length and direction given by $\mathbf{r}$, placed at random in the porous medium will have *both* its ends in the pore space.

For a Poisson porous medium of spheres, the probability that no sphere center is within a distance a from the end points of the line segment $\mathbf{r}$ thrown at random into the medium is

$$G(\mathbf{r}) = \exp(-nV_{\mathbf{r}}) = \begin{cases} \exp\left(-n\frac{4\pi a^3}{3}\left(1 + \frac{3r}{4a} - \frac{r^3}{16a^3}\right)\right), & r < 2a, \\ \exp\left(-n\frac{8\pi a^3}{3}\right), & r > 2a, \end{cases} \tag{2.22}$$

where $V_{\mathbf{r}}$ is the volume covered by two spheres of radius a separated by a distance $r = |\mathbf{r}|$. Thus $G(\mathbf{r})$ is the conditional probability (density) that given a point in pore space, the point at a distance $\mathbf{r}$ is also in pore space.

The void–void correlation function satisfies several general relations (Berryman, 1985; Berryman and Blair, 1986): From the definition it follows immediately that

$$G(0) = \frac{1}{V} \int_V d^3\mathbf{x}\, X(\mathbf{x})X(\mathbf{x}) = \frac{1}{V} \int_V d^3\mathbf{x}\, X(\mathbf{x}) = \phi, \qquad (2.23)$$

since $X(\mathbf{x})^2 = X(\mathbf{x})$.

In the limit $r \to \infty$, the probability for being in the pore space at $\mathbf{x}$ is uncorrelated with the probability for being in the pore space at $\mathbf{x} + \mathbf{r}$, and therefore we have

$$G(r \to \infty) = \phi^2. \qquad (2.24)$$

These relations are satisfied by $G(r)$ in Eq. (2.22) for the Poisson porous media.

There is an interesting relation between $G(r)$ and the *specific surface S*, the pore surface per unit volume

$$S = \frac{\text{Surface of pore space}}{\text{Total sample volume}}. \qquad (2.25)$$

It can be shown that the following relation holds in general (Berryman and Blair, 1986; Debye et al., 1957):

$$S = -4 \left.\frac{\partial G(r)}{\partial r}\right|_{r=0}. \qquad (2.26)$$

Note that Weissberg's expression, Eq. (2.22), satisfies the relations (2.23) and (2.24) and gives the following result for the specific surface for a *random packing of penetrable spheres*:

$$S = 4\pi a^2 n\phi = -\frac{3}{a}\phi \ln \phi, \qquad (2.27)$$

where we have used Eq. (2.17) to express n in terms of the porosity. This result is also obtained by the following argument (Strieder and Aris, 1973). The specific surface of the spheres is $S = 4\pi a^2 n$. But only a fraction of this surface is exposed due to the overlap of spheres. The probability that a point in the surface is a neighbor to the void space is ϕ, and Eq. (2.27) results.

Various results relating to the void–void correlation function are illustrated in Figure 2.8.

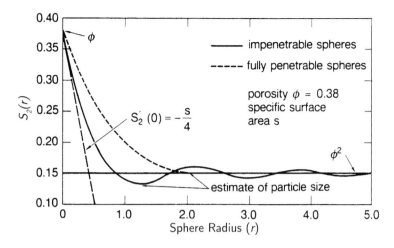

Figure 2.8 Two-point void–void correlation functions for model materials composed of penetrable and impenetrable spheres. Quantities that can be determined directly from the shape of the two-point correlation function include: (1) porosity, (2) specific surface area and (3) mean particle size (if the size distribution is narrow so that exclusion volume effects are important). (After Berryman and Blair (1986).)

2.3 Minkowski Functionals

What is the difference between *geometry* and *topology*? Geometry describes the shape of objects. Suppose these objects are made of rubber and we deform them without tearing. The geometry changes. However, some aspects describing the objects remain invariant under such deformations. For example, a cup with a handle has one hole. We may deform the cup into a doughnut. It looks completely different, but it still has one hole. This is an invariant. Topology is the description of the properties that remain invariant under such deformations.

The *Euler characteristic* is measure of the topology of a geometric object. If the object is a polyhedron, the Euler characteristic χ is given by

$$\chi = V - E + F, \tag{2.28}$$

where V is the number of vertices (i.e. corners), E the number of edges and F the number of faces. Table 2.2 shows the Euler characteristics for some three-dimensional convex polyhedra. They all have the same $\chi = 2$.

As we construct convex polyhedra with increasing numbers of faces, the number of edges and the number of corners grow in such a way that χ remains equal to two. This value survives in the limit of an infinite number of faces. Hence, $\chi = 2$ for the sphere. We may now deform the sphere. The Euler characteristic would remain the same as long as the topology of the object does not change. We have listed in Table 2.3 the Euler characteristic of a number of different objects. Any object

Table 2.2. *Euler characteristic* χ *for some three-dimensional convex polyhedra.*

Object	Vertices V	Edges E	Faces F	$\chi = V - E + F$
Tetrahedron	4	6	4	2
Cube	8	12	6	2
Octahedron	6	12	8	2
Dodecahedron	20	30	12	2
Icosahedron	12	30	20	2

Table 2.3. *Euler characteristic* χ *for the surface of some geometrical objects.*

Object	χ
Sphere	2
Torus	0
Möbius strip	0
N disjoint spheres	$2N$

that can be formed by deformation without tearing from these objects will have the same χ. The Euler characteristic is a number characterizing the *topology* of the object.

It turns out that the Euler characteristic can be calculated from an integration over the surface of the geometrical object through the *Gauss–Bonnet theorem* (Misner et al., 2017). Any point **s** on the two-dimensional surface S of the object at hand will have two principal radii of curvature $r_1(\mathbf{s})$ and $r_2(\mathbf{s})$ associated with it. If the surface S has no one-dimensional boundaries, the Gauss–Bonnet theorem states that

$$\int_S \frac{\mathrm{d}s}{r_1 r_2} = 2\pi \chi. \tag{2.29}$$

Let us consider a sphere with radius r. We parametrize the sphere using polar coordinates, that is, $\mathrm{d}s = r^2 \mathrm{d}\phi \mathrm{d}\theta$, where $0 \leq \phi < 2\pi$ and $0 \leq \theta \leq \pi$. We furthermore have that $r_1 = r_2 = r$. Hence, we find

$$\int_S \frac{\mathrm{d}s}{r_1 r_2} = \int_0^{2\pi} \int_0^{\pi} \frac{\mathrm{d}\phi \, \mathrm{d}\theta \, \sin(\theta) r^2}{r^2} = 4\pi, \tag{2.30}$$

giving $\chi = 2$ in accordance with Table 2.3.

Whereas the Euler characteristic describes the topology of the object, it does not describe the geometry fully. More information is needed for that. For a

d-dimensional object, one may define $d+1$ invariants; that is, quantities that depend on the shape of the object and not how it is oriented. These quantities are the *Minkowski functionals* (Armstrong et al., 2018; Mecke, 2000). Three-dimensional objects have four. They are the pore volume

$$M_0 = V_p, \tag{2.31}$$

the pore surface

$$M_1 = S_p, \tag{2.32}$$

the mean curvature of the surface,

$$M_2 = \int_S ds \left(\frac{1}{r_1} + \frac{1}{r_2} \right), \tag{2.33}$$

and the Euler characteristic

$$M_3 = 2\pi \chi. \tag{2.34}$$

The *Hadwiger theorem* states that Minkowski functionals form a complete basis set for all extensive – extensive means additive – functions F that are invariant with respect to any orientation of the object. Hence, we have that

$$F = \sum_{i=0}^{3} f_i M_i, \tag{2.35}$$

where f_i are coefficients.

Mecke and Arns (2005) demonstrate that the thermodynamic potentials of fluids in porous media may be expressed in the form of Eq. (2.35). More recently, Khanamiri et al. (2018) calculated the free energy of two immiscible fluids moving in a porous medium based on pore-scale images, comparing it to experimental measurements with good results.

2.4 Visualization of Porous Media

We have up to now discussed the geometry of a number of model porous media. But, what about real porous media? The techniques to image porous media at the pore scale has made tremendous headway over the last years, and it is now possible to follow the motion of immiscible fluids at this scale in real time. This field has become so large, however, that it goes far beyond what we may present here. It deserves books on its own. Luckily such books exist, and we refer the interested reader to Blunt (2017).

Exercises

2.1 **Euler characteristics of spheres and toruses:**
A cube may be gradually converted into a sphere by addition of new faces, and a torus produced from a cube with a hole through it. Start by drawing a rectangle and find V, E and F. Draw a line connecting two of the side edges, and show that $V \rightarrow V + 2$, $E \rightarrow E + 3$ and $F \rightarrow F + 1$, and hence that X is unchanged.

2.2 Draw a cube and find V, E and F. Show that the Euler characteristic $X = 2$ for a cube.

2.3 Add a hole through the cube and determine how the numbers V, E and F change. Show that the Euler characteristic $X = 0$ for a cube with such a hole, and argue that $X = 0$ for a torus and $X = 2$ for a sphere. Hint: Make sure that the polyhedron that you make when you introduce the hole is connected with the original one.

2.4 Duplicate the cube with the hole and merge one side of the copy with the original. Argue that this two-step process leads to the changes

$$V \rightarrow 2V \rightarrow 2V - 4 \qquad (2.36)$$

$$E \rightarrow 2E \rightarrow 2E - 4 \qquad (2.37)$$

$$F \rightarrow 2F \rightarrow 2F - 2 \qquad (2.38)$$

and so $X \rightarrow 2X - 2 = -2$.

2.5 A capillary fiber bundle model is a model for a porous medium that consists of a set of tubes. It can be made by drilling N holes in a cube. Explain that it will have $X = -2(N - 1)$.

2.6 If you make a sideways connection between two tubes in this model, how does that change X? Explain how this allows you to re-interpret N in the above expression.

3

Fractals

For more than 2000 years, ever since Euclid, geometry was simple. That is, the objects considered were simple: spheres, cubes, cones, etc. Nature, on the other hand, has shapes that are anything but simple. The photo in Figure 3.1 is an illustration of this. It shows cubistic trees made of concrete at the World Exhibition in Paris in 1925. The geometry of these trees consists of intersecting planes, and they are eminently analyzable using the geometry we all learned in school. However, they are anything but realistic. Real trees do not look like this. Birches do. But, try analyzing the common birch using school geometry.

This is where *fractals* enter. They have provided the tools making the geometry of trees like birches possible to analyze quantitatively. This section is meant to give the reader a command over this toolbox.

A fractal is a geomtrical structure that repeats itself on different scales, or in other words, where a subset looks like the whole in some way. Nature is full of examples where this is the case, like the cumulus cloud in Figure 3.2 or the cauliflower in Figure 3.3. Fractals may be both ordered and disordered; in either case, quantifying them represents a geometric description that has no classical counterpart (Feder, 1988), cf. the cubist trees in Figure 3.1.

In the previous sections, we have studied pore geometries with only one characteristic length scale, such as the diameter of the sphere in a sphere packing. For instance, Eq. (2.11) describes a distribution that becomes increasingly peaked around one single average particle number $\langle i \rangle$ as the sampling volume v is increased. This means that the packing density will look more and more homogeneous when it is calculated over volumes larger than v. However, it is not always the case that there is only one length scale present. Sometimes there are several relevant length scales. This may be the case for the geometry of the porous medium itself. It is

Figure 3.1 Cubist trees by Jan and Joël Martel at the *Exposition des Arts Decoratifs*, Paris, 1925.

Figure 3.2 A cumulus cloud.

very often the case for the displacement pattern that arises when one fluid invades a porous medium where there is already another fluid present. Often these patterns lack a typical length scale; they may in fact exhibit a whole range of scales, none of which is more pronounced than the others. And this is another way to define

Figure 3.3 A cauliflower.

a *fractal*: they do not have a single scale because they have all – that is, at least between some upper and lower bounds. In this chapter we will establish some basic concepts that define what fractals are and how to apply them. In particular, we will introduce the *fractal dimension* and show how it is measured.

We will proceed from ordered, mathematical fractals to fractals with disorder.

3.1 Box-Counting Dimension

Fractals are characterized by a non-integer fractal dimension. So, we need a robust procedure to measure the dimension of an object. We are familiar with the dimension of simple shapes: A line has Euclidian dimension $d = 1$, a plane $d = 2$ and a solid cube $d = 3$. If we now cover these objects with boxes of linear dimension δ, as is illustrated in Figure 3.4, count these boxes and record how the number varies with δ, the result will give us their dimension. For the line of extent L, the number of boxes N will depend on δ through the expression

$$N(\delta) = \frac{L}{\delta}, \tag{3.1}$$

while for the plane we need to put boxes in both directions of the plane and

$$N(\delta) = \left(\frac{L}{\delta}\right)^2, \tag{3.2}$$

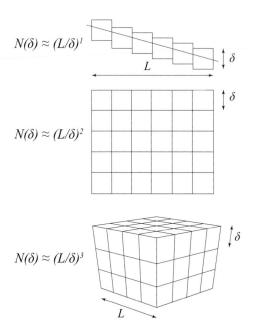

$N(\delta) \approx (L/\delta)^1$

$N(\delta) \approx (L/\delta)^2$

$N(\delta) \approx (L/\delta)^3$

Figure 3.4 Number of boxes required to cover a one-, two- and three-dimensional object.

while for the cube

$$N(\delta) = \left(\frac{L}{\delta}\right)^3. \tag{3.3}$$

Generally,

$$N(\delta) = \left(\frac{L}{\delta}\right)^d, \tag{3.4}$$

where d is the integer Euclidean dimension. This is the expression that may be generalized to give the fractal dimension D, that is,

$$N(\delta) = \left(\frac{L}{\delta}\right)^D. \tag{3.5}$$

A simple fractal where we may test this idea is the *Koch curve* shown in Figure 3.5. It is constructed by replacing a straight line by the $n = 1$ shape in the figure. This shape is called the *generator* of the fractal. To construct a new generation ($n \rightarrow n+1$), every straight line is replaced by the generator, as shown in

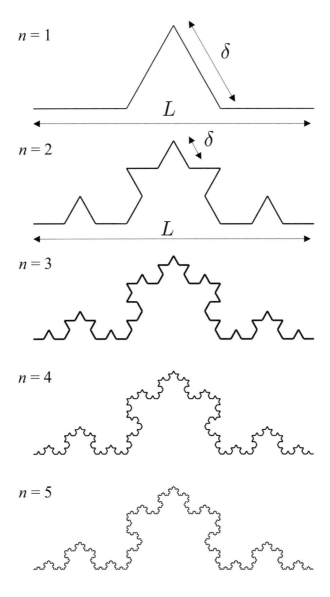

Figure 3.5 Generations n of the Koch curve showing the overall length L and the size of the box δ.

the $n = 2$ case. The same procedure is applied to reach $n = 3$, and so on forever. For any finite n, the shape is referred to as a *pre-fractal*, and the actual fractal results only as $n \to \infty$. For naturally occuring fractals, there are always some upper and lower n values. The cauliflower, for instance, has some smallest bud-size in its structure.

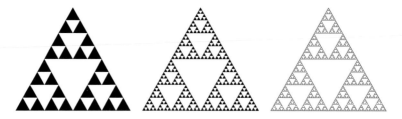

Figure 3.6 Construction of the Sierpinsky gasket with $n = 3, 4$ and 5.

For the Koch curve, we may find $N(\delta)$ by noting that when $n \to n+1$, the number of segments to be covered increases by a factor 4, $N \to 4N$, while the length $\delta \to \delta/3$. By iteration,

$$\delta = \frac{L}{3^n}, \tag{3.6}$$

$$N = 4^n. \tag{3.7}$$

Solving the first of these equations for n gives $n = \ln(L/\delta)/\ln 3$, which may be inserted in the last equation to give

$$N(\delta) = 4^{\ln(L/\delta)/\ln 3} = \exp\left(\frac{\ln 4}{\ln 3} \ln \frac{L}{\delta}\right) = \left(\frac{L}{\delta}\right)^D, \tag{3.8}$$

where

$$D = \frac{\ln 4}{\ln 3} \approx 1.262, \tag{3.9}$$

a number slightly larger than 1.

Another example is the Sierpiensky gasket, which is illustrated in Figure 3.6. The $n = 0$ case is just a black triangle, and to pass from n to $n+1$, a hole is cut in each black triangle as shown in the figure. One starts with the biggest hole, which gives three remaining black triangles, and for each new generation, each black triangle is replaced by three smaller ones.

In this case, the number of covering boxes equals the number of black triangles. When $n \to n+1$, the number of triangles to be covered increases by a factor 3, $N \to 3N$, while the length δ is only reduced by half, $\delta \to \delta/2$. By the exact same argument as for the Koch curve, we find that

$$N(\delta) = \left(\frac{L}{\delta}\right)^D, \tag{3.10}$$

with

$$D = \frac{\ln 3}{\ln 2} \approx 1.585. \tag{3.11}$$

Patterns where a part looks like the whole are called *self-similar*. The Sierpinsky gasket is an example of such a pattern. Note that if the Sierpinsky gasket was cut out in paper and weighed, one would find a mass $m \propto \delta^2 N(\delta)$ since every triangle has an area that is proportional to δ^2. Now, since $D < 2$, we find that $m(\delta) \to 0$ as $\delta \to 0$.

3.2 Mass Dimension

There are two alternative ways to construct a fractal: Either one may fix the overall size and create ever finer structure on the small scale, as in the last section, or one may fix the size of the smallest scale and include an ever increasing region of the surrounding structure. In Eq. (3.5), we fixed L and asked how the number of boxes N scaled with δ, but this equation is also valid if δ is considered fixed and L allowed to vary. If we consider the changes with n in this way, we are looking at ever bigger sections of the (pre-) fractal as n increases. With this in mind, Figure 3.5 still illustrates the changes, only now L increases. Also, Eq. (3.5) remains valid and may be used to determine D. Now, the smallest elements of size δ are fixed so that we may assign them a fixed mass m_0. The total mass will then depend only on L and take the form

$$M(L) = m_0 N(\delta) = m_0 \left(\frac{L}{\delta}\right)^D. \tag{3.12}$$

The mass density

$$\rho = \frac{M(L)}{L^d} = \rho_0 L^{D-d}, \tag{3.13}$$

where $\rho_0 = m_0/\delta^d$, is the constant density of the elementary particles that make up the fractal. It is important to note that the use of the word "mass" here is a bit sloppy as it often does not have the unit of mass, but rather the units of length, area or volume. But it will always be proportional to the number of elements or boxes $N(\delta, L)$, which is dimensionless.

The calculation of D will look a little different but give the same result. As an example, take the Koch curve of Figure 3.5, where now L is increased to include ever bigger sections of the curve while δ remains fixed. When $n \to n + 1$, we get that $L \to 3L$ and $N \to 4N$, as before. Then, instead of Eq. (3.7), we get $N = 4^n$ and $L = 3^n \delta$. Eliminating n between these equations still gives Eq. (3.5) with $D = \ln 4 / \ln 3$ as before. The reason is that the fraction L/δ changes the same way whether we keep L or δ constant.

3.2.1 Olbers' Paradox and Charlier's Fractal Universe Model

While the idea of fractals, pioneered by Mandelbrot (1982) and others, only attracted massive scientific attention in the 1980s, the idea was not entirely new. For instance, in astrophysics and physical cosmology, Olbers' paradox, named after the German astronomer Heinrich Wilhelm Olbers (1758–1840), found a solution based on the idea of fractals already in the 1920s. The paradox, which is also known as the German *dark night sky paradox*, is the argument that the darkness of the night sky conflicts with the assumption of an infinite and eternal static universe. If the universe were static, infinitely old and had a homogeneous distribution of stars, then any line of sight from the Earth would end at the (very bright) surface of a star, and hence the night sky should be completely bright. This clearly contradicts observations.

This paradox was resolved by the introduction of a fractal distribution of stars. This is most easily understood by first considering the two-dimensional case, which could be a model for a forest. In a forest too, where the trees might very well have a rather uniform distribution, there is an obstacle along every line of sight. How might this be different? If the trees were distributed in a fractal way, then one *could* be able look out of the forest.

In Figure 3.7, a two-dimensional Charlier fractal is illustrated and could be considered such a model for a fractal forest. Could one see out of it? To answer this we note that as $N \to 4N$, the length $\delta \to \delta/3$, and we find the fractal dimension $D = \ln 4 / \ln 3 = 1.26$, the same as for the Koch curve. According to Eq. (3.13), the mass density $\rho \sim L^{D-d}$. The probability for your line of sight hitting a tree beyond a certain distance L is the integral of the mass density along this line,

$$P(L) = \int_{L}^{\infty} dL' \, \rho(L'), \qquad (3.14)$$

which then gives

$$P(L) \sim L^{D-d+1}. \qquad (3.15)$$

Since $D - d + 1 = 1.26 - 2 + 1 > 0$, this means that $P(L)$ increases indefinitely with L, and one would not see out of the forest. However, if Figure 3.7 is taken as a model for a three-dimensional universe where every block is replaced by eight new blocks of $1/3$ the width, then D would become $D = \ln 8 / \ln 3 \approx 1.89$, while d would increase to 3. In this case $P(L) \to 0$ as L increases. This means that if you only pass the first few nearest stars that are closer than L, then the probability of

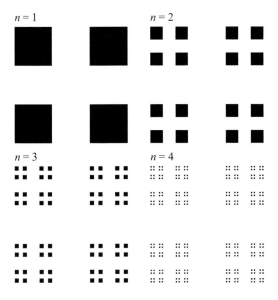

Figure 3.7 Construction of a two-dimensional Charlier fractal.

hitting another one beyond that drops to zero. This would explain the darkness of the night sky, as was first shown by Charlier.[1]

Of course, by now we know that the universe is neither static nor infinitely old, so the paradox does not appear in the first place. But this was not known in Charlier's time. In fact, the darkness of the night sky is one of the pieces of evidence for a dynamic universe, such as the Big Bang model.

3.3 Measuring the Fractal Dimension with Log–Log Plots

Often the problem is to actually measure the fractal dimension of a given object. In this case we need to use the mass dimension, that is, to record the mass of an increasing section of length L. Taking the logarithm on both sides of Eq. (3.12) gives

$$\log\left(\frac{M(L)}{m_0}\right) = D\log\left(\frac{L}{\delta}\right). \tag{3.16}$$

The means that in a plot of $\log M(L)$ versus $\log L$, the fractal dimension D can be read off as the slope. Normally the logarithm with base 10 is used, so that the plot shows the number of decades. For real objects, Eq. (3.16) is only satisfied for a limited range, $L_{min} < L < L_{max}$, where L_{min} is the lower and L_{max} the upper

---

[1] Carl Vilhelm Ludwig Charlier (1862–1934) was a Swedish astronomer.

cut-off length of the fractal behavior. For a cumulus cloud, the smallest length could be the length at which the turbulent flow of the air becomes laminar, and the largest the size of the cloud. For the cauliflower, the smallest length would that of a single flower bud, and the largest no bigger than the size of the vegetable. There will always be a physical reason, or mechanism, that sets the upper and lower cut-offs.

The measurement of D means establishing a power law, and this will always go hand-in-hand with the question of its validity. Any differentiable curve looks like a straight line if one only looks at a small enough segment of it. So, if the range between L_{min} and L_{max} is sufficiently small, the plot may always look like a power law. For this reason, two factors become important: First the L_{min} to L_{max} range must be large enough. As a rule of thumb, it has become conventional to require three decades between the two. Second, one needs to justify the cross-over values. If a change of behavior kicks in at a surprising value of L_{min} or L_{max}, it is likely that the power law is not a single power law after all.

3.4 Disordered Fractals

In Figure 3.8, we have generated a Koch curve where there is some variation added to the size of the generator as one passes from one generation to the next.

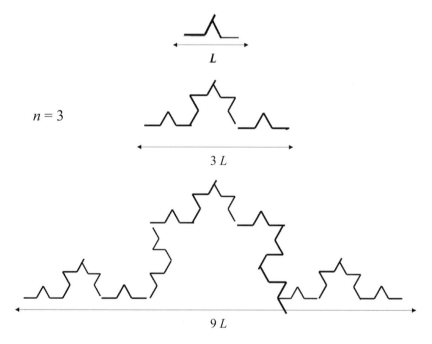

Figure 3.8 A Koch curve with some noise added to the length factor that is used to generate each new generation.

However, on the average, the overall mass depends on the length the same way as before, that is,

$$M(L) \rightarrow 4M(L) \qquad (3.17)$$

when $L \rightarrow 3L$. This means that that a log–log plot based on Eq. (3.16) will give the same fractal dimension as the Koch curve without such variations. The only difference will be that there might be some fluctuations on this curve.

Figure 3.9 A digitized image taken from an atlas showing the coast of southern Norway. Here $\delta = 50$ km.

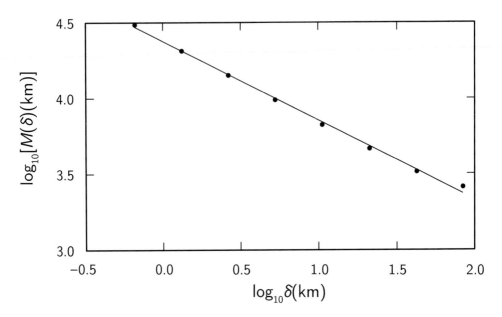

Figure 3.10 Log–log plot of the mass (length) of the coast of Norway shown in Figure 3.9 as a function of the yard stick δ.

3.4.1 Coast of Norway

A famous example of a real, disordered fractal is the coastline of Norway. In Figure 3.9, a digitized image of this coast is shown. The log–log plot of the coast length as a function of the length of the yard stick is shown in Figure 3.10, where the curve has a slope $D = 1.52 \pm 0.01$. The fact that this slope is so different from 1 means that the length of the coast will vary strongly with the resolution. A ship that chooses the shortest path between Norwegian cities will cover a very different distance compared to a pedestrian who follows the coast as closely as possible.

3.4.2 Fast Drainage

Figure 3.11 shows an example of a fractal that has been produced by fast drainage (Zhao et al., 2019). The model porous medium is a circular Hele–Shaw cell with a hole in the center and an open boundary. The Hele–Shaw cell, with an opening of 100 μm, is filled with fixed glass beads. The model is initially filled with a fluid that wets the glass beads and the walls of the cell. A second fluid, immiscible with respect to the first, is then injected through the central hole. This fluid has a wetting angle of 150° with respect to the beads. With a injection rate corresponding to a capillary number of 2.9, this is a very fast process.

Figure 3.11 Fast drainage, capillary number = 2.9 and a wetting angle of 150°
for the injected fluid in a circular Hele–Shaw cell filled with fixed glass beads.
(Adapted from Zhao et al. (2019).) (A black and white version of this figure will
appear in some formats. For the colour version, refer to the plate section.)

The authors find by using the box-counting method that the injected fluid body
forms a fractal with dimension D around 1.57. This value is then used as a quantitative measure to assess the accuracy of the some 10 different numerical models
used to replicate the experiment – some doing well, some less well.

3.5 Multifractals*

Fractals describe geometrical objects. A geometrical object is a shape of some sort.
Intrinsically, this is a binary concept. We answer "yes" or "no" to the question "does
this point belong to the object?"

Suppose we now give each point in space belonging to the fractal a property
characterized by a value h. Clearly all points having a value in the interval h to
$h + dh$ form a subset of all points belonging to the fractal. This subset of points
may themselves form a fractal with the same fractal dimension as the full set of
points, but not necessarily. The dimension may be different. If it is different, the
fractal dimension of the subset must be smaller.

3.5.1 Nested Fractals*

We first discuss the case when h takes on discrete values. We are then dealing with
nested fractals rather than *multifractals*.

We may illustrate this using the Charlier fractal in Figure 3.7. Suppose we give the squares in the upper left and lower right corners of each generation the value $h = h_1$ and the squares in the lower left and upper right corners the value $h = h_2$. The fractal dimension of the squares with value h_1 is $D_1 = \ln 2/\ln 3$, as is the set with value $h = h_2$, $D_2 = \ln 2/\ln 3$.

Suppose now we decide that only the square in the upper left corner has the value $h = h_1$, whereas the remaining three corners have the value $h = h_2$. The fractal dimension of the subset for which $h = h_1$ is $D_1 = \ln 1/\ln 3 = 0$, and the fractal dimension of the subset with value $h = h_2$ is $D_2 = \ln 3/\ln 3 = 1$.

This is a curious result. The fractal dimension of the Charlier fractal in Figure 3.7 is $D = \ln 4/\ln 3$, whereas the fractal dimensions of the two subsets are either $D_1 = D_2 = \ln 2/\ln 3$ in the first example or $D_1 = 0$ and $D_2 = 1$ in the second example. Perhaps one would have expected that the fractal dimensions of the subsets somehow should have "added up" to $D = \ln 4/\ln 3$, but they do not.

Let us first concentrate on the second example, that is, we have split the Charlier fractal into two subsets, one having the value $h = h_1$ associated with its points, and a fractal dimension $D_1 = 0$. The other subset has h_2 associated with its points, and a fractal dimension $D_2 = 1$. The two subsets combined without regard for the value of h have a fractal dimension equal to $D = \ln 4/\ln 3$.

We define the kth moment

$$\langle h^k \rangle = \sum_i h(i)^k, \tag{3.18}$$

where $h(i)$ is the value of h of element i in the fractal. k may be any real number.

We set $k = 0$, and the moment simply counts the number of points, that is, it measures the mass of the fractal. Hence, we find for the Charlier fractal

$$\langle h^0 \rangle = M \propto L^D = L^{\ln 4/\ln 3}. \tag{3.19}$$

However, for $k \neq 0$, we find

$$\langle h^k \rangle \propto h_1^k L^{D_1} + h_2^k L^{D_2}. \tag{3.20}$$

In the limit of $L \to \infty$, we thus find

$$\langle h^k \rangle \sim \begin{cases} L^D & \text{if } k = 0, \\ h_2^k L^{D_2} & \text{if } k \neq 0, \end{cases} \tag{3.21}$$

since $D_2 = 1 > D_1 = 0$.

But, what about the other Charlier fractal example, where we split the set into two equally large subsets, each having a fractal dimension $D_1 = D_2 = \ln 2/\ln 3$, but having the values $h = h_1$ or $h = h_2$ associated with them? A moment's reflection gives

$$\langle h^k \rangle \sim \begin{cases} L^D & \text{if } k = 0, \\ (h_1^k + h_2^k)L^{D_2} & \text{if } k \neq 0, \end{cases} \tag{3.22}$$

in the limit $L \to \infty$.

We may generalize this discussion to any fractal with fractal dimension D. We divide the points belonging to the fractal into K non-intersecting subsets, each characterized by a value h_i and a fractal dimension D_i, where $1 \leq i \leq K$. The kth moment will behave as L^{y_k}, where we have defined the moment scaling exponents y_k. We find

$$y_k = \begin{cases} D & \text{if } k = 0, \\ D_{\max} & \text{if } k \neq 0, \end{cases} \tag{3.23}$$

where

$$D_{\max} = \max(D_1, \ldots, D_K) \tag{3.24}$$

and $h_{\max}$ being the value of h associated with this subset.

But, the reader may ask, why bother with the moments $\langle h^k \rangle$ when they can only detect the full set with its fractal dimension D or the subset with the largest fractal dimension $D_{\max}$? We are losing all the information pertaining to all the other subsets, that is, $\{(h_1, D_1), \ldots, (h_K, D_K)\}$. We are, however, asking the reader for patience. We will need to make one more, seemingly strange detour before we get to the core of this section, namely multifractals. And there, it will be clear why we introduced the moments in Eq. (3.18) – and why information was lost in the case of nested fractals.

3.5.2 Power Law Distributions*

Suppose we have a fractal with fractal dimension D. Each point belonging to this fractal we give a value h *drawn from a probability distribution*, also known as a probability density $p(h)$. Didn't we just do this, using a bimodal statistical distribution with values h_1 and h_2 in the case of the Charlier fractal we discussed in the previous section? No, we did not. There was nothing statistical about how we placed the h-values there.

In the following, we will follow the discussion of Hansen et al. (1991). We define the *cumulative probability* associated with the statistical distribution $p(h)$ as

$$P(h) = \int_0^h dh'\, p(h'), \tag{3.25}$$

where we have assumed h to be positive. The cumulative probability is the probability of finding a value h or smaller when we draw a number from the $p(h)$ distribution.

There is, by the way, another very useful way to interpret the cumulative probability $P = P(h)$. We may also think of $P = P(h)$ as a new random variable that takes a new value each time a value h is picked from the $p(h)$ distribution. Then the distribution $p'(P)$ for P is given by $p'(P)dP = p(h)dh$. Since $p = dP/dh$, we immediately see that $p' = 1$, or, in other words, P is uniformely distributed on the unit interval. So, once more: $P = P(h)$ is a transformation of h to a new variable P that is uniformely distributed on the unit interval.

Before we proceed, we will need to invoke some basic results from *order statistics* (Gumbel, 2004): Suppose we draw M_N sequences of N numbers using the probability distribution $p(h)$. That is, we draw a first sequence of N numbers, then second sequence of N numbers, then a third, a fourth, etc. up to sequence number M_N.

In each of the M_N sequences, each containing N numbers, we order the h-values we have found so that $h_{(1)} \leq h_{(2)} \leq \cdots \leq h_{(N)}$. Hence, $h_{(n)}$ is the nth-smallest element in this sequence. Hence, there are $n - 1$ elements that are smaller and $N - n + 1$ elements that are larger. We now pose the question: What is the probability distribution for the n smallest element, $h_{(n)}$, out of N elements when the number of sequences of length N, $M_N \to \infty$? That is, what is the probability density for the nth smallest element $h_{(n)}$ to have the value h? It is

$$p_{(n)}(h) = n \binom{N}{n} P(h)^{n-1}[1 - P(h)]^{N-n} \, p(h), \tag{3.26}$$

based on there being $n - 1$ even smaller elements than the nth and $N - n$ elements that are larger. The binomial coefficient times n takes care of the number of ways this can be arranged.

The average of the nth smallest element is given by

$$\langle h_{(n)} \rangle = \int_0^\infty dh \, h \, p_{(n)}(h). \tag{3.27}$$

It is not possible to do this integral without knowing the initial probability distribution $p(h)$. However, in the limit of large N, there is a way around this. $P(h)$ is the cumulative probablity for h. However, it is also a function that maps a number h to a another number $P = P(h)$ on the interval $[0, 1]$. So, why not calculate the average of $P(h_{(n)})$? This is easy. By using Eq. (3.26), we find

$$\langle P(h_{(n)}) \rangle = \int_0^\infty dh \, P(h) p_{(n)}(h)$$

$$= \int_0^1 dP \, P \, n \binom{N}{n} P^{n-1}[1 - P]^{N-n} = \frac{n}{N+1}, \tag{3.28}$$

using repeated partial integration. We note here that underlying this calculation is that we have interpreted $P = P(h)$ as a new variable that is uniformly distributed on the unit interval as described above. In the limit of large N, fluctuations in $P(\langle h_{(n)} \rangle)$ will decay, and we have that

$$P(\langle h_{(n)} \rangle) = \langle P(h_{(n)}) \rangle = \frac{n}{N+1}, \tag{3.29}$$

or

$$\boxed{\langle h_{(n)} \rangle = P^{-1}\left(\frac{n}{N+1}\right),} \tag{3.30}$$

where P^{-1} is the functional inverse of the cumulative probability $P = P(h)$. This is so since the distribution $p_{(n)}(h)$ becomes more and more peaked with increasing N.

Let us now consider a specific statistical distribution, $p(h)$. We limit the values of h to be in the range $[0, 1]$ and let the cumulative probability be

$$P(h) = h^{\phi_0}, \tag{3.31}$$

where ϕ_0 is a positive exponent. Hence, $p(h) = \phi_0 h^{\phi_0 - 1}$.

We now combine Eqs (3.30) and (3.31) to find

$$\langle h_{(n)} \rangle^{\phi_0} = \frac{n}{L^D}, \tag{3.32}$$

where we have used that the denominator $N + 1$ in Eq. (3.30) behaves as L^D when we are assigning the values to a fractal with fractal dimension D.

Let us now *define*

$$\langle h_{(n)} \rangle = L^{-\alpha}, \tag{3.33}$$

or

$$\alpha = -\frac{\ln \langle h_{(n)} \rangle}{\ln L}. \tag{3.34}$$

We furthermore define a function f so that

$$n = L^f, \tag{3.35}$$

so that

$$f = \frac{\ln n}{\ln L}. \tag{3.36}$$

We now express Eq. (3.32) using the new variables α and f, finding

$$f = D - \phi_0 \alpha. \tag{3.37}$$

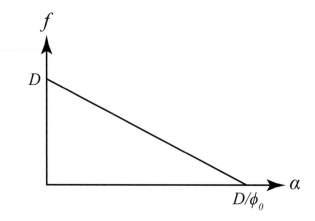

Figure 3.12 This is the f–α curve for the cumulative probability $P(h) = h^{\phi_0}$ for $0 \le h \le 1$ supported by a fractal with fractal dimension D.

The smallest value n can have is 1 and the largest is $N + 1 = L^D$. Hence, we have that $0 \le (\ln n / \ln L) \le D$, and as a consequence

$$0 \le \alpha \le \frac{D}{\phi_0}. \tag{3.38}$$

We show in Figure 3.12 f vs. α based on the cumulative probability given by Eq. (3.31). Here is a central message: f, defined in Eq. (3.36), vs. α, defined in Eq. (3.34), contains precisely the same information as 1. the fractal dimension of the set acting as support for the distribution of h and 2. the cumulative probability given in Eq. (3.31), but with one important difference: the variables f and α *do not depend on the size of the system*, L. We see this easily in Eq. (3.37): D and ϕ_0 do not depend on the size of the system L, and therefore f and α do not depend on the system size either. They are scale free. This is in contrast to the variables $\langle h_{(n)} \rangle$ and $P(\langle h_{(n)} \rangle)$. Said another way, the variables f and α are *intensive*, as opposed to being *extensive*.

Let us now make the probability distribution for h, $p(h)$, a bit more complex. So far, we have considered a power law tail toward $h \to 0$. Let us now consider a power law tail for $h \to \infty$, that is, using a cumulative probability

$$P(h) = 1 - h^{-\phi_\infty}, \tag{3.39}$$

for $1 \le h < \infty$. We now combine this cumulative probability with the order statistics expression (3.30) and retrace the steps (3.32)–(3.38), with the modification that we replace Eq. (3.35) by

$$L^D - n = L^f, \tag{3.40}$$

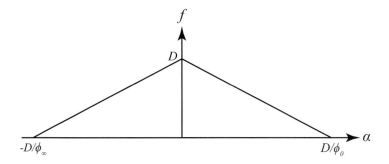

Figure 3.13 This is the f–α curve for the cumulative probability (3.44) for $0 \leq h < \infty$ supported by a fractal with fractal dimension D.

after noting that Eq. (3.32) now reads

$$\langle h_{(n)} \rangle^{\phi_0} = \frac{L^D - n}{L^D}. \tag{3.41}$$

We find that

$$f = D + \phi_\infty \alpha, \tag{3.42}$$

where

$$-\frac{D}{\phi_\infty} \leq \alpha \leq 0. \tag{3.43}$$

Lastly, we combine the two power law distributions (3.31) and (3.39) into one cumulative probability

$$P(h) = \begin{cases} (1-a)h^{\phi_0} & \text{if } 0 \leq h \leq 1, \\ (1-a) + a(1 - h^{-\phi_\infty}) & \text{if } h > 1, \end{cases} \tag{3.44}$$

where $0 < a < 1$. We show in Figure 3.13 the f–α curve for this distribution.

The variables f and α are, as we stated above, scale free. That is, they contain no intrinsic scales. Power law distributions are scale free. We have considered two, given by Eqs (3.31) and (3.39). Let us demonstrate this for the second one, leaving the first one as a (trivial) exercise. The distribution corresponding to (3.39) is

$$p(h) = \frac{dP(h)}{dh} = \phi_\infty h^{-\phi_\infty - 1}, \tag{3.45}$$

on the interval $[1, \infty)$. Let us now scale $h \to \tilde{h} = h/h_0$. We find

$$\begin{aligned} p(h)dh &= \phi_\infty h^{-\phi_\infty - 1} dh \\ &= \phi_\infty h_0^{-\phi_\infty} \tilde{h}^{-\phi_\infty - 1} d\tilde{h} \\ &= h_0^{-\phi_\infty} p(\tilde{h}) d\tilde{h}, \end{aligned} \tag{3.46}$$

on the interval $[h_0, \infty)$. We see that apart from an overall scaling factor, $h_0^{-\phi_\infty}$ coupled to the change of the lower cutoff $1 \to h_0$, the distribution is unchanged.

Let us now try the same, that is, $h \to \tilde{h} = h/h_0$, for the exponential distribution $p(h) = \exp(-h)$ on the interval $[0, \infty)$. We find

$$p(h)dh = e^{-h}dh = h_0 e^{-h_0\tilde{h}}d\tilde{h}$$
$$\neq h_0 p(\tilde{h})d\tilde{h}. \tag{3.47}$$

The scaling of h has changed the distribution. The exponential distribution is *not* scale free. Only power law distributions are scale free – with one exception, namely *mulifractals*. But, we will get to this.

We now make no demands on the shape of the probability distribution $p(h)$ on $h \in [0, \infty)$ and its corresponding cumulative probability $P(h)$ associated with the points belonging to the fractal characterized by the fractal dimension D. Equation (3.30) then takes the form

$$P(\langle h_{(n)} \rangle) = \frac{n}{L^D}. \tag{3.48}$$

By expressing n by f, Eq. (3.35) and $\langle h_{(n)} \rangle$ by α, Eq. (3.33), we find

$$f = D + \frac{\ln \left[P\left(L^{-\alpha} \right) \right]}{\ln L}. \tag{3.49}$$

Let us now assume that the values of h associated with the points belonging to the D-dimensional fractal are given by a cumulative probability according to

$$P(h) = h^{\phi_0} e^h, \tag{3.50}$$

where $0 < h < 1$. We assume $\phi_0 > 0$. What is the $f-\alpha$ curve for this distribution? We use Eq. (3.49) to find

$$f = D - \phi_0\alpha + \frac{L^{-\alpha}}{\ln L}, \tag{3.51}$$

where $0 \leq f \leq D$ as argued just before Eq. (3.38). Hence, also here we have that $0 \leq \alpha \leq D/\phi_0$. The last term in Eq. (3.51) then disappears as L becomes large. Only the scale-invariant component of the probability distribution survives in this limit.

What if we turn the cumulative probability in Eq. (3.50) around and consider

$$P(h) = 1 - h^{-\phi_\infty} e^{-h}, \tag{3.52}$$

on the interval $[1, \infty)$, and $\phi_\infty > 0$? In this case, we find

$$f = D + \phi_\infty\alpha + \frac{L^{-\alpha}}{\ln L}, \tag{3.53}$$

where $-D/\phi_\infty \leq \alpha \leq 0$. Also in this case, the last term disappears in the large-L limit. Only the scale-invariant component survives.

But what happens if there is no scale-invariant component in the probability distribution? A concrete example is the cumulative probability

$$P(h) = 1 - e^{-h} \tag{3.54}$$

on the interval $h \in [0, \infty)$. We find that

$$\langle h_{(n)} \rangle = L^{-\alpha} = \ln\left(\frac{L^D - n}{L^D}\right), \tag{3.55}$$

making $\alpha = 0$ as $L \to \infty$. Hence, the f–α curve reduces to a point, $\alpha = 0$ and $f = D$ in this case.

This is general. Only the power laws (and multifractals, which have not been defined yet... but we will get to that) show up in the f–α curves. All other distributions reduce to a point: $f = D$ for $\alpha = 0$.

We now consider the moments $\langle h^k \rangle$, first encountered in Section 3.5.1, see Eq. (3.18). In the context of the present section, they are

$$\langle h^k \rangle = \sum_{n=1}^{L^D} \langle h_{(n)} \rangle^k. \tag{3.56}$$

We now assume the power law cumulative probability (3.31) and use Eq. (3.30) to find

$$\langle h^k \rangle = \sum_{n=1}^{L^D} \left(\frac{n}{L^D}\right)^{k/\phi_0 - 1} \frac{n}{L^D} \approx \int_{1/L^D}^{1} dh\, h^{k/\phi_0 - 1}, \tag{3.57}$$

where the last expression is a good approximation for large L. We do the integral and find

$$\langle h^k \rangle = \frac{\phi_0}{k}\left(1 - L^{-kD/\phi_0}\right). \tag{3.58}$$

We see that the moment converges to a constant independent of L for large *positive* k. For negative k, on the other hand, we find that the moments scale as

$$\langle h^k \rangle \sim L^{-kD/\phi_0} = L^{y_k}, \tag{3.59}$$

where we have used the moment scaling exponents y_k, first encountered in Section 3.5.1. They show *constant gap scaling* – a term borrowed from the theory of critical phenomena. Constant gap scaling means that y_k is proportional to k.

We now turn to the power law distribution for $h \to \infty$, whose cumulative probability is given by Eq. (3.39). By the same arguments, we find that $\langle h^k \rangle$ is independent of L for negative k, whereas for positive k, we find

$$\langle h^k \rangle \sim L^{kD/\phi_\infty} = L^{y_k}, \tag{3.60}$$

so that $y_k = kD/\phi_\infty$ in this case, that is, constant gap scaling once again.

The zeroth moment behaves, as expected, as $\langle h^0 \rangle \sim L^D$. If we now collate the results and assume the combined power law, Eq. (3.44), we find that the moment exponents are given by

$$y_k = \begin{cases} -\frac{kD}{\phi_0} & \text{if } k < 0, \\ D & \text{if } k = 0, \\ \frac{kD}{\phi_\infty} & \text{if } k > 0. \end{cases} \tag{3.61}$$

The reader should compare the behavior of the moments studied in this section with those found for nested fractals, see Eq. (3.23). The behavior is substantially different.

3.5.3 Multifractal Distributions*

Now, finally, we get to multifractals. They are generalizations of the power law distributions studied in the previous section. We may define multifractals as follows:

A multifractal is a scale-invariant, spatially correlated and statistically homogeneous distribution.

A power law distribution, for comparison, is a scale-invariant spatially *uncorrelated* and statistically homogeneous distribution. We used this in the previous section. The h-values were distributed at random on the fractal; all points belonging to the fractal were treated equally. This is what we mean by "spatially uncorrelated." When we say "spatially correlated," we mean that the h values are not randomly placed. They obey some correlation function.

What about the next qualifier, "statistically homogeneous"? Let us first look at the opposite: The electric potential outside a point charge follows a power law, that is, it is inversely proportional to the distance from the charge. The point at which the charge is sitting stands out as special. "Statistically homogeneous" means that no point stands out; all points are equally important.

So, how do we recognize a multifractal? This is where the f–α curve becomes of the uttermost importance. We have already worked it out for the power law distributions, see Figure 3.13. We now give the general prescription to measure this curve. If $\tilde{p}(h)$ is the normalized histogram of measured h values – normalized meaning that it sums up to one – we define

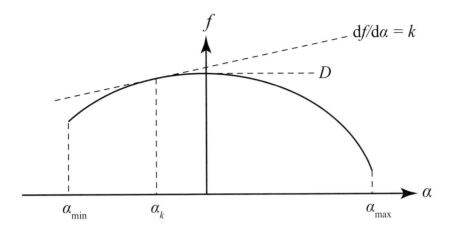

Figure 3.14 A generic multifractal f–α curve. We also illustrate Eq. (3.66) here.

$$f = \frac{\ln L^D \tilde{p}(h)}{\ln L} \tag{3.62}$$

and

$$\alpha = -\frac{\ln h}{\ln L}. \tag{3.63}$$

(In the previous section, we used order statistics to work out the histogram for the power law distributions.) When the f–α curves are as in Figure 3.13, we are dealing with spatially uncorrelated power law distributions. When they do *not* consist of straight lines as in this figure, we are dealing with multifractals. For example, they may look like the sketch in Figure 3.14.

Let us now calculate the kth moment for a multifractal using the f–α curve. We have

$$\langle h^k \rangle = \int_{h_{\min}}^{h_{\max}} dh\, h^k \frac{d\tilde{p}(h)}{dh}. \tag{3.64}$$

By using the definitions (3.62) and (3.63) in addition to a partial integration to get rid of the derivative $d\tilde{p}/dh$, we have

$$\langle h^k \rangle \sim L^{yk} \sim \int_{\alpha_{\min}}^{\alpha_{\max}} d\alpha\, L^{f-\alpha k}. \tag{3.65}$$

We do this integral using the saddle point approximation. This approximation assumes that the major contribution to the integral comes from the region where the integrand is stationary. This happens for the value of $\alpha = \alpha_k$ that solves the equation $d(f - \alpha k)/d\alpha = 0$ or

$$\frac{\mathrm{d}f}{\mathrm{d}\alpha} = k \quad \Rightarrow \quad \alpha = \alpha_k, \qquad (3.66)$$

giving the final result

$$y_k = f(\alpha_k) - k\alpha_k, \qquad (3.67)$$

where α_k is the solution of Eq. (3.66).

This is the way to detect multifractality: Measure the moments of the variable you are interested in. If you find 1. that the moments scale as L^{y_k} and 2. the moment scaling exponents y_k do *not* show constant gap scaling, that is, they behave as

$$y_k \neq ak + b, \qquad (3.68)$$

where a and b are constants, in addition to there being spatial homogeneity, you are dealing with a multifractal.

The support of a multifractal does not have to be a fractal. In other words, the fractal dimension D can be the dimension of the embedding space itself.

3.5.4 Relevance of Multifractals in Porous Media

We show in Figure 3.15 the velocity distribution in a two-phase immiscible dynamic network model (Aker et al., 1998; Sinha et al., 2021). It models a two-dimensional square lattice where each link represents an hourglass-shaped pore. All links have the same length, but the radius at the narrowest point along the links vary, introducing disorder into the model. It uses biperiodic boundary conditions – that is, it has the topology of a torus. Through a trick, a pressure difference is set up along a line cutting the lattice at 45°. This forces the fluids to move around the torus, thus setting up a steady state. This pressure difference is adjusted continuously so that the volumetric flow rate Q is kept constant. In the figure, the model is filled 70 volume percent wetting fluid, the rest being non-wetting fluid; that is, a wetting saturation of 0.7. The capillary number Ca was 3.4×10^{-4} in the upper illustration and 3.4×10^{-3} in the lower. The figure shows snapshots of the model. Each link is shaded according to the logarithm of the flow velocity it carried at the time of the snapshot – the darker the color, the higher the velocity.

It is evident from the figure that the velocity distribution is highly correlated. There is a "highway system" of high-velocity paths that criss-cross the network; between these "highways" there are patches of low-velocity areas. When the capillary number is increased, the network of "highways" becomes denser.

The network has size $L_N \times L_N$, where L_N is number of nodes N in each direction times the link length divided by the square root of two. We pick a square sublattice

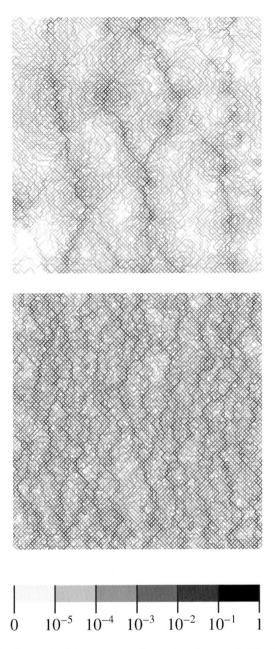

Figure 3.15 The flow velocities in a two-dimensional network filled with immiscible fluids at 70% wetting fluid saturation. The capillary number was 3.4×10^{-4} in the upper figure and 3.4×10^{-3} in the lower figure. The links are shaded according to the logarithm of the flow velocity in the links: the darker the shade, the faster the flow. The figure is based on a two-dimensional network simulator (Aker et al., 1998; Sinha et al., 2021). The two-dimensional network has been implemented on a torus and forced to circulate at constant average pressure gradient. This generates a steady-state flow. (Figure courtesy of Santanu Sinha.)

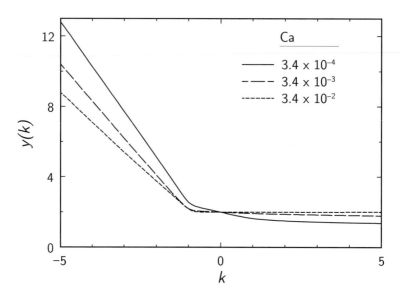

Figure 3.16 Scaling exponents $y(k)$ for the moments of the velocity distribution, Eq. (3.70) for three different capillary numbers. (Courtesy of Santanu Sinha.)

of size $L \times L$ and measure the statistical moments of the velocity distribution within this sublattice. If v_i is the instantaneous velocity in link i within the sublattice, we calculate

$$\sum_i |v_i|^k. \tag{3.69}$$

Next, we average this sum both over time and over position. That is, we calculate the sum at different times and diffent positions of the sublattice. This gives us the kth moment, $\langle |v|^k \rangle$. Plotting this moment against L for different values of L on a logarithmic scale gives us the scaling exponents $y(k)$,

$$\langle |v|^k \rangle \sim L^{y(k)}. \tag{3.70}$$

We show the resulting scaling exponents $y(k)$ for three values of the capillary number Ca in Figure 3.16.

By using Eqs (3.65)–(3.67), the f–α curve may be constructed. The result is shown in Figure 3.17. As is evident from the figure, the curves do not consist of two straight lines as in Figure 3.13; rather, they have the generic form shown in Figure 3.14. This signals that the velocity distribution is multifractal. We see that the curvatures in the f–α curves are larger for the Ca $= 3.2 \times 10^{-4}$ case compared to the other two cases. We also see that the curvature in $y(k)$ as a function of $k > 0$ is larger for the Ca $= 3.2 \times 10^{-4}$ case. It seems that the multifractal aspects of the velocity distribution develop with decreasing capillary number.

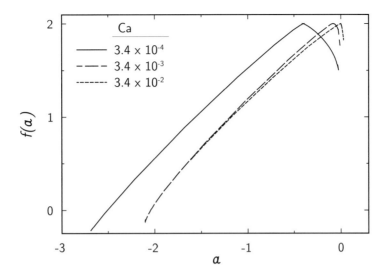

Figure 3.17 The f–α curves for three different capillary numbers. (Courtesy of Santanu Sinha.)

A word of caution before we move on the next topic, self-affinity. There are many examples in the literature of the use of multifractals in connection with various aspects of porous media. Not all of this work is well founded. Perhaps a problem is that people use the literature on *dynamical systems theory* to learn the topic, an approach that is very different from the one we have described here. Multifractals in the context of *percolation theory* is the appropriate one. We discuss this further in Section 4.6.

3.6 Self-Affine Surfaces

Dip a piece of paper halfway into water and watch the imbibition front rise due to capillarity and stabilized by gravity. This front is an example of a (one-dimensional) *self-affine surface*; a self-affine trace, see Figure 3.18. Here a paper towel is dipped into water colored by ink. The details behind the capillary forces causing this imbibition process will be discussed in Section 9.4.3. We will discuss the shape of the imbibition front further in Section 4.5, where we introduce directed percolation.

Here is an example of a two-dimensional self-affine surface: break something brittle, and watch the fracture surface.[2] That surface will be self affine. In fact, as fractals, self-affine surfaces are ubiquitous in nature, and the concept belongs to the

---

[2] If you break glass, you need an electron microscope to see the roughness of the fracture surface.

Figure 3.18 Imbibition: A paper towel dipped into a cup of water colored by ink producing a self-affine invasion front. (A black and white version of this figure will appear in some formats. For the colour version, refer to the plate section.)

new geometry toolbox that comes in addition to the old one of Euclid. We show in Figure 3.19 an artificially generated self-affine trace.

The properties of self-affine surfaces are related to those of fractals. However, they are different – even though the older literature did not always make a clear distinction. And, we have to warn the reader. If (s)he felt that it got complicated with fractals, here we go again. This ride is unfortunately no less bumpy.

Let us start the ride by considering the simplest possible model that generates a self-affine trace, that is, a one-dimensional self-affine surface. We will return to this model, but in a very different context, in Chapter 6 where we discuss dispersion.

Consider a small sphere with a diameter of a couple of millimeters rolling down an inclined plane – perhaps made of plexiglass – which has been made rough by sand blasting. We place a Cartesian coordinate system along the plane. We let the origin ($x = 0, y = 0$) be at the middle of the upper rim and let the x axis point along the plane in the downwards direction. The y is orthogonal to it and is placed along the upper rim. The motion of the sphere along this plane is in fact quite complex as Riguidel et al. (1994) show, but the following simple model will do fine: The sphere will obey the equation of motion

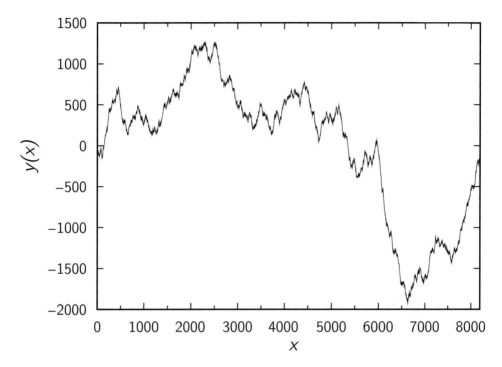

Figure 3.19 A self-affine trace with Hurst exponent $H = 0.75$. We have used the Fourier method to generate it, see Section 3.6.4, as with the other self-affine traces shown in this section.

$$\frac{dy(x)}{dx} = \eta(x), \tag{3.71}$$

where $\eta(x)$ is a noise term and represent the "kicks" the sphere gets from the rough asperities along the rough plane. There is no acceleration term as we assume the friction felt by the sphere makes it reach a velocity component along the x axis that has a well defined, constant average. Such an equation of motion is an example of a Langevin equation. We assume the noise to be Gaussian – that is, distributed according to a Gaussian distribution and averaging to zero,

$$\langle \eta(x) \rangle = 0, \tag{3.72}$$

and to be uncorrelated,

$$\langle \eta(x)\eta(x') \rangle = 2D\delta(x - x'), \tag{3.73}$$

where $\delta(x)$ is the Dirac delta-function. This is defined as $\delta(x) = 0$ for $x \neq 0$ and $\int_{-\infty}^{\infty} dx\,\delta(x) = 1$. By the averaging in Eqs (3.72) and (3.73), we mean that we average over an ensemble of many spheres rolling down the plane; that is, we do the experiment over and over.

Equation (3.71) is easy to solve. We assume as initial condition that $y(x = 0) = 0$. We then find

$$y(x) = \int_0^x dx'\, \eta(x').\qquad(3.74)$$

We average this solution,

$$\langle y(x) \rangle = \int_0^x dx'\, \langle \eta(x') \rangle = 0.\qquad(3.75)$$

This is no surprise; there is no bias in the motion of the sphere. The sphere does not prefer $y > 0$ to $y < 0$. What about the average spread w? We define this as $w^2 = \langle y(x)^2 \rangle - \langle y(x) \rangle^2 = \langle y(x)^2 \rangle$. The right-hand side of this equation comes from Eq. (3.75). We find that w is

$$\begin{aligned}
w^2 = \langle y(x)^2 \rangle &= \int_0^x dx' \int_0^x dx'' \langle \eta(x')\eta(x'') \rangle \\
&= \int_0^x dx' \int_0^x dx'' 2D\delta(x' - x'') \qquad(3.76)\\
&= \int_0^x dx'\, 2D = 2Dx.
\end{aligned}$$

The fact that the average spread w, more commonly referred to as the *roughness*, follows a power law in x, that is, $w \sim x^{1/2}$, is a defining property of self-affine surfaces.

Let us, however, be entirely precise as to what we mean by self-affinity. Suppose we focus on the point (x, y) on the inclined plane. We know that the sphere will pass the distance x at some value of y. What is the probability $p(x, y)dy$ that it passes within dy of the point y at distance x from the upper edge? It turns out – as we shall see in Chapter 8 – that this probability is given by a Gaussian,

$$p(x, y) = \frac{1}{\sqrt{4\pi Dx}}\, e^{-y^2/4Dx},\qquad(3.77)$$

see Eq. (8.15). We see that it obeys the scaling relation

$$\lambda^{1/2} p(\lambda x, \lambda^{1/2} y) = p(x, y).\qquad(3.78)$$

This scaling is the defining property making the trace of the sphere self affine. The scaling exponent $1/2$ is the Hurst exponent, or roughness exponent, which is as important for self-affine surfaces as the fractal dimension is for fractals.

So far we have used the term "surface," but only discussed traces. When dealing with a surface that has higher dimensionality than one, that is, we must describe a point on the surface by $(\mathbf{x}, y)$ rather than (x, y), we may still reduce the analysis to that of a trace by making a cut through the self-affine surface. The cut will be a one-dimensional trace. We will in the following assume this procedure: we consider cuts even if the surface is two-dimensional.[3]

A self-affine trace – which is perhaps a cut through a two-dimensional surface – is defined by the scaling relation

$$\lambda^H p(\lambda x, \lambda^H y) = p(x, y) \quad \text{for} \quad 0 < H < 1, \tag{3.79}$$

where H is the Hurst exponent.

As a consequence of this definition, when assuming $\langle y \rangle = 0$, the spread, or roughness, is

$$
\begin{aligned}
w^2 = \langle y^2 \rangle &= \int_{-\infty}^{+\infty} \mathrm{d}y \, y^2 p(x, y) \\
&= \int_{-\infty}^{+\infty} \mathrm{d}y \, \lambda^H y^2 p(\lambda x, \lambda^H y) \\
&= \int_{-\infty}^{+\infty} \mathrm{d}y \, x^H y^2 p(1, x^{-H} y) \\
&= x^{2H} \int_{-\infty}^{+\infty} \mathrm{d}(x^{-H} y) \, (x^{-H} y)^2 p(1, x^{-H} y) \\
&= x^{2H} \int_{-\infty}^{+\infty} \mathrm{d}y \, y^2 p(1, y).
\end{aligned}
\tag{3.80}
$$

Hence, we have that $w \sim x^H$. We note that the trace or surface is asymptotically flat since $w/x \to 0$ when $x \to \infty$, as $H < 1$.

We note that when calculating w^2 in Eq. (3.80), we assumed that $y(x = 0) = 0$. If we now add the assumption that *the probability distribution only depends on intervals*, that is, $p(x - x_0, y - y(x_0))$, then the roughness w^2 may in fact be written as a height–height correlation function

$$C_2(x) = \langle (y(x + x_0) - y(x_0))^2 \rangle. \tag{3.81}$$

A subtle point here is that the averaging now is also done over x_0. This additional assumption is in fact a defining property of self-affine surfaces, and we find

$$C_2(x) \sim x^{2H}. \tag{3.82}$$

We will in the next section see why this extra assumption is an essential ingredient.

---

[3] However, in some special cases, this may lead to problems, see Hansen et al. (2001).

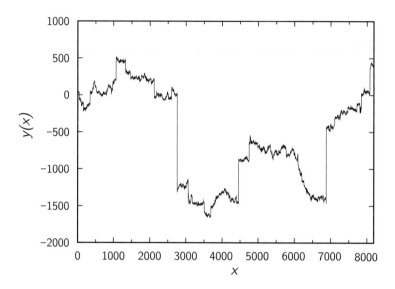

Figure 3.20 A Lévy α-stable trace with an apparent Hurst exponent $H_a = 1/\alpha = 0.75$. Compare this trace to Figure 3.19.

3.6.1 Lévy α-Stable Distributions*

The self-affine trace we show in Figure 3.19 has an apparent Hurst exponent $H = 0.75$. Now consider the trace shown in Figure 3.20. That too obeys the defining scaling relation (3.79) with the same $H_a = H = 0.75$. But, they look completely different!

They are indeed different. Figure 3.20 shows the trace based on a Lévy α-stable distribution. A stable distribution in a variable y is the following. Suppose we generate a sequence of elements $y_1, \ldots, y_N$ from the distribution. We form the sum $Y = \sum_{i=1}^{N} y_i$. If the distribution for Y has the *same functional shape* as the one for y, then the distribution is stable. The Lévy α-stable distributions are of this kind. Here is what they look like:

$$p_\alpha(x, y) = \frac{1}{\sqrt{\pi}} \int_{-\infty}^{+\infty} d\omega \exp\left(i\omega y - x|\omega|^\alpha\right), \qquad (3.83)$$

where $0 < \alpha < 2$. The value $\alpha = 2$ is special. The Fourier transform in Eq. (3.83) is in this case of the Gaussian distribution in Eq. (3.77). For $0 < \alpha < 2$, however, we find that

$$p_\alpha(x, y) \sim |y|^{-\alpha-1} \quad \text{for } |y| \to \infty. \qquad (3.84)$$

We return to the sphere rolling on the inclined, sand blasted plane. We assume the sphere still obeys the equation of motion (3.71) with the noise term still being unbiased and uncorrelated, that is, obeying Eqs (3.72) and (3.73). However,

Figure 3.21 The length of the trace is L. We consider the height–height correlation function over a distance x averaged over all starting positions x_0.

the noise itself is now not Gaussian as earlier but distributed according to a Lévy α-stable distribution. Due to the stability, we then know that Eq. (3.74) is also a Lévy α-stable distribution, with the same α as the noise itself. This is what Figure 3.20 shows. Note the large jumps. They are due to the power-law distribution of the noise.

It is easy to demonstrate from Eq. (3.83) that we have the scaling relation

$$\lambda^{1/\alpha} p(\lambda x, \lambda^{1/\alpha} y) = p(x, y) \quad \text{for } 0 < \alpha < 2. \tag{3.85}$$

Hence, the Lévy α-stable distributions obeys the same scaling relation as *self-affine traces* with an apparent Hurst exponent

$$H_a = \frac{1}{\alpha} \quad \text{for } 1 < \alpha \le 2, \tag{3.86}$$

where the lower limit on α comes from the constraint that a self-affine trace should be asymptotically flat. We furthermore see that the range of apparent Hurst exponents H_a accessible using the Lévy α-stable distributions is $1/2 \le H_a < 1$.

Let us now turn to the roughness w^2 – or rather the height–height correlation function $C_2(x)$ defined in Eq. (3.81). Remember the innocent-looking comment after this this equation? "A subtle point here is that the averaging now is also done over x_0"? We will now focus on this. We assume the trace to have a length L, so that $0 \le x_0 \le L$. We calculate the height–height correlation function over the interval x and average over all possible starting positions x_0, as shown in Figure 3.21. The correlation function will be dominated by the largest jump to be expected over the entire length, L. We invoke order statistics, using Eq. (3.30). For large jumps, η, we have from Eq. (3.84) that the cumulative probability will behave as $P(\eta) = 1 - \eta^{-\alpha}$. With L samplings of the Lévy α-stable distribution, we expect the largest jump to be $\eta_{(L)} = L^{1/\alpha}$. The height–height correlation function will be dominated by this largest jump, and we have

$$C_2(x) \sim \eta_{(L)}^2 = L^{2/\alpha}. \tag{3.87}$$

Using the apparent Hurst exponent $H_a = 1/\alpha$, we may compare this result to $C_2(x)$ for a self-affine trace, Eq. (3.82). The scaling is the same, but in the self-affine case, the scaling parameter is x, whereas in the Lévy α-stable case, it is the system size. This is the difference between the two.

3.6.2 Dance of the Signs*

Comparing Figures 3.19 and 3.20 indicates that in the self-affine case, it is correlations in the direction of the trace – whether it is moving in the positive or negative y directions – that generate the correlations picked up by the height–height correlation function $C_2(x)$. In the Lévy α-stable case, it is the jumps that generate the scaling that leads us to allocate to it an apparent Hurst exponent $H_a = 1/\alpha$. We will now investigate the interplay between uncorrelated jumps and correlated changes in direction.

Let us rewrite Eq. (3.74) as

$$y(x) = \int_0^x dx'\, \eta(x') = \int_0^x dx'\, \text{sign}\, \eta(x')\, |\eta(x')|, \tag{3.88}$$

where we have defined the sign-function

$$\text{sign}\, \eta = \begin{cases} +1 & \text{if } \eta > 0, \\ -1 & \text{if } \eta < 0. \end{cases} \tag{3.89}$$

Let us now introduce a parameter m and define

$$y_m(x) = \int_0^x dx'\, \text{sign}\, \eta(x')\, |\eta(x')|^m. \tag{3.90}$$

When $m = 1$, we have that $y_{m=1}(x) = y(x)$. However, in the limit $m \to 0$, we have

$$y_0(x) = \int_0^x dx'\, \text{sign}\, \eta(x')\, |\eta(x')|^0 = \int_0^x dx'\, \text{sign}\, \eta(x'). \tag{3.91}$$

Hence, we see that in this limit, any self-affinity must come from the correlations in the sequence of signs. When there are no correlations, then $H = 1/2$ as we then are dealing with a random walk.

Now, to some numerical experiments. We show in Figure 3.22 a self-affine trace $y = y(x)$ with Hurst exponent $H = 0.80$. We generate the transformed trace $y = y_0(x)$ and show this in Figure 3.23. It is hard to see a difference between the two traces except the scale along the y axis. When measuring the Hurst exponent of the trace $y = y_0(x)$, H_0, we find it to have the same value as the original Hurst exponent $H = 0.80$. Hence, the self affinity is due to correlations in the sign-function, sign $\eta(x)$.

Now consider Figure 3.24 showing a self-affine trace $y = y(x)$ with Hurst exponent $H = 0.25$. Figure 3.25 shows the corresponding transformed trace $y = y_0(x)$.

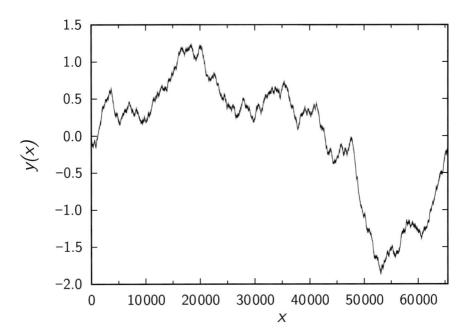

Figure 3.22 A self-affine trace $y = y(x)$ with Hurst exponent $H = 0.80$.

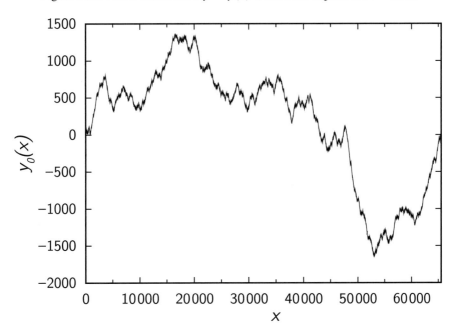

Figure 3.23 The transformed trace $y(x) \rightarrow y_0(x)$ based on the self-affine trace shown in Figure 3.22.

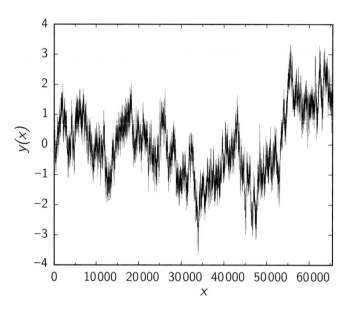

Figure 3.24 A self-affine trace $y = y(x)$ with Hurst exponent $H = 0.25$.

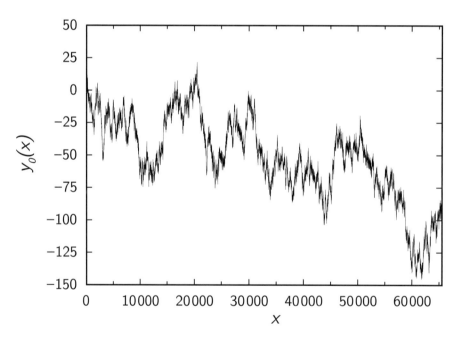

Figure 3.25 The transformed trace $y(x) \rightarrow y_0(x)$ based on the self-affine trace shown in Figure 3.24.

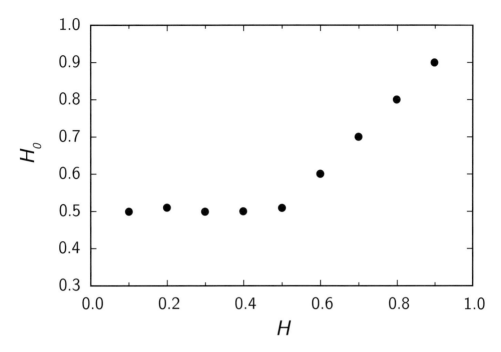

Figure 3.26 The Hurst exponent H_0 of the trace $y = y_0(x)$ vs. the Hurst exponent of the original trace $y = y(x)$. Each data point is averaged over 100 samples of length $L = 65\,536$.

This time the transformed trace looks very different from the original trace. When we measure the Hurst exponent of the transformed trace, we find $H_0 = 0.50$. There are no correlations in the sequence of sign changes! It behaves as a random walk!

Indeed, we can do a systematic study of the Hurst exponent of the transformed trace $y = y_0(x)$, H_0, versus the Hurst exponent H of the original trace $y = y(x)$. We show the result in Figure 3.26. We find

$$H_0 = \max\left(H, \frac{1}{2}\right). \tag{3.92}$$

This is a highly surprising result since it is "common knowledge" that self-affine traces with Hurst exponent larger than $1/2$ – the value for the uncorrelated random walk – are persistent, and traces with Hurst exponent less than $1/2$ are antipersistent. This is not correct. The sign is *uncorrelated* when $H < 1/2$. We have to conclude that the Hurst exponent H, when it is less than $1/2$, comes from correlations between the size of the jumps $\eta(x)$ and the sign. This has as of yet not been fully explored.

We will now discuss some methods to measure the Hurst exponent – or to generate self-affine traces or surfaces given a specified Hurst exponent. Many more details may, however, be found in Schmittbuhl et al. (1995) and Bakke and Hansen (2007).

3.6.3 Box Counting

We start by discussing the oldest method used to measure the Hurst exponent H, box counting, see Section 3.1. Even though it is still occasionally used, especially by the engineering community, we do not recommend it since it tries to pretend that the self-affine trace (or surface) is a fractal – which it is not. It assigns a fractal dimension to the trace, and it is often this quantity we find given in the publications based on this method. However, as we know from the discussion following Eq. (3.80), the trace is asymptotically flat, and hence it should be described as having a fractal dimension of one on large scales. It is only on *sufficiently* small scales that a non-trivial fractal dimension may be assigned as a proxy for the Hurst exponent.

We cover the self-affine trace $y(x)$ by boxes of size $\delta \times \delta$. The trace has length L. The boxes will form a structure of L/δ columns. Each column will have $|\Delta y|/\delta$ boxes stacked on top of each other, $|\Delta y|$ being the largest y value minus the smallest y value occurring in that column. If $|\Delta y| < \delta$, only one box will be placed in that column.

Since the trace is self-affine, we will on average have $|\Delta y| \sim \delta^H$, and hence, the number of boxes in each column on average will be $\max(\delta^H/\delta, 1)$ and the total number of boxes is therefore

$$N(\delta) \sim \frac{L}{\delta} \max\left(\delta^{H-1}, 1\right) \sim \delta^{-D}, \tag{3.93}$$

where the right-hand side has invoked the box-counting dimension defined in Eq. (3.5). Hence, we see that

$$D = \begin{cases} 2 - H & \text{for small } \delta, \\ 1 & \text{for large } \delta. \end{cases} \tag{3.94}$$

Hence, as already remarked, the effective fractal dimension $D = 2 - H$ is not scale invariant. Hence, it is not a good parameter for characterizing self affinity. H is scale invariant. The Hurst exponent H should be measured directly. We will present two such methods next.

3.6.4 Fourier Spectrum Method

This method handles two-dimensional self-affine surfaces as well as one-dimensional self-affine traces. We discuss the method considering one-dimensional traces and leave it to the reader to work out the two-dimensional case. Help for this may, for example, be found in Hansen et al. (2001).

The Fourier transform of a trace $y(x)$ is

$$\tilde{y}(\omega) = \frac{1}{\sqrt{2\pi}} \int_{-\infty}^{+\infty} dx \, e^{-i\omega x} y(x), \tag{3.95}$$

and its inverse is

$$y(x) = \frac{1}{\sqrt{2\pi}} \int_{-\infty}^{+\infty} d\omega \, e^{i\omega x} \tilde{y}(\omega).$$ (3.96)

We assume $y(x)$ to be a self-affine trace. Then, the height–height correlation function $C_2(x)$ will obey the scaling law given by Eq. (3.81). We express $C_2(x)$ in terms of $\tilde{y}(\omega)$, making the assumption that Fourier components with different wavelengths average to zero, that is,

$$\langle \tilde{y}(\omega)\tilde{y}(\omega') \rangle = \delta(\omega + \omega')P(\omega),$$ (3.97)

where we have introduced the Fourier power spectrum

$$P(\omega) = \langle |\tilde{y}(\omega)|^2 \rangle.$$ (3.98)

After some straightforward algebra, we find

$$C_2(x) = \langle (y(x_0 + x) - y(x_0))^2 \rangle$$
$$= \frac{1}{\pi} \int_{-\infty}^{\infty} d\omega \, (1 - \cos(\omega x)) \, P(\omega).$$ (3.99)

The scaling law (3.81) may be expressed as

$$C_2(\lambda x) = \lambda^{2H} C_2(x).$$ (3.100)

For Eq. (3.99) to fulfill this scaling relation, we must have that

$$P(\lambda\omega) = \lambda^{-2H-1} P(\omega).$$ (3.101)

Setting $\lambda = 1/\omega$ in this equation, we find

$$P(\omega) = \frac{P(1)}{\omega^{2H+1}}.$$ (3.102)

And this is it. Measure the Fourier power spectrum P of the signal $y(x)$, plot it against ω using logarithmic axes and you will find a straight line with slope $-(2H + 1)$. Press et al. (2007) explain how to calculate the power spectrum using fast Fourier transforms.

Before you go ahead and calculate the Fourier spectrum, however, it is very important that the signal is *detrended*. This means forcing $\langle y(x) \rangle = 0$. If this is not done, the measured Hurst exponent will be larger than it should be – a straight line $y(x) = ax$ will have an apparent Hurst exponent of one. Detrending is simple. If the signal starts at $x = 0$ and ends at $x = L$, then calculate

$$y(x) \to y(x) - \frac{y(L) - y(0)}{L} x.$$ (3.103)

The Fourier spectrum method may be used to generate artificially self-affine surfaces. If we introduce a noise $\tilde{\eta}(\omega)$ such that

$$\langle \tilde{\eta}(\omega) \rangle = 0, \tag{3.104}$$

and

$$\langle \tilde{\eta}(\omega)\tilde{\eta}(\omega') \rangle = 2D\delta(\omega + \omega'), \tag{3.105}$$

we may then form

$$\tilde{y}(\omega) = \frac{\tilde{\eta}(\omega)}{\omega^{H+1/2}}. \tag{3.106}$$

Fourier transforming $\tilde{y}(\omega)$ to form $y(x)$ according to Eq. (3.96) produces a self-affine trace.

3.6.5 Wavelet Method

The book by Press et al. (2007) is an excellent place to start if one is new to wavelets. Wavelets are localized functions that span function space. That is, any function may be written as a weighted sum over them. By "localized," we mean the wavelet functions are zero outside some range. We start by defining a mother wavelet $\psi = \psi(x)$. From this we generate a set of shifted and scaled versions of it,

$$\psi_{a,b} = \psi\left(\frac{x-b}{a}\right), \tag{3.107}$$

where a is the scale and b is the position of the wavelet. If we now pick a function $y(x)$, we may decompose it in its wavelet components

$$W_{a,b} = \frac{1}{\sqrt{a}} \int_{-\infty}^{+\infty} dx \, y(x)\psi_{a,b}^*(x), \tag{3.108}$$

where $\psi_{a,b}^*$ is the complex conjugate of $\psi_{a,b}$.

Wavelet coefficients are easy to relate to the initial function $y = y(x)$. $W_{a,b}$ contains the "average" of $y(x)$ over a range a at position b.[4]

If we now in a rather imprecise way describe a self-affine signal as statistically obeying the scaling rule $\lambda^{-H} y(\lambda x) = y(x)$, we see from Eq. (3.108) that the wavelet coefficients must obey the scaling rule $W_{\lambda a, \lambda b} = \lambda^{H+1/2} W_{a,b}$. If we now average over the *absolute value* of the wavelet coefficients at a given scale a, $W_a = \langle |W_{a,b}| \rangle$, we find the scaling law

$$W_{\lambda a} = \lambda^{H+1/2} W_a. \tag{3.109}$$

---

[4] Fourier coefficients mix the entire trace, and it is very difficult to intuitively connect the Fourier coefficients to the original trace.

Setting $\lambda = 1/a$ then gives

$$W_a = \frac{W_1}{a^{H+1/2}}. \tag{3.110}$$

Hence, we calculate the wavelet coefficients, average their absolute values scale by scale and plot this against the scale a on a log–log scale. We will find a straight line with slope $-(H + 1/2)$, see Mehrabi et al. (1997); Simonsen et al. (1998).

Our favorite wavelet basis for measuring Hurst exponents is the Daubechies wavelet basis, see Press et al. (2007). Interestingly, when using this basis, it is *not* necessary to detrend the trace $y(x)$ as this is done automatically. A drawback with the method, as we have presented it here, is that it only works for one-dimensional traces. This is unlike the Fourier spectrum method that works equally well for two-dimensional surfaces.

As with the Fourier spectrum method, we may also in this case "invert" and generate self-affine traces with prescribed Hurst exponents, see Simonsen and Hansen (2002). Here is how to do it. For each scale a, generate coefficients $W_{a,b} = \eta_b / a^{H+1/2}$, where η_b is a white noise symmetrically distributed around zero. Then inverse wavelet transform the signal to form $y(x)$, which then will be self-affine with Hurst exponent H.

3.7 Multiaffinity*

Self-affine surfaces may in addition be *multiaffine surfaces*. Does this have anything to do with *multifractals*, which we discussed in Section 3.5? The answer is yes and no. Here is what multiaffinity means. We generalize the height–height correlation function to

$$C_k(x) = \langle (y(x_0 + x) - y(x_0))^k \rangle, \tag{3.111}$$

where k is an even number. If we assume multiaffinity, we find that

$$C_k(x) \sim x^{kH_k}, \tag{3.112}$$

where kH_k is *not* of the form ak, where a is a constant independent of k. As with self-affine traces, the system size L does not appear in the expression.

This scaling resembles strongly that found in multifractals, see Eq. (3.68). This is where the two concepts overlap. Otherwise, they are very different. Nevertheless, Barabási and Vicsek (1991) present a construction that has precisely this behavior – and they call it "multifractality of self-affine fractals," a sign that this is an older paper.

The following construction presented by Mitchell (2005) is interesting, see Figure 3.27. We assume that the boxes of height η form a fractal along the x-axis

Fractals

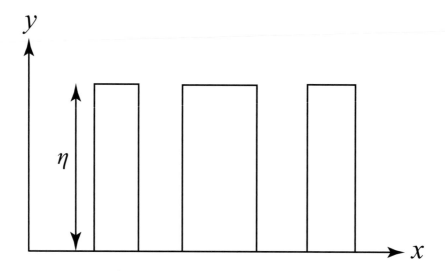

Figure 3.27 This trace, consisting of boxes of height η and lengths distributed as a fractal with fractal dimension D, has a height–height correlation function that behaves as $C_k(x) \sim \epsilon^k x^{1-D}$.

with fractal dimension D. Hence, over a distance x, we expect to find x^{1-D} jumps of size η. This gives us immediately that the height–height correlation function behaves as

$$C_k = \langle (y(x_0 + x) - y(x_0))^k \rangle = \eta^k x^{1-D} \sim x^{kH_k}, \qquad (3.113)$$

where we have used the definition for H_k in Eq. (3.112) on the right-hand side. Hence, we have

$$kH_k = 1 - D, \qquad (3.114)$$

making it multiaffine according to the definition, since $H_k = (1 - D)/k$.

Consider Figure 3.28. It shows a fractal curve described by a fractal dimension D_s. We consider the curve from above. This causes us only to see the top part of the curve. Hence, it will be full of jumps due to overhangs. The distribution of distances between the jumps Δ along the x axis is $p_x(\Delta) \sim \Delta^{-D_s}$, and the distribution of jump sizes η is $p_y(\eta) \sim |\eta|^{-D_s-1}$, see Furuberg et al. (1991) for a derivation. An example of such a trace relevant for flow in porous media is the front produced during very slow drainage in a tilted two-dimensional Hele–Shaw cell. This front behaves essentially as invasion percolation in a gradient, see Hansen et al. (2007).

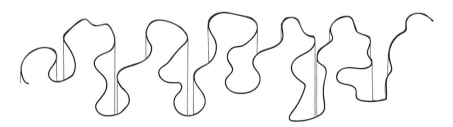

Figure 3.28 Here is a fractal curve characterized by a fractal dimension D_s. We look at it from above and see only the parts visible from that perspective. Due to the appearance of overhangs, the trace will contain jumps. These will be distributed according to a power law.

The structure of such a trace is a self-affine trace that is mixed with power-law distributed jumps. By transforming $y(x) \to y_0(x)$ as in Eq. (3.91), we may find the Hurst exponent of the self-affine part of the trace. We may furthermore analyze the apparent Hurst exponent due to the jumps, but we run into the same problems as with the Lévy α-stable distributions, see Eq. (3.87): the dependence of $C_k(x)$ is with respect to the system size, not x. In the invasion percolation problem was studied by Hansen et al. (2007), where $D_s = 7/4$, they found analytically that H_0, the Hurst exponent of $y_0(x)$, was 2/3 and $C_k(x) \sim L^{k/4+2/3}$, referring to this as "multiaffinity for small k."

To summarize this perhaps frustrating section, we have defined multiaffinity as scaling of the generalization of the height–height correlation function $C_k(x) \sim x^{kH_k}$, where $kH_k \neq ak$ in analogy with multifractality. When there are overhangs causing jumps as shown in Figure 3.28, we find $kH_k = ak + b$ where $b \neq 0$.

Exercises

3.1 The Sierpinsky gasket is shown in Figure 3.6. Suppose now that the edges of the triangular areas are resistors with resistance proportional to their length. Calculate the kth moment of the current distribution in this network given that current is injected and extracted at two of the three external corners of the gasket. Is the current distribution multifractal?

3.2 Generate self-affine traces with Hurst exponents $H = 0.8$ and $H = 0.2$ using the Fourier method. Then use the wavelet method to measure their Hurst exponents.

3.3 Generate self-affine traces with Hurst exponents $H = 0.8$ and $H = 0.2$ using the wavelet method. Then use the Fourier method to measure their Hurst exponents.

3.4 Prepare a number of paper sheets in sizes A3, A4, A5 and A6. Crumble them into little balls, and then flatten them as well as you can. Then place the paper sheets on a flat surface and measure the height of the highest point from the flat surface. Plot on a log–log diagram the value against the diagonal size of the paper sheets. What is the value of the Hurst exponent you find?

4

Percolation

When Broadbent and Hammersley (1957) first made *percolation theory* (Stauffer and Aharony, 1992) a subject of scientific investigation, they made the following introductory statement:

There are many physical phenomena in which a fluid spreads randomly through a medium. Here fluid and medium bear general interpretations: We may be concerned with a solute diffusing through a solvent, electrons migrating over an atomic lattice, molecules penetrating a porous solid or disease infecting a community. Besides the random mechanism, external forces may govern the process, as with water percolating through limestone under gravity. According to the nature of the problem, it may be natural to ascribe the random mechanism either to the fluid or to the medium.

The subject grew, and a vast literature was produced during the 1980s and beyond. *Percolation* is the phenomenon of making the connection through a medium. The medium may be a porous one, and the agent making the connection may be a liquid, but the there are many other possibilities. Perhaps the simplest way of understanding things is to consider a lattice of sites that may be occupied or not and where clusters connected via occupied nearest neighbors exist. The connectivity may also be defined via the bonds on the lattice rather than the sites, in which case it is known as *bond percolation*. The simplest two-dimensional case of site percolation on a square lattice is illustrated in Figure 4.1. In Figure 4.2, off-lattice percolation is illustrated. In this case, a neighbor is a particle within a distance small enough so that they overlap. In Figure 4.3, site percolation on a triangular lattice is illustrated. In what follows we shall focus on percolation on lattices.

Clusters may have any size, but when the occupation probability f exceeds a critical value f_c, a cluster that connects through the whole system will emerge. This cluster is sometimes refered to as the *incipient infinite cluster* (illustrated in Figure 4.4) since it would be infinite on an unlimited lattice. For site percolation on

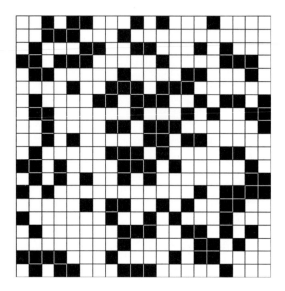

Figure 4.1 Percolation on a square lattice. The white lines are between neighbors. Note that there are clusters of size 1, 2 and 3, as well as larger ones.

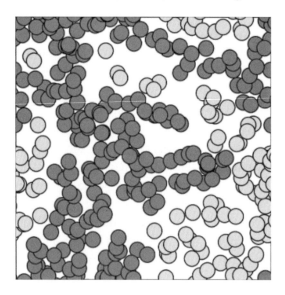

Figure 4.2 Off-lattice percolation where neighboring particles are those that overlap. (A black and white version of this figure will appear in some formats. For the colour version, refer to the plate section.)

the triangular lattice and bond percolation on the square lattice, $f_c = 1/2$, while in general no rational number is known for this value.

In Table 4.1, some different f_c values are tabulated. Note the exact value of $f_c = 1/2$ for site percolation on triangular lattices and bond percolation on square lattices.

Figure 4.3 Percolation on a triangular lattice. Neighbors are those that have common boundaries. The lattice that is defined by all boundaries is a honeycomb lattice. (A black and white version of this figure will appear in some formats. For the colour version, refer to the plate section.)

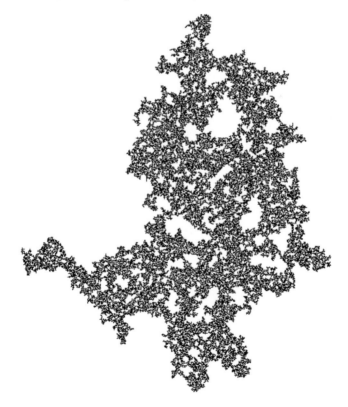

Figure 4.4 An incipient infinite percolation cluster on a square lattice.

Table 4.1. *Percolation thresholds f_c*
for some simple lattices in the cases
of either site or bond percolation.
Only nearest neighbors form clusters.

Lattice	Site	Bond
Honeycomb	0.6962	0.6527
Square	0.5927	0.5
Triangular	0.5	0.3472
Simple cubic	0.3116	0.2488
FCC	0.198	0.119

Percolation is a critical phenomenon where the phase transition is represented by the occurence of the infinite cluster, and f is the analog of temperature. While f_c will depend on the type of lattice, the critical exponents will only depend on dimension. In this chapter, we will introduce the most important of these exponents and the relations among them.

4.1 Statistical Description of Percolation Clusters

In order to arrive at a general description and make the definitions both concrete and precise, it is a good idea to start with the simplest possible model, which would be a lattice of N sites in one dimension with connectivity defined between nearest neighbors only. In one dimension, all sites must be occupied to percolate through the lattice, so

$$f_c = 1. \tag{4.1}$$

With this model, all distributions and exponents are exactly known, and we shall use these results as the basis for generalizations to higher dimensions. Taking the higher-dimensional results to have the same form as in one dimension, all that remains is to measure the critical f_c and the exponents. Although many of the exponent values are known by exact, or at least plausible, arguments in two dimensions as well, we will not derive these results here.

4.1.1 The Cluster Size Distribution

As f approaches f_c from below, the average cluster size starts to diverge. The building block of the theory is the *cluster size distribution*

$$n_s = \frac{\#s\text{-clusters}}{N}, \tag{4.2}$$

where s is the number of sites in the cluster. The fact that Nn_s is the number of s-clusters means that

$$Nn_s s = \#(\text{sites in } s\text{-clusters}) \tag{4.3}$$

and that

$$N \sum_s sn_s = \#(\text{sites in any cluster}) = Nf. \tag{4.4}$$

In one dimension, n_s may be written out explicitly. Note that the number of s-clusters is also the number of sites on the lattice that are at the leftmost position in such clusters. The probability of being such a site is simply

$$n_s = f^s(1 - f)^2, \tag{4.5}$$

that is, the probability of hitting s consecutive sites surrounded by two empty sites, hence the $(1 - f)^2$ term. It is then an easy check to verify that Eq. (4.4) is verified:

$$\begin{aligned}
\sum_s sn_s &= \sum_{s=0}^{\infty} sf^s(1 - f)^2 \\
&= (1 - f)^2 f \frac{\mathrm{d}}{\mathrm{d}f} \sum_{s=0}^{\infty} f^s \\
&= (1 - f)^2 f \frac{\mathrm{d}}{\mathrm{d}f} \frac{1}{1 - f} \\
&= f.
\end{aligned} \tag{4.6}$$

Now, we want to define an average cluster size, since this quantity will define the characteristic length scale of the clusters. For this purpose we note that there are two natural definitions: First, we may treat all clusters equally and pick them with the same probability when we are calculating their average size

$$\langle s \rangle' = \frac{\sum_s sn_s}{\sum_s n_s}. \tag{4.7}$$

This is a perfectly valid choice and would give $\langle s \rangle' = f/(1 - f)$ in one dimension.

However, the definition favored by convention is based on selecting the clusters by picking random *sites on clusters* rather than random clusters, when calculating the average. This procedure increases the probability of picking larger clusters. When a random occupied site is picked, the probability W_s of being on a cluster of size s is s times the probability of hitting the leftmost site in an s-cluster, or upon normalization,

$$W_s = \frac{sn_s}{\sum_s sn_s}. \tag{4.8}$$

Note the difference between W_s and sn_s: the former is normalized to 1 while the latter sums to f. The averaged cluster size now becomes

$$\langle s \rangle = \sum_s W_s s = \frac{\sum_s s^2 n_s}{\sum_s s n_s} = \frac{\sum_s s^2 n_s}{f}, \qquad (4.9)$$

which in one dimension becomes $\langle s \rangle = (1 + f)/(1 - f)$.

A way to picture the difference between $\langle s \rangle$ and $\langle s \rangle'$ is given by the following analogy: Say you are on a very large train station and wish to calculate the average of the train sizes s. A passenger might do this by means of a list over all the trains at the station, sampling randomly from all trains, picking them with equal probability regardless of their size. This procedure may be referred to as *passenger averaging*. A bird, on the other hand, that likes to land on train wagons and picks these with equal probability, will come to a different result if it computes the average of s from such visits. The *bird average* will favor longer trains as they have more wagons, and it is the bird procedure that corresponds to W_s. Note that the bird only lands on the trains, never on the tracks, so that the probability of landing on a train is $\sum_s W_s = 1$.

4.2 Critical Exponents and Fractal Dimension of Clusters

At this point it is natural to introduce our first critical exponent. The so-called Fisher-exponent comes into the picture by generalizing Eq. (4.5) to higher dimensions, that is,

$$n_s = s^{-\tau} e^{-cs}. \qquad (4.10)$$

In Eq. (4.5), there is an s-independent term, $(1 - p)^2$, and the term $f^s = e^{-cs}$, where

$$c = -\ln f. \qquad (4.11)$$

So, in one dimension $\tau = 0$, but in higher dimensions this exponent characterizes how clusters are distributed below the cut-off size $1/c$. This size enters through the exponential function in Eq. (4.10), as above this size it becomes exponentially unlikly to find clusters. Below it, however, there may be cluster sizes ranging over many orders of magnitude. Note that as $f \to f_c$, $1/c \to \infty$ so that in this limit the whole s-distribution is given by the power law determined by τ.

Note that near the percolation threshold, we may Taylor-expand the logarithm in Eq. (4.11), which may then be written

$$c = -\ln(1 - (1 - f)) \approx 1 - f = |f - f_c|. \qquad (4.12)$$

The general form of Eq. (4.11) is then

$$c \propto |f - f_c|^{1/\sigma}, \qquad (4.13)$$

with σ a positive exponent, and in one dimension $\sigma = 1$.

4.2.1 The Correlation Length and the Incipient Infinite Network

The correlation function is, as usual, defined as the conditional probability that if a site is occupied, it is connected via a cluster to another occupied site a distance r away. In one dimension this means that

$$g(r) = f^r = e^{-r/\xi},$$

(4.14)

where we have introduced the correlation length

$$\xi = -\frac{1}{\ln f}$$

(4.15)

in the conventional way, that is, as the length over which $g(r)$ decays. Note that close to $f_c = 1$, this length may be written approximately as

$$\xi = -\frac{1}{\ln f} = -\frac{1}{\ln(1 - (1 - f))} \approx \frac{1}{(1 - f)},$$

(4.16)

which diverges as $f \to f_c$.

The average number of sites connected to an arbitrary occupied site is thus $\sum_r g(r)$, where the sum runs over all sites at distances r. But the average number of sites connected to an arbitrary site is just $\langle s \rangle$, so that

$$\langle s \rangle = \sum_r g(r) = \frac{1}{f} \sum_s s^2 n_s.$$

(4.17)

This may be checked in one dimension, where the sum over $g(r)$ is

$$\sum_r g(r) = \frac{2}{1 - f} \approx \langle s \rangle$$

(4.18)

when $f \approx f_c$. The factor of two in the numerator comes from the fact that r must be summed over both positive and negative values.

The general definition of the correlation length is

$$\boxed{\xi^2 = \frac{\sum_r r^2 g(r)}{\sum_r g(r)}.}$$

(4.19)

However, we may also get ξ^2 as the average separation Δr_s^2 between sites in a cluster of size s: This average is calculated by a sum that must run over all s sites in each cluster. Since a random site has a probability proportional to $s n_s$ of belonging to a cluster of size s, we must multiply Δr_s^2 both by this probability and by s, so that, by normalization,

$$\boxed{\xi^2 = \frac{\sum_s s^2 n_s \Delta r_s^2}{\sum_s s^2 n_s} \propto |f - f_c|^{-2\nu},}$$

(4.20)

where we have also introduced the *correlation length exponent* $\nu > 0$, which describes the expected divergence of ξ near the critical point. The average separation between cluster sites may be written

$$\Delta r_s^2 = \frac{\sum_{ij} |\mathbf{r}_i - \mathbf{r}_j|^2}{s^2}. \tag{4.21}$$

4.2.2 Fractal Dimension

The fractal dimension D characterizes clusters that are below the critical size $c^{-1} \sim |f - f_c|^{-1/\sigma}$. Below this size, the clusters sizes will scale with linear size essentially in the same way as at the critical point. The cluster size may also be considered the cluster mass, so that

$$s \propto \Delta r_s^D, \tag{4.22}$$

where D is the mass fractal dimension.

4.2.3 Exponent Relations

Taking D, σ and τ to be independent and as fundamental descriptions of the percolation cluster geometry, we would like to know how the other exponents characterizing ξ and $\langle s \rangle$ depend on D, σ and τ.

The correlation length exponent ν may be obtained from Eq. (4.20). Replacing the sum by an integral and using Eq. (4.10), the denominator in this equation may be written

$$\sum_s s^2 n_s \approx \int ds \, s^{2-\tau} e^{-cs} = c^{\tau-3} \int dx \, x^{2-\tau} e^{-x}, \tag{4.23}$$

where we have used the substitution $x = cs$. Note that the x-integral is only a constant factor. Now, if we make use of the fractal nature of the clusters $\Delta r_s^2 \sim s^{2/D}$, the numerator may be written

$$\sum_s s^2 n_s \Delta r_s^2 \approx \int ds \, s^{2/D+2-\tau} e^{-cs} \sim c^{\tau-3-2/D}. \tag{4.24}$$

Inserting these two results in Eq. (4.20) and using Eq. (4.13) to eliminate c, we find that $\xi^2 = |f - f_c|^{-2/(D\sigma)}$, or in other words, that

$$\boxed{\nu = \frac{1}{D\sigma}.} \tag{4.25}$$

Finally, we inquire about the behavior of the spanning cluster that arises at $f = f_c$ and the probability P of a given random site belonging to that cluster. At the critical point $f = f_c$, clusters of all sizes exist, while on the other side, $f > f_c$, the clusters that are not part of the infinite cluster will start to decrease in size again, as more and more of them will simply be absorbed in the infinite cluster.

We take Nn_s to be the number of *finite* clusters of size s, so that the total probability of belonging to *any* cluster (which is f) is the sum

$$f = P(f) + \sum_s sn_s. \tag{4.26}$$

In the $N \to \infty$ limit, $P(f) = 0$ for all $f < f_c$, so that

$$\lim_{f \to f_c^-} P(f) = P(f_c) = 0. \tag{4.27}$$

So, right at $f = f_c$, the probability of being on the infinite spanning cluster is zero, so

$$\sum_s n_s(f_c)s = f_c. \tag{4.28}$$

Subtracting Eq. (4.28) from Eq. (4.26) gives

$$P = \sum_s s\,(n_s(f_c) - n_s(f)), \tag{4.29}$$

where we have neglected the linear term in $f - f_c$ as this term is dominated by the sum, as will become clear in the following. By Eq. (4.10) we now get

$$P = \sum_s s^{1-\tau}(1 - e^{-cs}). \tag{4.30}$$

Replacing the sum by an integral again gives

$$P \approx \int ds\, s^{1-\tau}(1 - e^{-cs}) = c^{\tau-2} \int dx\, x^{1-\tau}(1 - e^{-x})$$

$$\propto |f - f_c|^{\frac{\tau-2}{\sigma}}. \tag{4.31}$$

Since P is often written $P \propto |f - f_c|^\beta$, we have just shown the exponent relation

$$\beta = \frac{\tau - 2}{\sigma}. \tag{4.32}$$

The presumably exact results in two dimensions are $\tau = 187/91$ and $\sigma = 36/91$, so that $\beta = 5/36$, which is much smaller than 1. Hence, the sum in Eq. (4.29) indeed dominates the neglected $(f - f_c)$ term.

In Table 4.2, we summarize the exponent results that exist in the literature. Note that Eqs (4.32) and (4.25) are indeed in agreement with these results.

Table 4.2. *Critical exponents for*
percolation. The fractions signify
that the results are assumed to be
exact, whereas the digits signal that
they are measured in simulations.

	d = 2	d = 3
β	5/36	0.41
ν	4/3	0.88
σ	36/91	0.45
τ	187/91	2.18
D	91/48	2.53

4.3 Renormalization Group Derivation of ν on the Triangular Lattice

Right at the critical point, $\xi \to \infty$, and the cluster sizes s are distributed as a power law

$$P(s) \sim s^{-\tau}. \tag{4.33}$$

Power laws have the unique property of scale invariance in the following sense: The probability of finding a cluster of size s is related to the probability of finding a cluster of size λs in the same way, regardless of the value of s. This follows immediately from the fact that s cancels in the ratio $P(\lambda s)/P(s)$. If, say, a cluster of size $s/2$ is four times as likely as a cluster of size s, this statement holds for all cluster sizes, at least up to the cut-off values of the power law. This is really a symptom of a self-similar structure.

There is an elegant way to derive the correlation length exponent that is based on a *renormalization group* procedure. This procedure is based on the self-similarity at the critical point $f = f_c$. At this point we may scale all the clusters such that $s \to s' = \lambda s$ without changing the probability distribution; that is, it remains invariant. This may be used to rescale the lattice onto a coarser lattice with the same cluster distribution. On this new *superlattice*, the percolation cluster will look the same when $f = f_c$.

When f is slightly smaller than f_c, everything will look the same only up to the cut-off length. This (characteristic) cut-off length is simply the correlation length of Eq. (4.20). The fact that ξ diverges as $f \to f_c$ ($\nu > 0$) means that this length is a measure of the largest cluster in the system. The only thing that is needed now is to determine how the connectivity is inherited from the original lattice to the superlattice.

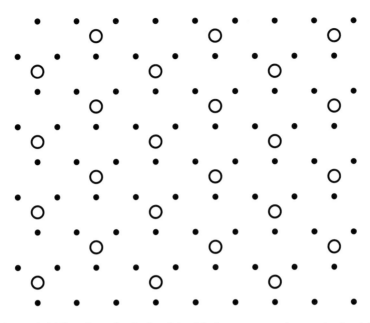

Figure 4.5 The triangular lattice (•) with the superposed super-lattice (○).

Figure 4.5 shows an illustration of a renormalization procedure on the triangular lattice. Each of the open symbols represent three of the closed ones, and they too form a triangular lattice, only with a 30° rotation. A supersite will be taken to be occupied if the original sites, which form triangles around the supersite, have a connected path through the triangle. This happens when two or three of the sites are occupied, and the probability of this happenning is

$$f' = f^3 + 3f^2(1 - f),\qquad(4.34)$$

where the $(1 - f)$ factor is the probability that one site is *not* occupied.

It is clear from Figure 4.5 that the superlattice has a lattice constant of $b' = \sqrt{3}b$. The rescaling will thus leave the cluster size distribution invariant up to a cut-off length that becomes shorter by a factor b'/b. In mathematical terms, this means that

$$\xi' = \frac{b}{b'}\xi.\qquad(4.35)$$

The scale invariance means that the functional form of $\xi(f)$ remains the same, so

$$\frac{(f' - f_c)^{-\nu}}{b} = \frac{(f - f_c)^{-\nu}}{b'},\qquad(4.36)$$

which may be solved for ν to give

$$\frac{1}{\nu} = \frac{\ln\left((f' - f_c)/(f - f_c)\right)}{\ln(b'/b)}.\qquad(4.37)$$

In the limit when $f \to f_c$ this becomes

$$\frac{1}{\nu} = \frac{\ln\left(\mathrm{d}f'/\mathrm{d}f|_{f_c}\right)}{\ln(b'/b)}. \qquad (4.38)$$

From Eq. (4.34), we calculate the derivative at $f = f_c = 1/2$ to be $\mathrm{d}f'/\mathrm{d}f\big|_{f_c} = 3/2$, so we get

$$\nu = \frac{\ln\sqrt{3}}{\ln(3/2)} \approx 1.355, \qquad (4.39)$$

which is in good agreement with the assumed exact value of $\nu = 4/3$ for two-dimensional percolation.

4.4 Invasion Percolation

The *invasion percolation* algorithm is one of the simplest ways of representing a complex displacement process in a porous medium and also a good example of how an algorithmic model may work by keeping only the essential features of a process. The algorithm assumes that in a sufficiently slow drainage process, the pores are invaded sequentially, and at every step, the pore with the smallest *capillary threshold pressure* p_t is invaded first. Generally, such a threshold pressure is determined by the pore width (wider pores have smaller p_t) and wetting properties.

In a simulation, the pores are represented as sites on a lattice (bonds are also possible representations), the simplest case being the square lattice in two dimensions and cubic lattice in three dimensions. Every site on the lattice is assigned a random number p_t that gives the invasion criterion on that site: A step in the simulation consists of localizing and invading the site of the smallest p_t along the perimeter of the cluster. This defines a new perimeter and new sites that are candidates for invasion. If the front closes off a region of fluid, this region must be recognized as a *trapped cluster*. Within such a cluster, fluid incompressibility prevents further growth, so the perimeter of this cluster is excluded from the growing front. In a simulation, a trapped cluster is localized by a search along the front, following each invasion step.

Invasion percolation was first studied by Wilkinson and Willemsen (1983). Simulations by Furuberg et al. (1996) gave insight into the order at which sites grow, a study that was only recently reproduced by experiments and understood theoretically by Moura et al. (2017).

The invasion percolation model should be contrasted to normal percolation, where the clusters are found by the occupation of all available sites with a probability f. First, in invasion percolation there is no analog to the occupation probability f, and there are no finite clusters, only the "infinite," spanning one. Second, invasion percolation is characterized by a sequential process where the invasion starts at one side. Third, trapped clusters may continue growing in normal percolation.

Yet, when the invasion percolation algorithm is modified to exclude trapping, the incipient infinite cluster and the invasion percolation cluster are very similar. Without trapping, Wilkinson and Willemsen found a fractal dimension of $D = 1.89$ for the invasion cluster, a value very close to that of the incipient infinite cluster at f_c. This fact has implications for the growing front in invasion percolation where there are as yet no trapped regions. In particular, one may there expect the correlation length to be given by the exponent as in the previous section.

4.4.1 The Introduction of Gravity in the Models

The introduction of gravity in two-dimensional models is achieved by the modification of threshold pressures p_t^0 to

$$p_t = p_t^0 - \Delta \rho g h. \tag{4.40}$$

Here, p_0^t is the threshold capillary pressure needed to invade a pore inside a horizontal model, $\Delta \rho$ is the difference in fluid mass densities, g is the gravitational acceleration and h the local height of the pore. When $\Delta \rho > 0$, the modification favors the invasion of higher pores over lower ones.

4.5 Directed Percolation*

Percolation theory concerns transport in disordered media. The essential question that percolation theory tries to answer is the following: "Is there an open path between point A and point B in the medium?" where "open path" is to be interpreted as transport along it being possible. There are no restrictions as to how contorted these paths may be, and at the percolation threshold, they will indeed be very contorted. Suppose now that we draw a straight line between A and B as shown in Figure 4.6. We add a stack of lines orthogonal to this line, also shown in the figure. We then draw two paths between A and B, $\mathcal{P}_\alpha$ and $\mathcal{P}_\beta$. Path $\mathcal{P}_\alpha$ has *backbends*, whereas path $\mathcal{P}_\beta$ does not. The backbends are here defined with respect to the stack of lines: a backbend occurs when the path crosses a line in the stack more than once. Path $\mathcal{P}_\beta$ is *directed*, as it has no backbends.

If we allow only *directed paths*, that is, paths without backbends, we are dealing with *directed percolation* (Kinzel, 1983). By removing the non-directed paths, we change the character of the percolation problem in significant ways. Before

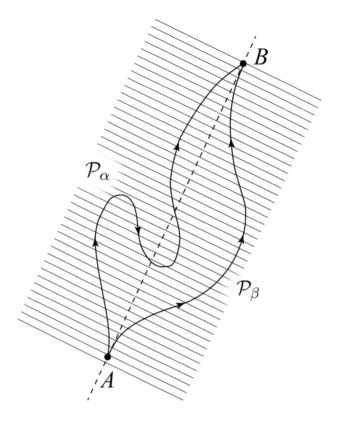

Figure 4.6 Two paths $\mathcal{P}_\alpha$ and $\mathcal{P}_\beta$ between points A and B in a disordered medium. Path $\mathcal{P}_\alpha$ has backbends, that is, the path crosses the same tranversal lines more than once; path $\mathcal{P}_\beta$ is directed as it has no backbends.

describing these changes, let us formulate the directed bond percolation problem on a lattice. To do so, we choose a square lattice, orienting it at 45° compared to the edges, see Figure 4.7. We address a given node by the coordinates (i, j), where $\leq i \, le L_\perp$ and $1 \leq j \leq_\parallel$. We assume periodic boundary conditions in the i-direction. That is, node $(i = L_\perp, j)$ is the left neighbor of node $(1, j)$. Each link between neighboring nodes takes on the value one with probability f or zero with probability $1 - f$. Hence, the link between node $(i - 1, j - 1)$ and node (i, j) has a variable $c(i - 1, j - 1; i, j)$ assigned to it that is either one or zero. We show in Figure 4.8 a possible configuration. The links assigned the value one are shown as solid lines; those assigned the value zero are shown as dotted lines. There is a percolating path of solid links from the bottom edge to the upper edge. The links belonging to this path are highlighted. We have indicated the directedness of the problem by drawing arrows along the links. All arrows point from the lower edge toward the upper edge. There are no backbends.

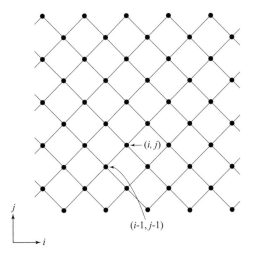

Figure 4.7 A square lattice oriented at 45° with respect to the edges. We have introduced a coordinate system (i, j) as shown.

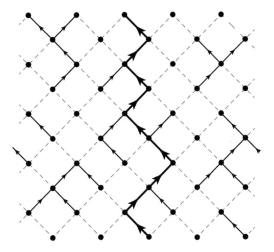

Figure 4.8 A configuration of links in the lattice in Figure 4.7. The links assigned a value $c = 1$ are shown as lines; those with $c = 0$ as dotted lines. We may interpret the $c = 1$ links as "present" and those with $c = 0$ as "absent." Furthermore, we have associated each "present" link with an arrow. This is to remind the reader that this is a directed problem – all links point from the lower toward the upper edge.

Let us now describe an operational way to determine whether there is directed percolation in this lattice. That is, is there at least one open path (i.e., with links having the value $c = 1$) from the lower edge to the upper edge without backbends? To do so, we assign to each node (i, j) a variable $C(i, j)$ taking the value zero or

Table 4.3. *Critical exponents*
for directed percolation.

	d = 2	d = 3
β	0.2765	0.583
$\nu_\parallel$	1.7338	0.733
$\nu_\perp$	1.0969	1.11

one. We initialize $C(i, j) = 0$ for all the nodes, except those along the lower edge, where we set $C(i, 1) = 1$. We now iterate row by row from $j = 2$ to $j = L_\perp$:

$$C(i, j) = \text{OR} \left[\text{AND} \left[C(i - 1, j - 1), c(i - 1, j - 1; i, j) \right], \right.$$
$$\left. \text{AND} \left[C(i + 1, j - 1), c(i + 1, j - 1; i, j) \right] \right]. \tag{4.41}$$

We have used logical operators here: We have that $\text{AND}[1, 1] = 1$, $\text{AND}[1, 0] = \text{AND}[0, 1] = \text{AND}[0, 0] = 0$, $\text{OR}[0, 0] = 0$ and $\text{OR}[1, 0] = \text{OR}[0, 1] = \text{OR}[1, 1] = 1$. If at least one of the variables along the $j = L_\perp$ has the value $C(i, L_\perp) = 1$, then the system percolates.

As in ordinary percolation, there will be a critical probability f_c, below which there will be no directed spanning clusters, whereas above, there will be such clusters in the limit of large systems. For the directed bond percolation problem on the square lattice – the one described here – $f_c = 0.6448$, whereas it is $f_c = 0.7056$ for the directed site percolation problem on the square lattice (Roux and Hansen, 1987). There are also in this case unversal critical exponents that only depend on the dimensionality of the system. We show some values in Table 4.3. There is one large difference between this table and the one for ordinary percolation, Table 4.2: There are *two* correlation length exponents, $\nu_\parallel$ and $\nu_\perp$, rather than the single one defined in Eq. (4.20) for ordinary percolation. Hence, the correlation length in the direction parallel to the "allowed" direction diverges as

$$\xi_\parallel \sim |f - f_c|^{-\nu_\parallel}, \tag{4.42}$$

whereas the correlation length in the direction perpendicular to the "allowed" direction diverges as

$$\xi_\perp \sim |f - f_c|^{-\nu_\perp}. \tag{4.43}$$

There is, on the other hand, only one β exponent and so forth in directed percolation.

4.5.1 Directed Percolation Depinning Model*

Figure 3.18 in the previous chapter shows ink rising into a vertically held piece of paper due to capillary forces – a process known as imbibition. We will in this section

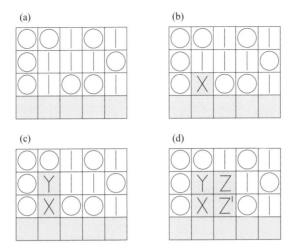

Figure 4.9 The imbibition model introduced by Buldyrev et al. (1992). Each unshaded square is assigned a value one with probability f or zero with probability $1 - f$. The squares assigned zero are blocked with respect to the fluid front entering them, whereas those assigned one may be invaded. The fluid front is assumed to cover the lower shaded row in (a). A square with value one neighboring the fluid front is then chosen (marked X), which then is filled with fluid (b). In (c), another neighbor to the fluid front is chosen among the neighbors having the value one. This new square is marked Y. In (d), the square next to the one marked Y is chosen and invaded by fluid, the square being marked Z. But, in addition to the square Z being set to invaded, so are all squares beneath it, irrespective of whether their value had been set to one or zero; in this case, the square marked Z'. (Adapted from Buldyrev et al. (1992).)

show how directed percolation may offer a quantitative description of this process. This discussion is based on Buldyrev et al. (1992), see also Barabási and Stanley (1995). The imbibition front, that is, where the paper goes from dry to wet, is a self-affine trace characterized by a Hurst exponent $H = 0.63 \pm 0.04$ (Buldyrev et al., 1992). A simple invasion model is capable of reproducing this behavior, see the caption of Figure 4.9 for an explanation of the model. Note in particular panel (c) in the figure. The step that fills in all uninvaded squares below the one that just got invaded guarantees that there will be no overhangs in the fluid front. It turns out that there is a critical $f = f_c$ below which the fluid interface gets pinned, that is, stops evolving. When $f > f_c$, growth goes on indefinitely. We show in Figure 4.10 a configuration where the fluid front has been stopped by a spanning cluster of blocking squares. This is in fact a directed percolation spanning cluster. It is directed since there are no overhangs along the front. This means that there are no backbends in the direction orthogonal to the flow direction.

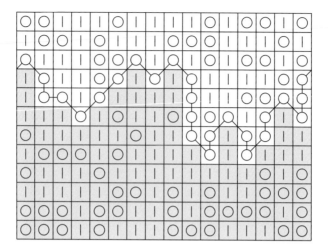

Figure 4.10 A spanning cluster of blocked squares (i.e., having the value zero) is here preventing the fluid front from developing further. (Adapted from Buldyrev et al. (1992).)

Armed with this insight, we may now relate the Hurst exponent describing the roughness of the front to the correlation length exponents in direction percolation. There will then be two correlation lengths, Eqs (4.42) and (4.43). We now assume that $\xi_\perp$ is proportional to the width of the fluid front w, and $\xi_\parallel$ is proportional to the width of the system, L. We then find by eliminating $|f - fc|$ in Eqs (4.42) and (4.43) that

$$w \sim L^{\nu_{perp}/\nu_\parallel}. \tag{4.44}$$

Since we have that $w \sim L^H$, see Eq. (3.80), we find that

$$H = \frac{\nu_{perp}}{\nu_\parallel} \approx 0.63, \tag{4.45}$$

from Table 4.3.

4.6 Transport Properties and Multifractality*

We now return to ordinary percolation. So far we have only considered *geometric* aspects, even when we considered dynamic processes drainage in Section 4.4 and imbition in Section 4.5. We will here address the critical properties associated with transport at the percolation threshold. That is, we are going beyond geometry. The properies we are after concern the flow field of whatever is transported.

The standard theoretical tool to study these transport properties is the *random resistor network*. Consider a lattice where each link is either an electrical conductor with finite conductance σ or an electrical insulator with zero conductance. The probability that the link has conductance σ is f, and the probability that it is an insulator is $1 - f$. There are no spatial correlations between the conductances of the links. It turns out that the conductance G of this network behaves as

$$G \sim \begin{cases} (f - f_c)^t & \text{if } f \geq f_c, \\ 0 & \text{if } f < f_c, \end{cases} \tag{4.46}$$

where t is the *conductivity exponent* (Kirkpatrick, 1971). Grassberger (1999) finds the value $t/\nu = 0.9825 \pm 0.0008$ in two dimensions and Batrouni et al. (1996) find the value $t/\nu = 1.282 \pm 0.005$ in three dimensions; ν is the correlation length exponent, see Eq. (4.20). The conductivity exponent is not known exactly. This is in contrast to the geometrical critical exponents, which are all known exactly in two dimensions (but not in three).

Suppose the random resistor network has size $L \times L$ in two dimensions or $L \times L \times L$ in three dimensions. Now, assume that f is close enough to the critical value f_c so that the percolation correlation length ξ defined in Eq. (4.20) is larger than L. Since L is the size of the entire system, the correlation length cannot be larger than L, so rather than being given by Eq. (4.20), it is L. In other words, L acts as an *effective correlation length*.

This seems to imply that Eq. (4.20) is not valid in a finite system. It is, however, as the percolation threshold in a finite system is different from that in a finite system; the percolation threshold in a finite system is an *effective* percolation threshold, $f_c \to f_{\text{eff}}$. So, when we measure the average percolation threshold in a system of size L, we are measuring f_{eff}. This effective percolation threshold needs to match L as the correlation length. Hence, we have

$$L \sim |f_{\text{eff}} - f_c|^{-\nu}. \tag{4.47}$$

We may rewrite this expression as

$$f_{\text{eff}} = f_c + AL^{-1/\nu}, \tag{4.48}$$

where A is a constant. This provides the standard way to determine f_c: Plot f_{eff} as a function of $L^{-\nu}$. This gives a straight line, from which we determine A and f_c.

If we now combine Eq. (4.47) with the expression for the conductivity, (4.46), we have

$$G \sim (f_{\text{eff}} - f_c)^t \sim L^{-t/\nu}. \tag{4.49}$$

Hence, the conductance G scales with the system size L to the power t/ν. This is an example of *finite size scaling analysis,* which is the most common procedure to determine the value of critical exponents numerically. This is why we gave the values of the conductance exponent t in terms of ν following Eq. (4.46).

Let us set up a voltage difference ΔV across the random resistor network. This leads to an electric current I flowing through it. The total *dissipation* in the network is then $\Delta V\,I$. This dissipation must be equal to the sum of the dissipation in each conducting link, that is,

$$\Delta V\,I = \sum_n \Delta v_n\, i_n, \tag{4.50}$$

where Δv_n is the voltage drop and i_n is the current flowing through link n. Since, $I = G\Delta V$ and $i_n = \sigma \Delta v_n$, we have

$$G = \sigma \sum_n \left(\frac{\Delta v_n}{\Delta V}\right)^2. \tag{4.51}$$

The sum on the right-hand side of this equation is the *second moment* of the relative voltage difference distribution in the network,

$$\left\langle \left(\frac{\Delta v_n}{\Delta V}\right)^2 \right\rangle = \sum_n \left(\frac{\Delta v_n}{\Delta V}\right)^2, \tag{4.52}$$

and we have from Eq. (4.49) that

$$\left\langle \left(\frac{\Delta v_n}{\Delta V}\right)^2 \right\rangle \sim L^{-t/\nu}. \tag{4.53}$$

What about the kth moment? This question was first addressed by de Arcangelis et al. (1985). There is scaling with respect to L,

$$\left\langle \left|\frac{\Delta v_n}{\Delta V}\right|^k \right\rangle \sim L^{y_k}, \tag{4.54}$$

where $y_2 = -t/\nu$. The zeroth moment is simply the number of links in the network that carry a current. This is *not* the number of bonds belonging to the percolating cluster since the percolating cluster also contains the *dead ends*. The fractal dimension of the current-carrying backbone of the percolating cluster is 1.64 ± 0.02 in two dimensions and 1.87 ± 0.03 in three dimensions (Porto et al., 1997).

The exponents y_k are given in Table 4.4 for two and three dimensions. They are *not* of the form $ak + b$ (see Eq. (3.68)): The voltage distribution is *multifractal*.

We discussed the velocity distribution in immiscible two-phase flow in porous media in Section 3.5.4, showing evidence that it too is multifractal. However, there is a huge difference between that example and the random resistor network at the percolation threshold. In the latter example, we need to be at (or close to) the critical probability f_c for there to be multifractality. No such fine tuning has been done in the example in Section 3.5.4. In that case, the multifractality must be a result of self-organization. Clearly, there are many lingering open questions.

Table 4.4. *Multifractal exponent y_k for
the moments of the voltage difference in
the random resistor network at the
percolation threshold. Data from
de Arcangelis et al. (1997), Batrouni
et al. (1996) and Porto et al. (1997).*

	d = 2	d = 3
y_0	1.64	1.87
y_2	−0.9825	−1.282
y_4	−3.12	−3.920
y_6	−5.15	−6.477

Exercises

4.1 The Hoshen–Kopelman Algorithm and the Fisher Exponent:
The Hoshen–Kopelman algorithm represents a systematic way of labeling percolation clusters. It is illustrated in Figure 4.11 and proceeds as follows: Scan
the lattice row by row, going right. If you arrive on an occupied site, then check
to the left and below for occipied sites:

- If no (occupied neighbor-) site, then make a new label.
- If occupied site above, then assign the label above to new site.
- If occupied site to the left, then assign left label to the new site.
- If occupied sites above and to the left, then (a) give new site label above
  and (b) give all sites in the left cluster the label of the cluster above.

Apply your favorite programming language to carry out the following tasks:

4.2 Set up a square lattice with with the critical occupation probability
$f_c = 0.5927$.

4.3 Program the Hoshen–Kopelman algorithm to label clusters.

4.4 Use the labels to compute the distribution n_s, and find the Fisher exponent τ.

4.5 The directed percolation threshold:*
We return here to the directed bond percolation problem on a square lattice.
Consider the lattice shown in Figure 4.7. Assign to each link a random number
$r \in [0,1]$. Choose a probability f. If the random number r is larger than f,
set the value c associated with the link to zero; otherwise set it to one. Then
follow the prescription given in Eq. (4.41) and determine whether the lattice
percolates or not.

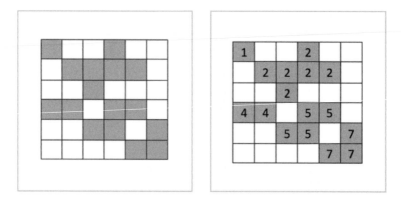

Figure 4.11 The labeling procedure of the Hoshen–Kopelman algorithm.

By repeating this for different values of f, we will be able to determine the directed percolation threshold. But, this is a slow, tedious method with low precision. There is a much better way.

Exchange the logical operator AND by min and the logical operator OR by max in Eq. (4.41), writing it

$$R(i, j) = \min\left[\max\left[R(i - 1, j - 1), r(i - 1, j - 1; i, j)\right],\right.$$
$$\left.\max\left[R(i + 1, j - 1), r(i + 1, j - 1; i, j)\right]\right], \quad (4.55)$$

where $R(i, j)$ takes on any value on the unit interval. The values r associated with the links are the random numbers described above. Initialize by setting all R-values to zero. Then iterate Eq. (4.55) layer by layer from $j = 2$ to $j = L_\parallel$. As a last step, determine

$$R = \min_{i=1}^{L_\perp} R(i, L_\parallel). \quad (4.56)$$

The value R is the directed percolation threshold for that lattice (Roux and Hansen, 1987). Averaging this value over many samples and extrapolating to infinite lattice sizes determines the directed percolation threshold for the square lattice, f_c.

4.6 Explain how Eq. (4.55) works. The logic behind this method to determine the (directed) percolation threshold is very important and we will meet it later on, for example, in connection with the Katz–Thompson model for permability, see Section 7.6.

4.7 Determine the directed *site* percolation threshold for the square lattice. Do this by using finite size scaling analysis. That is, determine the effective directed percolation threshold for different system sizes $L_\perp$, and extrapolate $L_\perp \to \infty$.

PLATES

Figure 3.11 Fast drainage, capillary number = 2.9, and a wetting angle of 150°
for the injected fluid in a circular Hele–Shaw cell filled with fixed glass beads.
(Adapted from Zhao et al. (2019).)

Figure 3.18 Imbibition: A paper towel dipped into a cup of water colored by ink producing a self-affine invasion front.

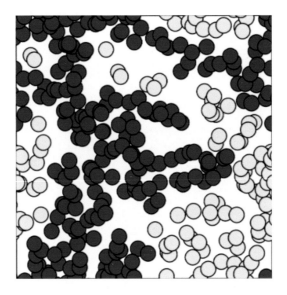

Figure 4.2 Off-lattice percolation where neighboring particles are those that overlap.

Figure 4.3 Percolation on a triangular lattice. Neighbors are those that have common boundaries. The lattice that is defined by all boundaries is a honeycomb lattice.

Figure 7.11 This bottle of French dressing contains a great example of a fluid possessing a yield threshold. The oil contains small pieces of carrot, onion, lettuce, parsley, dill etc. These vegetables do not have the same density. Still, they stay exactly where they are in the fluid. If the oil had been Newtonian, only one of the vegetable species could have floated neutrally, the others would have ended up at the bottom or at the top. What is happening is that the buoyancy is not large enough to create shear forces large enough to exceed the yield threshold. So, the pieces remain trapped where they are in a solid. Untill we shake the bottle. When vegetable pieces stop moving again after the shaking of the bottle, notice that they oscillate a few times before stopping completely. This signals that the fluid is acting as an elastic solid.

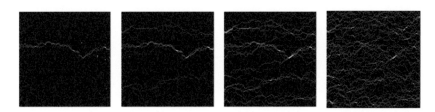

Figure 7.14 Flow regimes when a Bingham fluid moves in a porous medium with increasing pressure difference Δp (from Talon and Bauer (2013)). In the left most picture, the pressure difference is low and there is only a single channel that is open, i.e., the fluid in it moves. The next pictures show increasing pressure difference, from left to right. The increasing pressure difference results in more and more channels opening up.

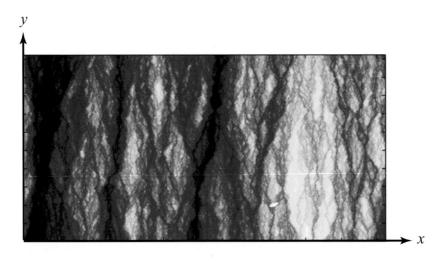

Figure 8.8 The minimal path passing through every link in this square lattice placed at $45°$ compared the x-axis. All paths start somewhere along the lower edge and end somewhere along the upper edge. The distance between the lower and upper edge is L. Darker color means smaller $T_{\mathcal{P}}$. (Adapted from Talon et al. (2013).)

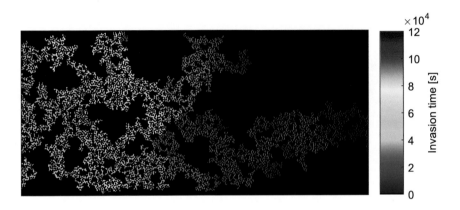

Figure 11.4 Spatio-temporal map of the invasion up to breakthrough (the average flow direction is from left to right). The color map shows the elapsed time for the invasion of a given pore (in seconds). The experiment lasts ∼ 33 h. (Figure courtesy of Marcel Moura.)

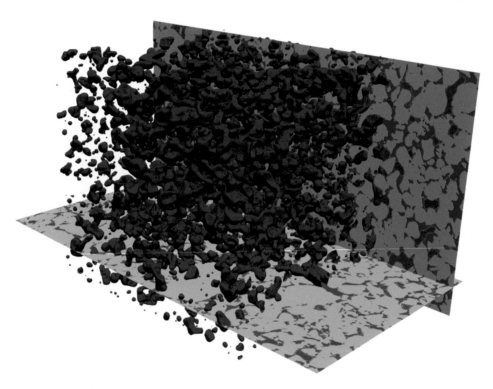

Figure 13.1 Pore scale simulation of two-phase flow, directly on a digital image of a sandstone sample, using a lattice Boltzmann method. The picture shows residual non-wetting fluid phase distribution in the pore space after an imbibition process. The non-wetting fluid clusters occupy the larger pore bodies of the model. (Figure courtesy Thomas Ramstad.)

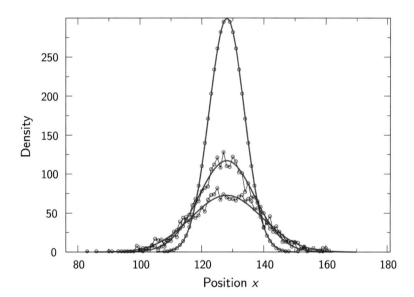

Figure 13.3 Density profiles corresponding to the same random walker algorithm as in Figure 13.2. However, the random walkers are initialized as a strip with translational symmetry in the y-direction and a Gaussian density profile in the x-direction. The density profiles are averaged in the y-direction, which extends 200 sites. The full lines show the solution of the one-dimensional diffusion equation. Note the noise in the measured graphs.

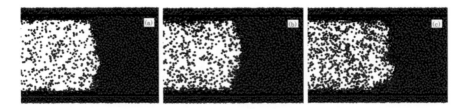

Figure 13.5 The liquid–vapor interface of a droplet confined in a nanochannel. The wall particles are red, the fluid particles blue. The liquid-to-vapor density ratio is about 37 in (a), 18 in (b) and 11 in (c), corresponding to the liquid-vapor coexistence temperatures $0.80\ \epsilon/k_B$, $0.85\ \epsilon/k_B$ and $0.90\ \epsilon/k_B$, respectively. Simulations by Wu et al. (2013).

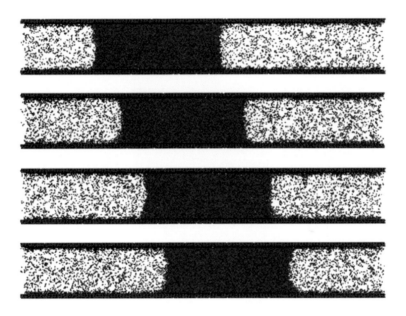

Figure 13.6 Snapshots of the liquid–vapor interface of a moving droplet confined in a nanochannel with a thermal gradient along it. (Wu et al., 2013).

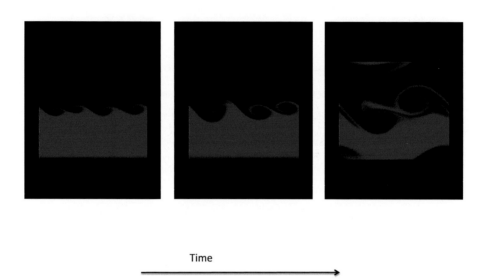

Time

Figure 13.13 Evolution of the Kelvin–Helholtz instability where the initial conditions are set up with velocity shear across the interface. The simulation is based on Eqs (13.86) and (13.90).

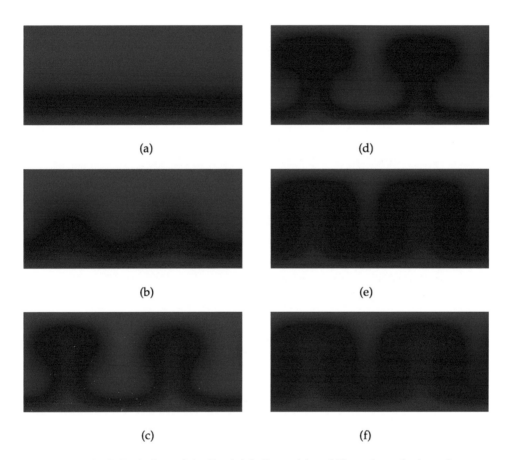

(a)

(d)

(b)

(e)

(c)

(f)

Figure 13.14 Evolution of the Rayleigh–Benard instability where the boundary condition is a larger temperature C at the bottom than at the top, and buoyancy is created by Eq. (13.95). The color represents temperature, red being hot and blue cold. Simulations are on a 128×64 lattice, $\alpha g = 0.0005$, $\nu = 0.25$ and $D = 0.25$. The timesteps are (a) $t = 6000$, (b) $t = 10\,000$, (c) $t = 12\,000$, (d) $t = 13\,000$, (e) $t = 70\,000$ and (f) $t = 154\,000$.

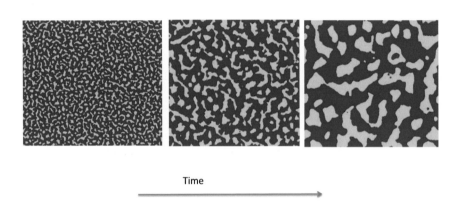

Time

Figure 13.15 Evolution of the phase separation process that results from an initial state with completely mixed phases. The simulations are done on a 128×128 lattice and are based on Eq. (13.98) and the subsequent steps of the imiscible lattice BGK model.

5

Laminar Flow in Channels and Pipes

The Navier–Stokes equations that describe hydrodynamics have a relatively simple form. However, they are non-linear, a fact that gives rise to serious mathematical difficulties that tend to obscure the simple underlying physics. In this chapter we therefore discuss some simple examples of fluid flow that may be understood without the full mathematical apparatus. The simplification arises for very slow flows.

5.1 Laminar Flow in a Channel

One of the simplest hydrodynamic flow problems is the low velocity flow of a fluid through a channel of thickness $2a$ and a width so large that it may be considered to be infinite. In such a case, the flow, as indicated in Figure 5.1, will be one-dimensional in the sense that the flow velocity $\mathbf{u} = (u, 0, 0)$ has a component only in the x-direction, and this component is a function of z only, $\mathbf{u} = \mathbf{u}(z)$.

We expect the velocity of the fluid to have its largest value at the center of the channel and decrease to zero at the walls. This expectation is based on our experience that fluids have a dissipation mechanism – the *viscosity* – that opposes velocity gradients. Qualitatively speaking, the faster fluid ($z > z_0$) drags the slower fluid ($z < z_0$) along (see Figure 5.2) so that there is a force σ per unit area at z_0, acting in the direction of the flow. Similarly, the slow fluid acts on the fast flow with an equal force but in the opposite direction.

It was Newton's assumption[1] that the stress, that is, (force/unit area), is given by

$$\sigma = \mu \frac{\partial u_x}{\partial z} \equiv \sigma_{xz}, \tag{5.1}$$

---

[1] Newton (1867), in section IX of Book II: Circular motion of Fluids, opens with a "Hypothesis," which reads: "... The resistance arising from the want of lubricity in the parts of a fluid is, other things being equal, proportional to the velocity with which the parts of the fluid are separated from one another...." Newton's *defectus lubricitatis* is what we today call internal friction or viscosity. As discussed by Dowson (1979), viscosity is a word with interesting roots: The Greek word ὀιξός for misteltoe is related to viscosity since misteltoe berries contain a sticky substance, *viscin*, that was used as glue on twigs to catch birds. The word came to the modern form though the Latin *viscositias*. Newton demonstrated that the idea that a viscous fluid in space could not explain the observed motion of the planets.

85

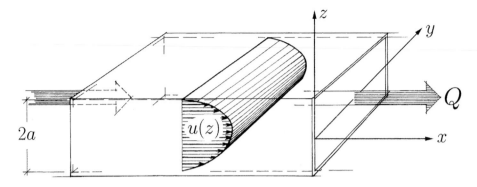

Figure 5.1 Flow in a channel. A flux Q per unit channel width gives rise to a parabolic velocity profile $u(z)$ of the fluid along the x-direction.

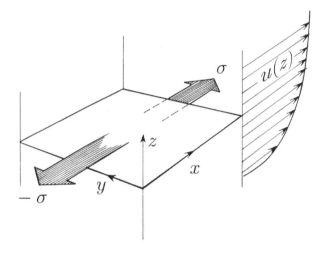

Figure 5.2 The shear forces per unit area, σ, due to a velocity gradient.

where we have introduced the notation σ_{xz} to indicate that the force (per unit area) acts in the x-direction on a surface that has a surface normal in the z-direction. Fluids that do not satisfy this *assumption* are called non-Newtonian fluids. The constant of proportionality, μ, in Eq. (5.1) is called the *viscosity* of the fluid. From the equation we see that μ has the dimension

$$[\mu] = \frac{\text{force/area}}{\text{velocity/distance}} = \text{N s m}^{-2} = \text{Pa s} \qquad (5.2)$$

in the SI system. The corresponding unit in the cgs system is poise $=$ dyn s/cm^2 $=$ 0.1 Pa s. The viscosity of water at 20°C is 0.01 poise $=$ 1 cP $=$ 10^{-3} Pa s. In many situations, the *kinematic viscosity* ν defined by $\nu = \mu/\rho$ with the dimension $[\nu] = \text{m}^2/\text{s}$ is the more useful quantity.

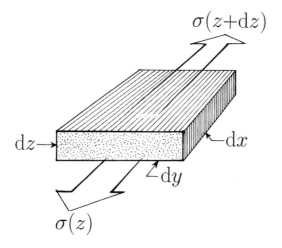

Figure 5.3 The shear forces acting on a volume element.

Now consider the forces on a volume element as indicated in Figure 5.3. The net force in the x-direction due to the viscosity terms is[2]

$$\left[\mu\left(\frac{\partial u}{\partial z}\right)_{z+dz} - \mu\left(\frac{\partial u}{\partial z}\right)_z\right] dx\, dy = \mu\left(\frac{\partial^2 u}{\partial z^2}\right) dx\, dy\, dz. \quad (5.3)$$

The fluid is driven through the channel by a difference in the pressures at the entrance and at the exit of the channel. Supposing that the pressure varies only in the x-direction, the forces due to the pressure gradient act as indicated in Figure 5.4. The net force on the fluid element due to the pressure gradient is then

$$[p(x) - p(x+dx)]\, dy\, dz = -\left(\frac{\partial p}{\partial x}\right) dx\, dy\, dz. \quad (5.4)$$

With stationary flow, each volume element flows without acceleration so that there is no net force on the volume element. Therefore, we must require the following balance equation:

$$\mu\left(\frac{\partial^2 u}{\partial z^2}\right) - \frac{\partial p}{\partial x} = 0. \quad (5.5)$$

Introducing the pressure gradient G by

$$G = -\frac{\partial p}{\partial x} = \frac{p(0) - p(L)}{L} = \frac{\Delta p}{L}, \quad (5.6)$$

---

[2] There should also be forces in the z-direction, otherwise the fluid element will rotate. We will give a more complete derivation in the following sections.

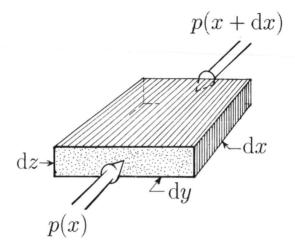

Figure 5.4 The forces on a fluid element due to a pressure gradient.

and assuming that u is independent of x and y, we see that the solution of Eq. (5.5) gives the velocity profile:

$$u(z) = \frac{G}{2\mu}(a^2 - z^2).$$ (5.7)

This is the classical parabolic velocity profile of flow in a channel, and we see that it is consistent with the boundary condition $\mathbf{u}(z = \pm a) = 0$, at the surface of the channel.

The average speed of the flow is easily found by the expression

$$U = \frac{1}{2a} \int_{-a}^{a} dz\, u(z) = \frac{a^2}{3\mu} G = \frac{2}{3} u_{max}.$$ (5.8)

The volume of fluid passing through the channel per unit time and per unit length in the y-direction is then

$$Q = \int_{-a}^{a} dz\, u = \frac{2a^3}{3\mu} \frac{\Delta p}{L},$$ (5.9)

or, in terms of the average velocity,

$$U = \frac{Q}{2a} = \frac{a^2}{3\mu} \frac{\Delta p}{L}.$$ (5.10)

This is known as the Hagen–Poiseuille law for a channel.

The average flow velocity, that is, the velocity $\langle u \rangle$ that would give the same Q but with a constant velocity over the cross section of the pipe, is determined from the relation

$$Q = \pi a^2 \langle u \rangle, \tag{5.20}$$

with the result

$$\langle u \rangle = \frac{G}{8\mu} a^2 = \frac{1}{2} u_{max}, \tag{5.21}$$

where u_{max} is the maximum velocity in the pipe, that is, $u(r = 0)$. Note that this result differs from the result in Eq. (5.8) for channel flow.

For a pipe with elliptical cross section, the flow equations may also be solved exactly. The resulting velocity profile (Landau and Lifshitz, 1987a, p. 53)

$$u = \frac{(p_2 - p_1)}{2\mu L} \frac{a^2 b^2}{a^2 + b^2} \left(1 - \frac{y^2}{a^2} - \frac{z^2}{b^2} \right). \tag{5.22}$$

Here a and b are the minor and major axes of the ellipsoidal cross section of the pipe. The discharge is found to be

$$Q = \frac{\pi (p_2 - p_1)}{4\mu L} \frac{a^3 b^3}{a^2 + b^2}. \tag{5.23}$$

It therefore follows that the average flow velocity U is given in terms of the maximum velocity

$$u_{max} = \frac{(p_2 - p_1)}{2\mu L} \frac{a^2 b^2}{a^2 + b^2} \tag{5.24}$$

by the expression

$$\langle u \rangle = \frac{1}{2} u_{max}. \tag{5.25}$$

It is remarkable that this relation is *independent* of the ellipticity of the channel.

5.3 Lubrication*

In a classical paper, Osborne Reynolds (1886) introduced the central ideas that form the foundation of the theory of lubrication (Batchelor, 1967). Reynolds first gave a qualitative discussion, with very clear illustrations, before he proceeded with the analytical calculations using Stokes' equations. Reynolds' illustrations require detailed explanations that are best given in his own words. The problem he addressed

Figure 5.6 Two nearly parallel surfaces separated by a viscous fluid. AB and CD are perpendicular sections of great extent compared with the distance h separating the surfaces, both surfaces being of unlimited length in the direction perpendicular to the paper (Reynolds, 1886).

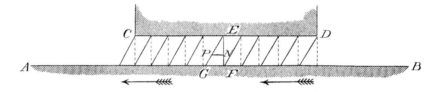

Figure 5.7 Case 1: Two parallel surfaces separated by a viscous fluid. AB moves to the left with a velocity U (Reynolds, 1886).

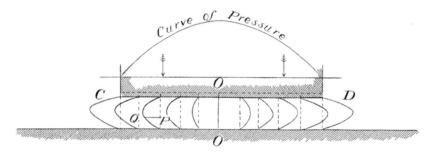

Figure 5.8 Case 2: Two parallel surfaces separated by a viscous fluid approaching with no tangential motion (Reynolds, 1886).

is defined in Figure 5.6. There are several special cases. First let the plates be parallel and move AB to the left with a velocity U, see Figure 5.7. Reynolds explained:

"...there will be a tangential resistance

$$F = \mu \frac{U}{h}$$

per unit area, and the tangential velocity will vary uniformly from U at AB to zero at CD. Thus if FG be taken to represent U, then PN will represent the velocity in the fluid at P. The slope of the line EG therefore may be taken to represent the force F, and the direction of the tangential force on either side is the same as if EG were in tension. The sloping lines therefore represent the condition of motion and stress throughout the film."

He then went on to discuss another special case, see Figure 5.8.

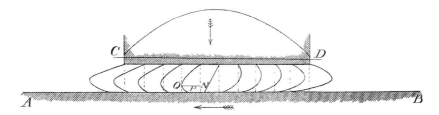

Figure 5.9 Case 3: Two parallel surfaces separated by a viscous fluid approaching with tangential motion (Reynolds, 1886).

"The fluid has to be squeezed out between the surfaces, and since there is no motion at the surface, the horizontal velocity outward will be greatest half-way between the surfaces, nothing at O, the middle of CD, and greatest at the ends. If at a certain stage of motion (shown by the dotted line, Figure 5.8) the space between AB and CD be divided in 10 equal parts by vertical lines (Figure 5.8, dashed vertical lines), and these lines supposed to move with the fluid, they will shortly after assume the positions of curved lines (Figure 5.8), in which the areas included between each pair of curved lines is the same as in the dotted figure. In this case, as in Case 1, the distance QP will represent the motion at any point P, and the slope of lines will represent the tangential forces in the fluid as if the lines were stretched elastic strings. It is immediately seen that the fluid will be pulled toward the middle of CD by the viscosity as though by the stretched elastic lines, and hence that the pressure will be greatest at O and fall of towards the ends C and D, and would be approximately represented by the curve at the top of the diagram".

The next case Reynolds discussed was a *superposition* of the two previous cases, that is, parallel surfaces approaching with tangential motion. Since Stokes' equations, which are the hydrodynamic equations for lubrication, are linear equations, the solutions satisfy the superposition principle. This third case is illustrated in Figure 5.9.

"The lines representing the motion in Cases 1 and 2 may be superimposed by adding the distances PQ in Figure 5.8 to the distances in Figure 5.7.
    The result will be as shown in Figure 5.9, in which the lines represent in the same way as before the stresses in the fluid where the surfaces are approaching with tangential motion.
    In this case the distribution of pressure over CD is nearly the same as in Case 2, and the mean tangential forces will be the same as in Case 1. The distribution of the friction over CD will, however, be different. This is shown by the inclination of the curves at the points where they meet the surface. Thus on CD the slope is greater on the left and less on the right, which shows that the friction will be greater on the left and less on the right than in Case 1. On AB the slope is greater on the right and less on the left, as is also the friction."

Now we come to the case of practical interest, that is, the situation illustrated in Figure 5.10 "Surfaces inclined with tangential movement only."

"AB is in motion as in Case 1, and CD is inclined as in Figure 5.10. The effect will be nearly the same as in the compound movement (Case 3, Figure 5.9).

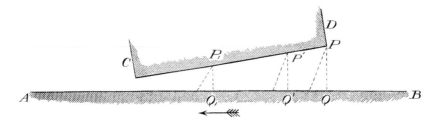

Figure 5.10 Case 4: Two inclined surfaces separated by a viscous fluid with tangential motion (Reynolds, 1886).

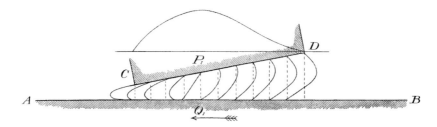

Figure 5.11 Case 4: Two inclined surfaces separated by a viscous fluid with tangential motion. The top curve represents the pressure, whereas the flow lines represent fluid velocities and stresses (Reynolds, 1886).

For if corresponding to the uniform movement U of AB the velocity of the fluid varied uniformly from the surface AB to CD, then the quantity carried across any section PQ would be $PQ \times \frac{1}{2}U$, and consequently would be proportional to PQ; but the quantities carried across all sections must be the same, as the surfaces do not change their relative distances; therefore there must be a general outflow from any vertical section PQ,P'Q' given by $\frac{1}{2}U(PQ-P'Q')$. This outflow will take place to the right and left of the section of greatest pressure. Let this be P_1Q_1, then the flow past any other section PQ, $\frac{1}{2}U(PQ - P_1Q_1)$, to the right or left according as PQ is to the right or left of P_1Q_1. Hence at this section the motion will be one of uniform variation, and to the right and left lines showing the motion will be nearly as in Figure 5.9. This is shown in Figure 5.11. The pressure of the intervening film of fluid would cause a force tending to separate the surfaces. The mean line or resultant of this force would act through some point O. This point O does not necessarily coincide with P_1, the point of maximum pressure. For equilibrium of the surface AB, O will be in the line of the resultant external force urging the surfaces together, otherwise the surface ACD would change its inclination.

The resultant pressure must also be equal to the resultant external force perpendicular to AB (neglecting the obliquity of CD). If the surfaces were free to approach the pressure would adjust itself to the load, for nearer the surfaces the greater would be the friction and consequent pressure for the same velocity, so that the surfaces would approach until pressure balanced the load.

As the distance between the surfaces diminished O would change its position, and therefore, to prevent an alteration of inclination, the surface CD must be constrained so that it could not turn around.

It is to be noticed that continuous lubrication between the plane surfaces can only take place with continuous motion in one direction, which is the direction of continuous incli-nation of the surfaces.

With reciprocating motion, in order that there may be continuous lubrication, the surfaces must be other than plane."

Reynolds then discussed the lubrication of revolving cylindrical surfaces as used for railroad wheel bearing. This will, however, take us too far from the subject of simple hydrodynamics for the purposes of flow in porous media. Suffice to say that Reynolds' paper defined a new scientific field of great practical importance. We have quoted Reynolds' qualitative discussion to show that complicated hydrody-namics may sometimes be understood without the complicated equations.

5.3.1 Lubrication: An Analysis*

As an example of the analytical treatment of lubrication, let us consider the situa-tion illustrated in Figure 5.7. We follow Batchelor's (1967) treatment, and take the positive x-axis to be to the left in Figure 5.11 with the z-axis vertical.

Let the bottom plate move with velocity U in the negative x-direction relative to the fixed top plate, that is, to the left in Figure 5.11. The flow again satisfies Eq. (5.5), and the solution is

$$u = \frac{z-h}{h} U - \frac{z(h-z)}{2\mu} \left(\frac{\partial p}{\partial x} \right).$$

(5.26)

Here, h is the plate separation. This solution satisfies the *no-slip boundary condition* which sets $\mathbf{u} = 0$ at the stationary wall and $\mathbf{u} = (U, 0, 0)$ at the moving wall AB.

The volume flow rate through this channel is (per unit channel length)

$$Q = \int_0^h dz\, u(z) = -\frac{h}{2} U - \frac{h^3}{12\mu} \left(\frac{\partial p}{\partial x} \right).$$

(5.27)

Now consider the situation illustrated in Figure 5.11, where the fixed block has a slight angle $\alpha = (h_D - h_C)/L$ relative to the moving bottom plate, that is, $dh = \alpha dx$. Here h_C and h_D are the heights of the block CD at C and D, respectively, and $h_D > h_C$. The pressure, $p(x)$, is now a non-linear function of position, which must vanish at the entry and at the exit of the narrow lubricating channel, that is, $p(0) = p(L) = 0$. At some position x_0 in the channel, the pressure has an extreme value $p_0 = p(x_0)$. For incompressible flow, the volume flow rate of the fluid is constant, so it follows from Eq. (5.27) that

$$Q_0 = -\frac{h_0}{2} U - \frac{h_0^3}{12\mu} \frac{\partial p}{\partial x} \bigg|_{x_0} = -\frac{h}{2} U - \frac{h^3}{12\mu} \frac{\partial p}{\partial x},$$

(5.28)

where $h_0 = h(x_0)$. It further follows that

$$h(x)^3 \left.\frac{\partial p}{\partial x}\right|_x = 6\mu U \left(h(x_0) - h(x)\right), \tag{5.29}$$

since $(\partial p/\partial x)|_{x_0} = 0$, by definition. We may integrate this equation from h to h_D using that $dp/dx = \alpha \, dp/dh$ to obtain the result

$$p(h_D) - p(h) = \frac{6\mu U}{\alpha} \int_h^{h_D} dh \, \frac{h(x_0) - h(x)}{h^3}, \tag{5.30}$$

or

$$p(h) = \frac{6\mu U}{\alpha} \left(\frac{1}{h_D} - \frac{1}{h} - \frac{h_0}{2h_D^2} + \frac{h_0}{2h^2}\right). \tag{5.31}$$

Here $h_0 = h(x_0)$ may be determined by noting that $p(h_C) = 0$, which when inserted into Eq. (5.31) gives

$$h_0 = \frac{2h_C h_D}{(h_C + h_D)}. \tag{5.32}$$

We may now calculate the vertical force on the block per unit width of the block:

$$F_\perp = \int_0^L dx \, p(x) = \frac{1}{\alpha} \int_{h_D}^{h_C} dh \, p(h). \tag{5.33}$$

By partial integration, this integral may be transformed into

$$F_\perp = -\frac{1}{\alpha} \int_{h_D}^{h_C} dh \, h \frac{dp}{dh}. \tag{5.34}$$

Using Eq. (5.29), the vertical force, on both surfaces, is found to be

$$F_\perp = \frac{6\mu U}{\alpha^2} \left[\ln\left(\frac{h_D}{h_C}\right) - 2\left(\frac{h_D - h_C}{h_D + h_C}\right)\right]. \tag{5.35}$$

This expression is seen to give a force lifting the fixed block provided that $h_D > h_C$, that is, $\alpha > 0$. Let $\Delta h = h_D - h_C$ and $\langle h \rangle = (h_D + h_C)/2$. Then for $(\Delta h/\langle h \rangle) \ll 1$, the logarithm in Eq. (5.35) may be expanded to the third power in $\Delta h/\langle h \rangle$ (the lower powers cancel) and the lifting force expressed by

$$F_\perp \simeq \frac{\mu U}{2\alpha^2} \left(\frac{\Delta h}{\langle h \rangle}\right)^3, \quad \text{for } \frac{|\Delta h|}{\langle h \rangle} \ll 1. \tag{5.36}$$

Thus if a heavy load is supported by a lubricating film on a moving surface, the film must be thin. Increasing the viscosity of the lubricant and increasing the velocity also increases the load carrying capability, see Batchelor (1967), p. 220.

Often the load and α may be given, and $\langle h \rangle$ is left to adjust. In this case, setting $\Delta h = \alpha L$ gives

$$F_\perp \simeq \frac{\mu U}{2} \alpha \left(\frac{L}{\langle h \rangle} \right)^3, \tag{5.37}$$

which goes to zero as $\alpha \to 0$, as one would expect. Solving Eq. (5.37) for $\langle h \rangle$ gives

$$\langle g \rangle = \left(\frac{\mu U \alpha}{2 F_\perp} \right)^{1/3} L. \tag{5.38}$$

The drag force on the block may be calculated using Eq. (5.29) in Eq. (5.26) to evaluate the velocity gradient $(\partial u / \partial z)|_h$:

$$F_\parallel = \int_0^L dx\, \mu \left(\frac{\partial u}{\partial z} \right)_{z=h} = \frac{2\mu U}{\alpha} \left[-\ln \left(\frac{h_D}{h_C} \right) + 3 \left(\frac{h_D - h_C}{h_D + h_C} \right) \right]. \tag{5.39}$$

In the limit of small $\Delta h / \langle h \rangle$, we find that:

$$F_\parallel \simeq \frac{\mu U}{\alpha} \frac{\Delta h}{\langle h \rangle}. \tag{5.40}$$

Setting $\Delta h = \alpha L$ again, $F_\parallel$ reduces to the α-independent expression

$$F_\parallel \simeq \frac{\mu U L}{\langle h \rangle}, \tag{5.41}$$

which is just the viscous stress tensor times L.

The ratio of the drag force to the lifting force is an effective coefficient of friction:

$$\mu_{\text{friction}} = \frac{F_\parallel}{F_\perp} \simeq 2\alpha \left(\frac{\langle h \rangle}{\Delta h} \right)^2, \tag{5.42}$$

in the $\Delta h / \langle h \rangle \ll 1$ limit. With $\Delta h = \alpha L$ this becomes

$$\mu_{\text{friction}} \simeq \frac{2}{\alpha} \left(\frac{\langle h \rangle}{L} \right)^2, \tag{5.43}$$

or, if we use Eq. (5.38) to write things in terms of the load, we find

$$\mu_{\text{friction}} = \frac{2}{\alpha^{1/3}} \left(\frac{\mu U}{2 F_\perp} \right)^{2/3}, \tag{5.44}$$

which has the somewhat surprising feature that the friction coefficient decreases as the load is increased. The total force required to push a given load increases very weakly with the load, that is, $F_\parallel \sim F_\perp^{1/3}$. Also, the friction decreases with α through the $1/3$ exponent, which means that the dependence is quite weak. Since $F_\parallel \sim U^{2/3}$, the friction may in principle be made arbitrarily small by reducing U. However, in practice the surfaces are not atomically smooth but have a self-affine fractal surface roughness, and contact between the surfaces on the atomic level will eventually arise as $\langle h \rangle$ is decreased.

Exercises

5.1 **Sliding Bearrings and Train Wheels:**
Older *train wheel designs* used *sliding bearings* instead of the more modern ball bearings. In these, the sliding parts were only separated by a layer of viscous lubricant oil, which could be described by Eq. (5.38). Taking the load in such a bearing to be the weight of 20 tons per wheel distributed over a width W of 10 cm, $F_\perp = 2 \cdot 10^6$ N/m. Taking also the viscosity of the lubricant fluid to be that of standard 15 W40 motor oil at 60°C, $\mu \simeq 4 \cdot 10^{-2}$ Pa s, $\alpha = 0.1$, $L = 10$ cm and the sliding velocity $U = 1$ m/s. Calculate $\langle h \rangle$

5.2 In order to avoid the wear that comes from solid–solid contacts, what requirements would you put on the smoothness of the shaft?

5.3 Estimate the power needed to overcome the friction at $U = 1$ m/s and $U = 4$ m/s, and discuss if such designs are realistic.

5.4 **Sphere Settling in a Viscous Fluid on a Plane:**
We want to know if, or how, a small smooth sphere of radius $a = 1$ mm and density $\rho + \Delta\rho$, where ρ is the mass density of the fluid, will displace the fluid and settle on a plane. Assume that as the smallest separation h_0 between the sphere and the plane decreases, the fluid forces will only come from a small circular section of radius $x \ll a$ inside which the vertical distance $h(x) = h_0 + \Delta h$ to the sphere is less than $2h_0$. Show by geometry that $\Delta h = x^2/(2a)$.

5.5 Explain why the vertical displacement flux caused by the moving sphere, $-\pi x^2 \dot{h}$, must equal the horizontal flux, $2\pi x h \langle u \rangle$. Use this along with the analog of Eq. (5.28) (with $U = 0$) to obtain an equation for $p(x)$.

5.6 Integrate p to obtain the total vertical force F on the sphere from the fluid, assuming that $p(2h_0) = 0$ (Hint: It is useful to change the integration variable from x to h).

5.7 Set $F = \Delta\rho V g$, where $V = (4\pi/3)a^3$, find the equation that determines $\dot{h}_0$ and solve for $h_0(t)$.

5.8 Determine on physical grounds if the sphere will ever touch the plane. Set $\mu = 10^{-3}$ Pa s, $\Delta\rho = 1000$ kg/m^3 and $a = 1$ μm, and calculate the characteristic settling time.

6

The Hydrodynamic Equations

The basic equations of hydrodynamics rely on a continuum description of the systems of interest. A central role is played by conserved variables. For instance, in single component fluids, the conservation of matter directly leads to the *continuity equation*. The conservation of energy as described by the first law of thermodynamics gives another equation. In more complicated systems, the rotational motion of molecules and the conservation of angular momentum give more equations. For fluid dynamics, however, the most important equation is *Newton's equation* relating the acceleration to the forces, that is, the conservation of momentum. In the following sections, we shall derive the simplest form of the hydrodynamic equations.

6.1 The Continuity Equation

Consider a volume element *fixed* in space that has volume V and a surface S as indicated in Figure 6.1. On every surface element of area dS we have an *outward* normal $\mathbf{n}$, so that we may write the surface element as $d\mathbf{S} = \mathbf{n} \, dS$. The flow of the fluid is described by giving the velocity $\mathbf{u} = \mathbf{u}(\mathbf{r}, t)$, as a function of position $\mathbf{r}$ and time t. This is the *Euler description*. A different description is obtained if one specifies the position, $\mathbf{r}$, of given fluid elements as a function of time by specifying $\mathbf{r}(\mathbf{r}_0, t)$ given that it started at $\mathbf{r}_0$ at time t_0. This is the *Lagrange description*, and it is directly related to the Euler description.

With the fluid density given by ρ, the *mass flux* is $\mathbf{j} = \rho \mathbf{u}$, which has the dimension $[j] = \mathrm{kg/m^2 s}$. The local mass flux out of the volume V is $\mathbf{j} \cdot d\mathbf{S}$. The rate of change of mass in the volume element is given by the difference between the mass flow into and out of the volume:

$$\underset{\substack{\text{Rate of increase of mass} \\ \text{of fluid within } V}}{\frac{d}{dt} \int_V \rho \, dV} \quad = \quad \underset{\substack{\text{Rate of addition of mass} \\ \text{across the surface } S}}{- \int_S \rho \mathbf{u} \cdot d\mathbf{S}.} \tag{6.1}$$

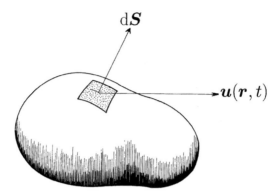

Figure 6.1 A fixed volume element in a velocity field **u**.

Here the fluid density is $\rho = \rho(\mathbf{r}, t)$. The volume is V and S its surface. d**S** is an infinitesimal surface element with a direction given by the surface normal. In the limit of an infinitesimal volume element, this equation becomes

$$\frac{\partial \rho}{\partial t} = -\lim_{V \to 0} \frac{1}{V} \int_S \rho \mathbf{u} \cdot \mathrm{d}\mathbf{S} = -\nabla \cdot (\rho \mathbf{u}), \tag{6.2}$$

where we have used the definition of the divergence operator $\nabla \cdot .$[1] Thus the conservation of matter, Eq. (6.1), is equivalent to the differential equation

$$\boxed{\frac{\partial \rho}{\partial t} + (\nabla \cdot \rho \mathbf{u}) = 0,} \tag{6.3}$$

which is the *continuity equation*.

6.2 Conservation of Momentum

The local volume flow of a fluid across the surface element dS is $(\mathbf{n} \cdot \mathbf{u})\mathrm{d}S$. The *momentum per unit volume* of the fluid is $\rho \mathbf{u}$, and therefore we conclude that $(\mathbf{n} \cdot \mathbf{u})\rho \mathbf{u}\,\mathrm{d}S$ is the rate at which momentum is carried across the surface dS because of the fluid flow. This expression may be rearranged as $[\mathbf{n} \cdot \rho \mathbf{u}\mathbf{u}]\mathrm{d}S$, and the quantity $\rho \mathbf{u}\mathbf{u}$ is the *momentum flux*, that is, the momentum per unit area per unit time convected with the fluid flow. The momentum flux is a tensor $\{\rho \mathbf{u}\mathbf{u}\}$ – a fact we emphasize with the notation[2] that for an arbitrary vector **x** gives $\{\mathbf{u}\mathbf{n}\} \cdot \mathbf{x} = \mathbf{u}(\mathbf{n} \cdot \mathbf{x})$. Here we have used the notation $\{\cdots\}$ to denote second-order tensors.

---

[1] *Gauss' divergence theorem* states that if V is a closed region with surface S, then $\int_V (\nabla \cdot \mathbf{u})\mathrm{d}V = \int_S (\mathbf{n} \cdot \mathbf{u})\mathrm{d}S$, where **n** is the outwardly directed unit normal vector.

[2] In this section, we follow Bird et al. (1987) and use ( ) to indicate scalars such as $(\mathbf{u} \cdot \mathbf{n})$. A vector result of a multiplication is indicated by square brackets [ ] as in $[\mathbf{n} \cdot \mathbf{u}\mathbf{u}]$, where **uu** is the *dyadic product*, that is, a second order *tensor*.

In addition to the momentum transport by flow, there will be momentum transfer due to molecular motion and interactions that will give rise to forces that change the momentum in the volume V.

In the simplest case of an *ideal fluid*, that is, for fluid flow where we may neglect *energy dissipation*, the force on the fluid element is simply given by the *pressure*, and the conservation of *momentum* corresponds to *Newton's law*:

$$\frac{d}{dt}\int_V \rho \mathbf{u}\, dV = -\int_S \{\rho \mathbf{u}\mathbf{u}\}\cdot d\mathbf{S} - \int_S p\, d\mathbf{S} + \int_V \rho \mathbf{g}\, dV.$$

| Rate of increase of momentum of fluid in V | Rate of addition of momentum across S by convection | Force on fluid in V by pressure | Force on fluid in V by gravity | (6.4) |

Here $\mathbf{g}$ is the acceleration due to gravity. The Gauss divergence theorem applies also to scalars and tensors, so that we have $\int_S p\, d\mathbf{S} = \int_V [\nabla p]dV$ and $\int_S \{\rho \mathbf{u}\mathbf{u}\}\cdot d\mathbf{S} = \int_V [\nabla\cdot \{\rho \mathbf{u}\mathbf{u}\}]dV$ so; thus we obtain an equation involving only volume integrals. In the limit $V \to 0$, we find *Euler's equation* for an ideal fluid (i.e. without viscosity):

$$\boxed{\frac{\partial}{\partial t}[\rho \mathbf{u}] + [\nabla\cdot\{\rho\mathbf{u}\mathbf{u}\}] = -[\nabla p] + [\rho\mathbf{g}].}$$

(6.5)

This equation was obtained by Euler (1755). Let us rewrite this equation into another form making use of the equation of continuity. First we remember the relation for the divergence of the dyadic **ab**:

$$[\nabla\cdot \mathbf{ab}] = [\mathbf{a}\cdot\nabla\mathbf{b}] + \mathbf{b}\,(\nabla\cdot\mathbf{a}),$$

(6.6)

which is easily derived by writing the expression in Cartesian coordinates defined by the *unit vectors* $\mathbf{e}_i$ with $i = 1, 2, 3$:

$$\mathbf{ab} = \sum_{jk} \mathbf{e}_j a_j b_k \mathbf{e}_k,$$

(6.7)

$$\nabla = \sum_i \frac{\partial}{\partial x_i}\mathbf{e}_i,$$

(6.8)

$$\begin{aligned}
[\nabla\cdot\mathbf{ab}] &= \sum_{ijk}\frac{\partial}{\partial x_i}\mathbf{e}_i\cdot\mathbf{e}_j\, a_j b_k\,\mathbf{e}_k \\
&= \sum_{ik}\left(\frac{\partial}{\partial x_i}a_i b_k\right)\mathbf{e}_k \\
&= \sum_i\left(\frac{\partial a_i}{\partial x_i}\right)\mathbf{b} + \sum_{ik} a_i\left(\frac{\partial b_k}{\partial x_i}\right)\mathbf{e}_k \\
&= \mathbf{b}\,(\nabla\cdot\mathbf{a}) + [\mathbf{a}\cdot\nabla\mathbf{b}].
\end{aligned}$$

(6.9)

Using this result, we find that

$$[\nabla \cdot \{\rho \mathbf{uu}\}] = [\rho \mathbf{u} \cdot \nabla \mathbf{u}] + \mathbf{u} (\nabla \cdot [\rho \mathbf{u}]) . \qquad (6.10)$$

Here we may now use the equation of continuity (6.3) and find that

$$[\nabla \cdot \{\rho \mathbf{uu}\}] = [\rho \mathbf{u} \cdot \nabla \mathbf{u}] - \mathbf{u} \frac{\partial \rho}{\partial t} . \qquad (6.11)$$

This result may now be reinserted into Eq. (6.5) to give the Euler equation in another form:

$$\boxed{\rho \frac{\partial}{\partial t} \mathbf{u} + \rho (\mathbf{u} \cdot \nabla) \mathbf{u} = -[\nabla p] + \rho \mathbf{g}, \qquad \text{(Euler)}} \qquad (6.12)$$

which is valid for ideal fluids, that is, fluid flows for which dissipation may be neglected, and thus the viscosity is ignored.

6.2.1 The Substantive Derivative

In order to discuss the effect of acceleration on fluid elements (the Lagrange point of view), we must remember that when we specify the velocity field $\mathbf{u}$, the particles in a fluid element may accelerate because the velocity field changes, that is, $\partial \mathbf{u} / \partial t \neq 0$, and also because the particles arrive at positions with a different velocity, that is, $\partial \mathbf{u} / \partial \mathbf{r} \neq 0$. For simplicity consider the variation of some scalar quantity such as the temperature T or density ρ. The change in temperature of a fluid element is given by

$$\delta T = \frac{\partial T}{\partial t} \delta t + \frac{\partial T}{\partial x} \delta x + \frac{\partial T}{\partial y} \delta y + \frac{\partial T}{\partial z} \delta z, \qquad (6.13)$$

and consequently we have

$$\frac{\delta T}{\delta t} = \frac{\partial T}{\partial t} + \frac{\partial T}{\partial x} \frac{\delta x}{\delta t} + \frac{\partial T}{\partial y} \frac{\delta y}{\delta t} + \frac{\partial T}{\partial z} \frac{\delta z}{\delta t} . \qquad (6.14)$$

Now we choose $(\partial x / \partial t, \partial y / \partial t, \partial z / \partial t)$ to be given by the velocity $\mathbf{u}$, that is, we evaluate the change in T at a position in space that is precisely the position occupied by the fluid element at a later time due to the *convection* prescribed by the velocity field $\mathbf{u}$ (the Lagrangian point of view). Since we get an equation of the same form as Eq. (6.14) for each and every component of a tensor, the general effect is described by introducing the *substantive derivative*

$$\frac{\mathrm{D}}{\mathrm{D}t} = \frac{\partial}{\partial t} + \mathbf{u} \cdot \nabla . \qquad (6.15)$$

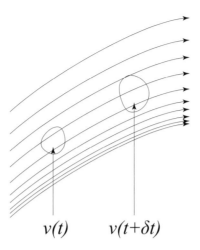

$v(t)$ $v(t+\delta t)$

Figure 6.2 A small fluid element of volume v is convected and deformed by the velocity field $\mathbf{u}(\mathbf{r}, t)$. Ignoring diffusion, the mass inside $v(t)$ does not change.

Newton's equation specifies that the rate of change of the momentum equals the force. The mass in a small (Lagrangian) volume v is $m = \rho v$, and the momentum of this mass is $m\mathbf{u}(\mathbf{r}, t)$. This mass is convected with the fluid; therefore, the rate of change of the momentum is $m\, D\mathbf{u}(\mathbf{r}, t)/Dt$. The choice of the small Lagrangian volume is arbitrary; therefore, the rate of change of momentum per unit volume is $\rho\, D\mathbf{u}(\mathbf{r}, t)/Dt$, which is given by

$$\boxed{\rho\frac{D\mathbf{u}}{Dt} = \rho\frac{\partial\mathbf{u}}{\partial t} + \rho(\mathbf{u}\cdot\nabla)\mathbf{u},} \tag{6.16}$$

which is also the left-hand side of Euler's equation (6.12). The force density is

$$\mathbf{f} = -\nabla p + \mathbf{f}_{\mu} + \mathbf{F}, \tag{6.17}$$

where $\mathbf{f}_{\mu}$ is the viscous force, to be discussed later. We conclude that Newton's equation of motion leads to:

$$\boxed{\rho\frac{\partial\mathbf{u}}{\partial t} + \rho(\mathbf{u}\cdot\nabla)\mathbf{u} = -[\nabla p] + \mathbf{f}_{\mu} + \rho\mathbf{g}.} \tag{6.18}$$

Here, the first term on the right-hand side is simply the effect of a pressure gradient discussed earlier. The last term represents forces due to fields such as gravity – $\mathbf{F} = m\mathbf{g}$. The middle term is due to viscosity. This term requires special attention, and we discuss it in the following section.

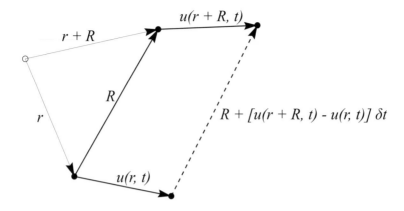

Figure 6.3 The velocity field at neighboring positions.

6.2.2 The Rate of Strain Tensor

Consider two points in a fluid separated by a vector distance **R**, as indicated in Figure 6.3. We expect the dissipation and the viscosity in a fluid to be related to the change in *relative* position of the fluid particles since there is no dissipation for a fluid at rest, in uniform translation or in rotation. The change in relative position can be expressed by the substantive derivative of the squared distance between neighboring fluid particles:

$$\frac{\mathrm{D}}{\mathrm{D}t}(R)^2 = \sum_{i=1}^{3} 2R_i \frac{\mathrm{D}R_i}{\mathrm{D}t}. \tag{6.19}$$

After a time increment δt, the particles originally at **r** and at **r** + **R** are separated by

$$\mathbf{R}(t + \delta t) = [(\mathbf{r} + \mathbf{R}(t)) + \delta t \, \mathbf{u}(\mathbf{r} + \mathbf{R}(t), t)] - [\mathbf{r} + \delta t \, u(\mathbf{r}, t)]. \tag{6.20}$$

Therefore, the change in **R** is

$$\delta \mathbf{R} = [\mathbf{u}(\mathbf{r} + \mathbf{R}, t) - \mathbf{u}(\mathbf{r}, t)] \, \delta t. \tag{6.21}$$

But we may write

$$u_i(\mathbf{r} + \mathbf{R}) = u_i + \sum_{j=1}^{3} \frac{\partial u_i}{\partial x_j} R_j, \tag{6.22}$$

and find

$$\frac{\mathrm{D}R_i}{\mathrm{D}t} = \frac{\delta R_i}{\delta t} = \sum_{j=1}^{3} \frac{\partial u_i}{\partial x_j} R_j. \tag{6.23}$$

We therefore conclude that

$$\frac{1}{2} \frac{1}{R^2} \frac{DR^2}{Dt} = \sum_{i,j=1}^{3} \left(\frac{R_i}{R}\right) \mathbf{e}_{ij} \left(\frac{R_j}{R}\right), \tag{6.24}$$

where we have introduced the *rate of strain tensor* **e** given by

$$e_{ij} = \frac{1}{2} \left(\frac{\partial u_i}{\partial x_j} + \frac{\partial u_j}{\partial x_i}\right). \tag{6.25}$$

This tensor has the same form as the strain tensor well known from the theory of elasticity. However, for hydrodynamics the local displacements are replaced by the local fluid velocities. Eq. (6.24) is a quadratic form, and we may take the limit $R \rightarrow 0$ without difficulty.

If the velocity is a constant, as is the case when the fluid in its container moves rigidly, then $\mathbf{e} = 0$, and no dissipation is expected. If instead, the container and the fluid rotate as a rigid body, we again expect no dissipation. In this case with a given angular velocity $\mathbf{\Omega}$, the velocity field is given by

$$\mathbf{u} = \mathbf{\Omega} \times \mathbf{r} = \Omega(-y, x, 0), \tag{6.26}$$

and we again find that all the components of $\mathbf{e} = 0$. We therefore conclude that we expect the dissipation to depend only on $\mathbf{e}$.

6.3 The Stress Tensor

In order to describe forces on fluid elements, we introduce the *stress tensor* in the same way as in the theory of elasticity. As illustrated in Figure 6.4, the stress tensor σ specifies the force σ_{ij} per unit area in the i-direction on a surface with a normal in the positive j-direction. Since tensors are invariant under rotations of the coordinate system, we can easily obtain the forces on a surface element oriented in any direction. The fundamental assumption is that for *Newtonian fluids*, the stress tensor σ is proportional to the rate of strain tensor $\mathbf{e}$:

$$\sigma_{ij} = \sum_{kl} \Lambda_{ijkl} e_{kl}. \tag{6.27}$$

This is Newton's assumption, and it has the same form for elastic solids, only that here the fourth order viscosity tensor $\mathbf{\Lambda}$ replaces the fourth order elastic constant tensor $\mathbf{C}$ in the theory of elasticity. The tensor $\mathbf{\Lambda}$ is characteristic of an *isotropic* fluid, and therefore it must have the symmetry of a fluid *at rest*. This condition greatly reduces the number of independent components of the $3^4 = 81$ component $\mathbf{\Lambda}$ tensor. There are only three independent coefficients:

$$\Lambda_{ijkl} = \lambda \delta_{ij} \delta_{kl} + \xi \delta_{ik} \delta_{jl} + \chi \delta_{il} \delta_{jk}. \tag{6.28}$$

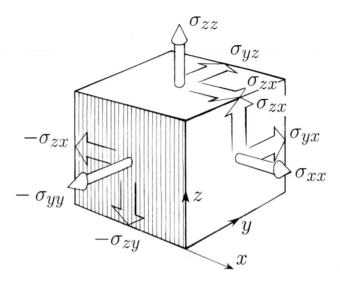

Figure 6.4 The various components of the stress tensor.

Therefore, we may write the stress tensor in the form

$$\sigma_{ij} = \lambda \delta_{ij}(e_{11} + e_{22} + e_{33}) + (\xi + \chi)e_{ij} = \lambda \delta_{ij} \nabla \cdot \mathbf{u} + 2\mu e_{ij}. \qquad (6.29)$$

Here we have expressed the viscosity μ in terms of the phenomenological constants ξ and χ by

$$\mu = \frac{1}{2}(\xi + \chi) \qquad (6.30)$$

and used that $\nabla \cdot \mathbf{u} = e_{11} + e_{22} + e_{33}$. The λ term is the *bulk viscosity*, and it is related to dissipation effects that arise when the density changes – as for instance in longitudinal sound waves.

Now consider the forces in the x-direction on a small volume element due to viscous stresses, as indicated in Figure 6.5. The pressure terms are written explicitly in Eq. (6.18), but could have been included in the stress terms. We find by summing the contributions of the viscous stresses on all the six sides of the volume element that the total viscous force in the x-direction is

$$\begin{aligned}
F_{\mu,x} &= (\sigma_{xx}(x + dx) - \sigma_{xx}(x))\, dy\, dz \\
&\quad + \left(\sigma_{xy}(y + dy) - \sigma_{xy}(y)\right) dx\, dz \qquad (6.31) \\
&\quad + (\sigma_{xz}(z + dz) - \sigma_{xz}(z))\, dx\, dy.
\end{aligned}$$

The force per unit volume is therefore

$$f_{\mu,x} = \frac{\partial \sigma_{xx}}{\partial x} + \frac{\partial \sigma_{xy}}{\partial y} + \frac{\partial \sigma_{xz}}{\partial z}. \qquad (6.32)$$

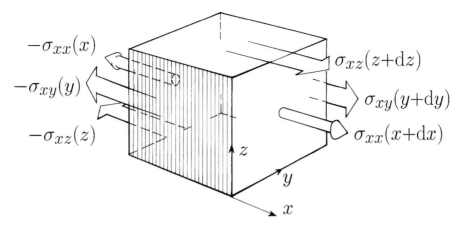

Figure 6.5 The stresses on a volume element.

By using Eqs (6.25) and (6.29), the above argument gives

$$
\begin{aligned}
f_{\mu,i} &= \sum_{j=1}^{3} \frac{\partial}{\partial x_j} \sigma_{ij} \\
&= \lambda \frac{\partial}{\partial x_i} \nabla \cdot \mathbf{u} + \mu \sum_{j=1}^{3} \frac{\partial}{\partial x_j} \left(\frac{\partial u_i}{\partial x_j} + \frac{\partial u_j}{\partial x_i} \right) \qquad (6.33) \\
&= (\lambda + \mu) \frac{\partial}{\partial x_j} \nabla \cdot \mathbf{u} + \mu \nabla^2 u_i
\end{aligned}
$$

for an arbitrary direction. The *viscous force* per unit fluid volume $\mathbf{f}_\mu$ is therefore given by

$$
\mathbf{f}_\mu = \mu \nabla^2 \mathbf{u} + (\mu + \lambda) \nabla (\nabla \cdot \mathbf{u}). \qquad (6.34)
$$

If the fluid is incompressible, it follows that $\partial \rho / \partial t = 0$, and then the continuity equation gives $\nabla \cdot \mathbf{u} = 0$. In this case, the viscous force is given by the simple expression

$$
\mathbf{f}_\mu = \mu \nabla^2 \mathbf{u}. \qquad (6.35)
$$

6.4 The Navier–Stokes Equation

The rate of change of momentum density, Eq. (6.16) is equal to the force per unit volume, Eq. (6.17), according to Newton's equation of motion. With the viscous force given by Eq. (6.34), the fluid equations of motion are

$$
\rho \frac{D\mathbf{u}}{Dt} = \rho \frac{\partial \mathbf{u}}{\partial t} + \rho \mathbf{u} \cdot \nabla \mathbf{u} = -\nabla p + \mu \nabla^2 \mathbf{u} + (\mu + \lambda) \nabla (\nabla \cdot \mathbf{u}) + \mathbf{F}, \qquad (6.36)
$$

with the continuity equation given by

$$\frac{\partial \rho}{\partial t} + \nabla \cdot \rho \mathbf{u} = 0. \tag{6.37}$$

These equations simplify to the *Navier–Stokes equations* if the fluid is incompressible (Navier, 1822; Stokes, 1845). In this case $\partial \rho / \partial t = 0$, and the continuity equation requires the velocity field to be divergence free:

$$\nabla \cdot \mathbf{u} = 0, \quad \text{for incompressible fluids.} \tag{6.38}$$

In this case, the dynamical equations simplify to

$$\rho \frac{\partial \mathbf{u}}{\partial t} + \rho (\mathbf{u} \cdot \nabla) \mathbf{u} = -\nabla p + \mu \nabla^2 \mathbf{u} + \mathbf{F}. \quad \text{(Navier–Stokes)} \tag{6.39}$$

This equation is a second order partial differential equation, and it has the major complication of being *non-linear* by the $(\mathbf{u} \cdot \nabla)\mathbf{u}$ term. In many problems, the external force $\mathbf{F} = 0$, and the fluid motion is caused by imposed pressure differences or the relative movement of boundaries. Generally fluid flow takes place in a field of gravity, which can not be ignored. If the density is uniform, the gravitational force is balanced by a vertical pressure gradient that is present whether or not the fluid is moving and that does not interact with any flow. This hydrostatic balance can be subtracted out of the dynamical equation and the problem reduced to one without body forces. This assumes, of course, that the fluid region is supported at the bottom and that there are no free surfaces at which waves can develop.

The Eqs (6.38) and (6.39) effectively are one scalar and one vector equation, which constitute a total of four equations for the determination of one scalar quantity, the pressure p, and one vector quantity, the velocity field $\mathbf{u}$. The number of unknowns is thus correctly matched by the number of equations. The pressure must necessarily be an intrinsic variable in any hydrodynamic problem for there to be enough variables to satisfy the basic laws of mechanics. In addition, the conservation of energy leads to one more equation that we do not discuss here.

6.5 Boundary Conditions

In order to obtain solutions of the Navier–Stokes equations, we need sufficient boundary conditions to specify the problem at hand. Boundary conditions cannot be derived from the continuum equations – they are microscopic in origin. The first condition that typically enters a fluid dynamics problem is the condition that the fluid does not penetrate the container wall (assuming the wall to be non-porous). Therefore, the fluid velocity normal to the wall moving with velocity $\mathbf{U}$ must satisfy

$$\mathbf{u} \cdot \mathbf{n} = \mathbf{U} \cdot \mathbf{n}, \tag{6.40}$$

where **n** is the unit normal to the surface. Often the frame of reference is chosen such that the wall is at rest; thus the boundary condition becomes

$$\mathbf{u} \cdot \mathbf{n} = 0. \tag{6.41}$$

Another condition often used is the *no-slip boundary condition*, stating that there is no relative tangential velocity between a rigid wall and the fluid immediately next to it. Formally this condition is written

$$\mathbf{u} \times \mathbf{n} = \mathbf{U} \times \mathbf{n}, \tag{6.42}$$

or with the walls at rest,

$$\mathbf{u} \times \mathbf{n} = 0. \tag{6.43}$$

This condition has no fundamental justification, and there are many instances where it does not apply. The question as to the correctness of the no-slip boundary condition must be decided by experiment for the case at hand. In most of our discussion we will treat problems for which the no-slip boundary condition is known to apply.

The stress on a surface is determined from the stress tensor σ, given in Eq. (6.29). Thus if we have a wall at $y = 0$ and the velocity has components $\mathbf{u} = (u, v, w)$, the force (per unit area) at the wall is

$$\mathbf{F}_{\text{wall}} = \mu \left\{ \left(\frac{\partial u}{\partial y} + \frac{\partial v}{\partial x} \right), 2 \left(\frac{\partial v}{\partial y} \right), \left(\frac{\partial w}{\partial y} + \frac{\partial v}{\partial z} \right) \right\}_{y=0}, \tag{6.44}$$

but the continuity equation gives the relation

$$\frac{\partial v}{\partial y} = -\frac{\partial u}{\partial x} - \frac{\partial w}{\partial z}, \tag{6.45}$$

so that if the no-slip condition, $u = v = w = 0$ at $y = 0$, applies,

$$\mathbf{F}_{\text{wall}} = \mu \left\{ \left(\frac{\partial u}{\partial y} \right), 0, \left(\frac{\partial w}{\partial y} \right) \right\}_{y=0}. \tag{6.46}$$

Thus the viscous force acts tangentially along the surface. The only normal force is due to the pressure.

Other commonly used boundary conditions refer to free surfaces where there is no stress from the other side. If there is no wall, so that the surface is free, the boundary condition must state that no forces can act across the free surface, and we must require the pressure to be continuous, $p_1 = p_2$, and the stress to vanish at the surface, $\sigma_{y=0} = 0$. In this case the hydrostatic pressure may vary because of waves on the surface, and therefore the hydrostatic pressure cannot be separated in the dynamical equations.

Often boundary conditions are applied at infinity. For instance, for flow around a single sphere, one will specify the uniform velocity at large distances:

$$\mathbf{u} \to \mathbf{u}_0 \text{ as } \mathbf{r} \to \infty. \tag{6.47}$$

6.6 The Reynolds Number and Scaling

The *Navier–Stokes equation* (6.39) may be written

$$\frac{\partial \mathbf{u}}{\partial t} + (\mathbf{u} \cdot \nabla)\mathbf{u} = -\frac{1}{\rho}\nabla p + \nu\nabla^2\mathbf{u} \tag{6.48}$$

for the situation that there are no body forces and the flow is given by the boundary conditions. Here we have introduced the *kinematic viscosity* ν defined by

$$\nu = \frac{\mu}{\rho}. \tag{6.49}$$

The kinematic viscosity has dimension $[\nu] = \text{m}^2/\text{s}$, which is the same as for a diffusion constant. The kinematic viscosity is $0.15\,\text{cm}^2/\text{s}$ for air and $0.01\,\text{cm}^2/\text{s}$ for water. The kinematic viscosity is the only *molecular* property of the fluid that enters. If the density is constant, as it is for incompressible fluids, then the only difference between the flow of water and the flow of air is the kinematic viscosity.

For a sphere of diameter ℓ moving through an infinite fluid with velocity U, the only variables that can affect the flow field are ν, ℓ and U. Therefore, we expect different flow regimes to be characterized by different values of some dimensionless combination of the relevant physical parameters. The conventional choice of such a dimensionless combination is the *Reynolds number* Re, given by

$$\boxed{\text{Re} = \frac{U\ell}{\nu}. \qquad \text{(Reynolds number)}} \tag{6.50}$$

The dimensions of the quantities used here are $[U] = \text{m/s}$, $[\ell] = \text{m}$ and $[\nu] = \text{m}^2/\text{s}$; thus it is easy to see that the Reynolds number is indeed a dimensionless quantity. For flow in a pipe of diameter d, the length scale in Re is replaced by d. In general ℓ is taken to be a characteristic length set by the boundary conditions of the problem. Of course, we could have taken the pipe length L instead of the diameter d to set the length scale. We could also have chosen ℓ to be the radius of curvature R if the pipe is bent. Different choices only change the numerical value of Re and do not alter the fact that we expect different types of flow depending on whether Re is very small or very large. In Figure 6.6 we see that at Re ~ 2300, there is a marked increase in the flow through a pipe for a given pressure gradient. This increase in

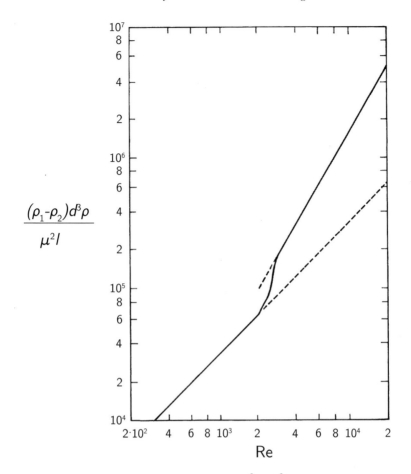

$$\frac{(p_1-p_2)d^3\rho}{\mu^2 l}$$

Figure 6.6 The normalized pressure drop $\Delta p \rho d^3/L\mu^2$ over a pipe of diameter d and length L as a function of Re. The transition from low Reynolds number laminar flow to turbulent flow at high Reynolds numbers near Re $\simeq 3000$ is clearly visible. (After Tritton (1977).).

flow means that the velocity profile changes drastically and becomes the turbulent flow characteristic of high Reynolds numbers.[3]

Another example of the effect of changing the Reynolds number is seen in Figure 6.7 where the flow past a cylinder is visualized by letting a dye flow out at the downstream point of the cylinder. For low flow velocities, Re < 30, the fluid

---

[3] The dimensionless pressure is $\hat{p} = p/\rho U^2$ (see Eq. (6.54)), and the Reynolds number is Re $= \rho U d/\mu$, where d is the pipe diameter. By multiplying the dimensionless pressure by Re^2, we find another form of the dimensionless pressure: $\tilde{p} = \hat{p}$Re$^2 = p\rho d^2/\mu^2$. The dimensionless pressure difference between the ends of the pipe may be written $(\tilde{p}_1 - \tilde{p}_2)d/L = (p_1 - p_2)\rho d^3/\mu^2 L$. Here the dimensionless factor d/L was used so that the dimensionless pressure difference will be independent of pipe length. This form of the dimensionless pressure difference consist only of easily measurable quantities and is the form used in Figure 6.6.

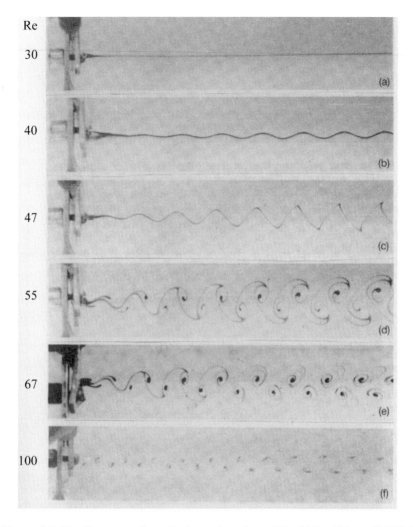

Figure 6.7 The flow around a cylinder as function of Re. (After Tritton (1977).)

streams nicely past the cylinder. As the flow rate is increased, the flow becomes unstable, and beautiful periodic oscillations in the flow field emerge. Finally, at high Reynolds numbers, the flow pattern becomes chaotic or turbulent.

Flows of biological interest are often at very *low* Reynolds numbers. For instance, bacteria have typical dimensions of less than 1μm, and consequently the Reynolds number of bacteria swimming in water must be very small indeed (with $U < 1$ mm/s, we have Re $< 10^{-3}$). On the other hand, flows on a geophysical scale, such as winds or ocean currents, are often characterized by *high* Reynolds numbers simply because the typical distances are very large – say hundreds of km – and therefore unsteady, and turbulent motion is to be expected. For a boat with a 15 m mast sailing in a (strong) $U = 10\,\text{m/s}$ wind, the Reynolds number is 10^6.

In the Navier–Stokes equation (6.48), let us measure lengths in terms of ℓ and velocities in terms of U, so that we introduce the dimensionless variables:

$$\hat{\mathbf{r}} = \mathbf{r}/\ell, \quad \hat{\mathbf{u}} = \mathbf{u}/U, \quad \hat{t} = tU/\ell. \tag{6.51}$$

Then the Navier–Stokes equation takes the form

$$\frac{\partial \hat{\mathbf{u}}}{\partial \hat{t}} + (\hat{\mathbf{u}} \cdot \hat{\nabla})\hat{\mathbf{u}} = -\hat{\nabla}\hat{p} + \frac{1}{\text{Re}}\hat{\nabla}^2\hat{\mathbf{u}}. \tag{6.52}$$

Here the reduced pressure is $\hat{p} = p/(\rho U^2)$. Let us return to the flow around a moving sphere. From Eq. (6.52), it follows that the velocity field can only depend on $\hat{\mathbf{r}}$ and the Reynolds number:

$$\mathbf{u}(\mathbf{r}, U, \ell) = U\mathbf{f}\left(\frac{\mathbf{r}}{\ell}, \text{Re}\right), \tag{6.53}$$

where $\mathbf{f}$ is a (vector) function that depends on the geometry of the boundary conditions. For a sphere in a flow of velocity U, the vector function $\mathbf{f}$ behaves in such a way that the sphere has no effect on the flow field when $|\mathbf{r}/\ell| \to \infty$ and $\text{Re} \to 0$. Eq. (6.53) is Reynolds' *law of similarity*, and it follows that the geometries of the flows for spheres of different diameters, at different velocities and at various kinematic viscosities are all the same if the Reynolds number is the same.

The reduced pressure, $\hat{p} = p/\rho U^2$, can only depend on $\mathbf{r}/\ell$ and Re:

$$p = \rho U^2 f\left(\frac{\mathbf{r}}{\ell}, \text{Re}\right), \tag{6.54}$$

where f is a scalar function of its arguments.

It is interesting to note that the *force* on the sphere moving with a velocity U depends *not* on $\mathbf{r}$ but only on Re and ℓ. The force must have the dimension of pressure times area, and it follows that the force on the sphere must have the following scaling:

$$F = \rho U^2 \ell^2 f\left(\frac{\mathbf{r}}{\ell}\bigg|_{|\mathbf{r}|=\ell}, \text{Re}\right) = \rho U^2 \ell^2 f(\text{Re}), \tag{6.55}$$

consistent with Stokes' law derived in the next section.

In the limit of very low Re, the last term in Eq. (6.52) dominates over the non-linear term, which can be ignored, and leads to the *Stokes equation*, valid for low Reynolds number *stationary* flow:

$$\boxed{\nabla p - \mu \nabla^2 \mathbf{u} = 0, \quad \text{Re} \ll 1. \quad \text{(Stokes)}} \tag{6.56}$$

This equation is used for *creeping flow* – mostly for systems where the characteristic length scale is *microscopic*, as it is for porous media and for the motion of micro-organisms.

In the limit of very high Reynolds numbers, viscosity is ignored, and the Navier–Stokes equation simplifies to

$$\frac{\partial \mathbf{u}}{t} + (\mathbf{u} \cdot \nabla)\mathbf{u} = -\frac{1}{\rho}\nabla p, \quad \text{Re} \gg 1, \quad \text{(Euler)} \tag{6.57}$$

which is *Euler's equation* first obtained by L. Euler (1755). This equation is used when viscosity can be ignored, as in fully developed turbulence. This equation is *scale-invariant* under a rescaling by an *arbitrary* positive factor b, so that

$$\mathbf{r} \to b \cdot \mathbf{r}, \quad \mathbf{u} \to b^H \cdot \mathbf{u},$$
$$t \to b^{1-H} \cdot t, \quad p/\rho \to b^{2H} p/\rho. \tag{6.58}$$

When these transformed variables are inserted into Euler's equation, it follows that the scaling factor b cancels in the resulting equation, and the untransformed equation is recovered. This transformation is valid for arbitrary H. In the famous Kolmogorov theory (1941a, b), global scale invariance (in a statistical sense) is implicitly assumed, and the exponent $H = \frac{1}{3}$ is singled out by an energy argument; see Kolmogorov (1962) for a more "refined" argument. The regime where this scaling is valid is called the *inertial range* of *fully developed turbulence*.

6.7 Two Theorems Based on the Steady Euler Equation

In steady state when $\partial \mathbf{u}/\partial t = 0$, the Euler equation reduces to

$$(\mathbf{u} \cdot \nabla)\mathbf{u} = -\frac{1}{\rho}\nabla p. \tag{6.59}$$

Now, the operator $\mathbf{u} \cdot \nabla$ is just the velocity times the derivative along a streamline. We can introduce a local coordinate l that measures the length along a given streamline and a unit vector $\mathbf{e}_l$ so that $(\mathbf{u} \cdot \nabla)\mathbf{u} = u\mathbf{e}_l \cdot \nabla\mathbf{e}_l u = u \, d(\mathbf{e}_l u)/dl$. Now, $d\mathbf{e}_l/dl \perp \mathbf{e}_l$, so that by taking the component of Eq. (6.59) along $\mathbf{e}_l$, we get

$$u\frac{du}{dl} = -\frac{1}{\rho}\frac{dP}{dl}, \tag{6.60}$$

or

$$\frac{d}{dl}\left(\frac{1}{2}\rho u^2 + P\right) = 0, \tag{6.61}$$

which means that $\frac{1}{2}\rho u^2 + P$ is constant along each streamline. This result is known as *Bernoulli's theorem*.

Since the Euler equation comes from ignoring viscous forces, it lacks a description of shear forces. The pressure only acts in the normal direction across any surface contained in the fluid. This fact means that the flow itself does not generate any internal rotation. Mathematically, this is expressed by *Kelvin's circulation theorem*, which we prove in the following using the steady Euler equation. The circulation is expressed as $\oint d\mathbf{l} \cdot \mathbf{u}$, and we are interested in showing that this integral is conserved along the flow. This means that one should think of the integral as a Riemann sum where the line segments $d\mathbf{l} \to \delta\mathbf{l}$ are small but finite and move along with the flow in the sense that the ends of the segments $\delta\mathbf{l}$ are anchored in the fluid. This means that the time rate of change

$$\frac{d(\delta\mathbf{l})}{dt} = \mathbf{u}(\mathbf{x} + \delta\mathbf{l}) - \mathbf{u}(\mathbf{x}) = \delta\mathbf{l} \cdot \nabla\mathbf{u}. \tag{6.62}$$

The substantial derivative of the circulation takes the form

$$\frac{d}{dt} \oint d\mathbf{l} \cdot \mathbf{u} = \oint \frac{d\delta\mathbf{l}}{dt} \cdot \mathbf{u} + \oint d\mathbf{l} \cdot \frac{d\mathbf{u}}{dt}, \tag{6.63}$$

where the last term $d\mathbf{u}/dt$ may be replaced by $-\nabla(P/\rho)$. In the first term $d(\delta\mathbf{l}/dt) \cdot \mathbf{u} = \delta\mathbf{l} \cdot \nabla\mathbf{u} \cdot \mathbf{u} = \frac{1}{2}\delta\mathbf{l} \cdot \nabla u^2$. This allows us to write Eq. (6.62) as

$$\frac{d}{dt} \oint d\mathbf{l} \cdot \mathbf{u} = \oint d\mathbf{l} \cdot \nabla \left(\frac{1}{2}u^2 - \frac{P}{\rho} \right)$$
$$= \int d\mathbf{S} \cdot \nabla \times \nabla \left(\frac{1}{2}u^2 - \frac{P}{\rho} \right) = 0. \tag{6.64}$$

In the last step, we have applied Stokes' theorem, so that the $d\mathbf{S}$ integral is taken over the surface limited by the moving contour. The integral vanishes because $\nabla \times \nabla f = 0$ for any vector field $\mathbf{f}$. This means that if the flow has no circulation at some large distance from an obstacle, it will not aquire any circulation as it passes it, as long as the Euler equation actually describes the flow. This, however, will not be the case very close to the surface, where viscous forces will start to become important. Such boundary layers exist everywhere in high Reynolds number flows past obstacles and define a transition region to the flow where viscous effects become important.

Kelvin's circulation theorem allows a stronger version of Bernoulli's theorem as we may now take $\nabla \times \mathbf{u} = 0$, provided the flow starts out without circulation. This means that there exists some field ϕ such that $\mathbf{u} = \nabla\phi$. Noting that

$$\nabla \frac{1}{2}u^2 = \nabla\phi \cdot \nabla^2\phi = (\mathbf{u} \cdot \nabla)\mathbf{u}, \tag{6.65}$$

the steady Euler equation then takes the form

$$\nabla \left(\frac{\rho}{2}\nabla\phi^2 + P \right) = 0, \tag{6.66}$$

or in other words

$$\frac{1}{2}\rho u^2 + P = \text{const.} \tag{6.67}$$

throughout space, and not just along a streamline.

6.8 The Stream Function and Moffatt Eddies*

When the flow is incompressible and two-dimensional, a particularly simple representation of the velocity field exists. This representation is in terms of the stream function ψ and comes about in the following way.

6.8.1 Stream Function*

In general, when $\nabla \cdot \mathbf{u} = 0$, there exists a field $\boldsymbol{\psi}$ such that $\mathbf{u} = \nabla \times \boldsymbol{\psi}$. Taking x and y to be Cartesian coordinates in the plane in which the flow varies, we can orient the z-direction transverse to that. Then we can take $\boldsymbol{\psi} = \mathbf{e}_z \psi(x, y)$ so that

$$\mathbf{u} = \nabla \times \boldsymbol{\psi} = -\mathbf{e}_z \times \nabla \psi. \tag{6.68}$$

Now, by construction $\nabla \cdot \mathbf{u} = \nabla \cdot (\nabla \times \boldsymbol{\psi}) = 0$, so the incompressibility condition is automatically fullfilled. Also, since the velocity $\mathbf{e}_z \times \nabla \psi \perp \nabla \psi$, the flow is in the normal direction to the gradient of ψ, or, in other words, the streamlines are contour lines of the stream function. Taking the curl of Eq. (6.56), we get

$$0 = \nabla \times (\nabla^2 \mathbf{u}) = \nabla^2 (\nabla \times \mathbf{u}) = \nabla^2 (\nabla \times \nabla \times \boldsymbol{\psi}). \tag{6.69}$$

Using

$$\nabla \times \nabla \times \boldsymbol{\psi} = \nabla(\nabla \cdot \boldsymbol{\psi}) - \nabla^2 \boldsymbol{\psi} = -\nabla^2 \boldsymbol{\psi}, \tag{6.70}$$

we get the biharmonic equation

$$\nabla^4 \psi = 0. \tag{6.71}$$

As an example we use this formalism to compute the structure of the viscous eddies that form in a sharp corner and that were first discovered by Moffatt (1964). Such flows inside a corner may be created by a far field flow pattern as is illustrated in Figure 6.8. The opening angle here is 2α.

We will use polar coordinates with unit vectors $\mathbf{e}_r$ and $\mathbf{e}_\theta$ so that

$$\mathbf{u} = \frac{\partial \psi}{\partial r} \mathbf{e}_\theta - \frac{1}{r} \frac{\partial \psi}{\partial \theta} \mathbf{e}_r \tag{6.72}$$

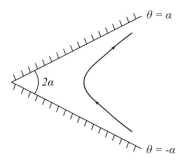

Figure 6.8 The flow past a sharp corner.

and apply no-slip boundary conditions so that $\mathbf{u}(\pm\alpha) = 0$. We look for flows that are antisymmetric around $\alpha = 0$ in the sense that $u_r(-\theta) = -u_r(\theta)$ and $u_\theta(-\theta) = u_\theta(\theta)$. This follows if $\psi(r,\theta)$ is symmetric in θ.

Now, Moffatt looked for solutions of the separated form

$$\psi = r^{\lambda+1} f_\lambda(\theta). \tag{6.73}$$

Note that in polar coordinates

$$\nabla^2\psi = \left(\frac{1}{r}\frac{\partial}{\partial r}r\frac{\partial}{\partial r} + \frac{1}{r^2}\frac{\partial^2}{\partial\theta^2}\right)r^{\lambda+1}f_\lambda(\theta)$$

$$= \left((\lambda+1)^2 + \frac{\partial^2}{\partial\theta^2}\right)r^{\lambda-1}f_\lambda(\theta), \tag{6.74}$$

so that applying ∇^2 once again only gives

$$\nabla^4\psi = \left((\lambda+1)^2 + \frac{\partial^2}{\partial\theta^2}\right)\left((\lambda-1)^2 + \frac{\partial^2}{\partial\theta^2}\right)r^{\lambda-3}f_\lambda(\theta). \tag{6.75}$$

Now, taking $f_\lambda(\theta) \sim e^{ik\theta}$, we see that either $k^2 = (\lambda+1)^2$ or $k^2 = (\lambda-1)^2$, which will produce four linearly independent solutions, provided $\lambda \neq 0$ or ± 1. The solutions corresponding to these special cases may be shown to be irrelevant of the boundary conditions that we apply (Moffatt, 1964). Since we are taking $f_\lambda(\theta)$ to be even, that leaves only

$$f_\lambda(\theta) = A\cos((\lambda+1)\theta) + B\cos((\lambda-1)\theta). \tag{6.76}$$

The boundary condition then reduces to $f_\lambda(\alpha) = f_\lambda'(\alpha) = 0$, so that

$$A\cos((\lambda+1)\theta) + B\cos((\lambda-1)\theta) = 0 \tag{6.77}$$

and

$$A(\lambda+1)\sin((\lambda+1)\theta) + B(\lambda-1)\sin((\lambda-1) = 0, \tag{6.78}$$

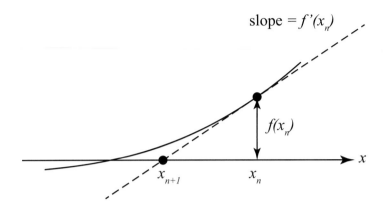

Figure 6.9 The Newton–Rapson iteration scheme.

which only has non-zero solutions for A and B if the coefficient matrix has a vanishing determinant. This condition may be written

$$\lambda \sin(2\alpha) + \sin(2\lambda\alpha) = 0, \qquad (6.79)$$

which turns out to have complex solutions for λ. It is necessary, however, to determine these numerically.

6.8.2 Newton–Rapson Method for Root Finding*

A simple way to solve Eq. (6.79) is to apply the Newton–Rapson method, which is a simple numerical scheme to find a root of an equation $f(x) = 0$ for which we know the derivative $f'(x)$. Taking the discrete approximation for the derivative using two consecutive points x_n and x_{n+1}, we may write

$$f'(x_n) = \frac{f(x_{n+1}) - f(x_n)}{x_{n+1} - x_n}. \qquad (6.80)$$

Now, taking $f(x_{n+1}) = 0$, we may solve this equation for x_{n+1} to get the iteration scheme

$$x_{n+1} = x_n - \frac{f(x_n)}{f'(x_n)}. \qquad (6.81)$$

The procedure is illustrated in Figure 6.9. As long as the derivative is not divergent or zero, it does not matter if x is complex, so that we may apply it to the function of $\lambda = \lambda_r + i\lambda_i$ (λ_r and λ_i being the real and imaginary parts of λ) on the left-hand side of Eq. (6.79).

There is an infinite sequence of roots to Eq. (6.79), but starting with the initial guess $\lambda_0 = 2\alpha = 2 + i$ gives the solution with the smallest real part. In the limit of small r, this is the solution that will dominate all the others. Figure 6.10 shows the

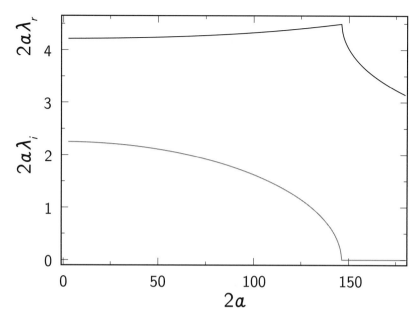

Figure 6.10 The real and imaginary part of λ multiplied by the opening angle 2α.

real and imaginary parts of λ. The fact that $\lambda_i \to 0$ at $2\alpha = 2\alpha_c = 146°$ means that at this critical angle, λ becomes real and the eddies disappear.

6.8.3 Self-Similar Streamline Pattern*

The angular component of the velocity depends on r as

$$u_\theta(r) \sim r^{\lambda_r} e^{i\lambda_i \ln r},$$ (6.82)

so it is seen to oscillate infinitely many times as $r \to 0$. In fact, $u_\theta = 0$ for values of r satisfying

$$\lambda_i \ln r_n + a = n\pi,$$ (6.83)

where n now labels the zeros, and a is some constant. Then

$$\frac{r_n}{r_{n+1}} = \frac{r_n - r_{n+1}}{r_{n+1} - r_{n+2}} = e^{\pi/\lambda_i},$$ (6.84)

so that the dimensions of successive eddies fall off in geometric progression with a common ratio e^{π/λ_i}.

In Figure 6.11, the infinite sequence of eddies is shown. Note that a magnification of a small section of the corner eddies will produce an identical image, and in this respect the pattern is self-similar. The relative strength of the eddies may be found

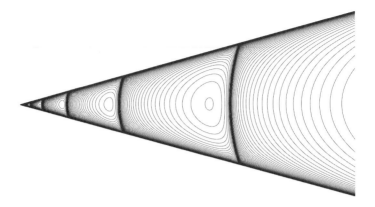

Figure 6.11 The streamlines in a wedge of opening angle $2\alpha = 30°$.

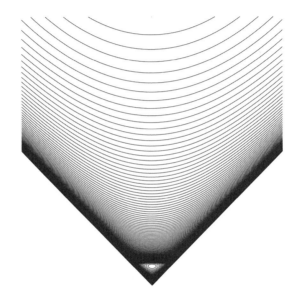

Figure 6.12 The streamlines in a wedge of opening angle $2\alpha = 90°$.

as follows. Taking the strength to be measured by $u_\theta(r) \sim r^{\lambda_r}$, we may use the eddie positions given in Eq. (6.84) to get

$$\frac{u_\theta(r_{n+1})}{u_\theta(r_n)} = \left(\frac{u_\theta(r_{n+1})}{r_n}\right)^{\lambda_r} = e^{\pi \frac{\lambda_r}{\lambda_i}} . \tag{6.85}$$

In Figure 6.10 it is seen that the ratio λ_r/λ_i does not vary very much for $\alpha < 75°$. For these values, the strength ratio is always large, around 300–400, but ony for larger values of α does it increase sharply.

In Figure 6.12 the eddies disappear much more quickly, though they remain self-similar in the same sense.

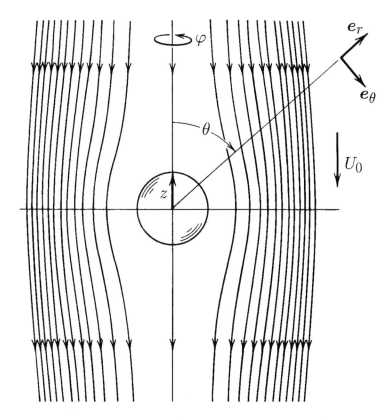

Figure 6.13 Flow around a sphere with fluid velocity $-U_0\mathbf{e}_z$ far from the sphere.

6.9 Stokes Flow Past a Sphere*

The creeping motion described by Eq. (6.56) was used by Stokes to calculate the drag force on a sphere moving in a viscous fluid. This calculation allows the measurement of fluid viscosity by observing the fall velocity of spheres in fluids. Consider a sphere of radius a fixed at the origin in a fluid moving downward that has the velocity $\mathbf{U}_0 = -U_0\mathbf{e}_z$ far from the sphere (see Figure 6.13).

The Stokes equation is

$$\nabla p - \mu\nabla^2\mathbf{u} = 0, \tag{6.86}$$

and it is natural to choose polar coordinates for this problem. In polar coordinates with unit vectors $\mathbf{e}_z$, $\mathbf{e}_\theta$, $\mathbf{e}_\varphi$, the uniform flow field is given by

$$\mathbf{U} = U_r\mathbf{e}_r + U_\theta\mathbf{e}_\theta + U_\varphi\mathbf{e}_\varphi, \tag{6.87}$$

where the orthogonal set of unit vectors $\mathbf{e}_r$, $\mathbf{e}_\theta$ and $\mathbf{e}_\varphi$ specify the directions for the field components given by

$$U_r = -U_0 \cos \theta,$$
$$U_\theta = U_0 \sin \theta, \tag{6.88}$$
$$U_\varphi = 0.$$

In polar coordinates, the pressure term in Eq. (6.86) is

$$\nabla p = \left(\frac{\partial p}{\partial r} \right) \mathbf{e}_r + \left(\frac{1}{r} \frac{\partial p}{\partial \theta} \right) \mathbf{e}_\theta + \left(\frac{1}{r \sin \theta} \frac{\partial p}{\partial \varphi} \right) \mathbf{e}_\varphi, \tag{6.89}$$

whereas the Laplace term in Eq. (6.86) is quite complicated:

$$[\nabla^2 \mathbf{u}]_r = \frac{\partial}{\partial r} \left(\frac{1}{r^2} \frac{\partial (r^2 u_r)}{\partial r} \right) + \frac{1}{r^2 \sin \theta} \frac{\partial}{\partial \theta} \left(\sin \theta \frac{\partial u_r}{\partial \theta} \right)$$
$$+ \frac{1}{r^2 \sin^2 \theta} \frac{\partial^2 u_r}{\partial \varphi^2} - \frac{2}{r^2 \sin \theta} \frac{\partial (u_\theta \sin \theta)}{\partial \theta} - \frac{2}{r^2 \sin \theta} \frac{\partial u_\varphi}{\partial \varphi},$$

$$[\nabla^2 \mathbf{u}]_\theta = \frac{1}{r^2} \frac{\partial}{\partial r} \left(\frac{1}{r^2} \frac{\partial u_\theta}{\partial r} \right) + \frac{1}{r^2} \frac{\partial}{\partial \theta} \left(\frac{1}{\sin \theta} \frac{\partial (u_\theta \sin \theta)}{\partial \theta} \right)$$
$$+ \frac{1}{r^2 \sin^2 \theta} \frac{\partial^2 u_\theta}{\partial \varphi^2} - \frac{2}{r^2} \frac{\partial u_r}{\partial \theta} - \frac{2 \cos \theta}{r^2 \sin^2 \theta} \frac{\partial u_\varphi}{\partial \varphi}, \tag{6.90}$$

$$[\nabla^2 \mathbf{u}]_\varphi = \frac{1}{r^2} \frac{\partial}{\partial r} \left(\frac{1}{r^2} \frac{\partial u_\varphi}{\partial r} \right) + \frac{1}{r^2} \frac{\partial}{\partial \theta} \left(\frac{1}{\sin \theta} \frac{\partial (u_\varphi \sin \theta)}{\partial \theta} \right)$$
$$+ \frac{1}{r^2 \sin^2 \theta} \frac{\partial^2 u_\varphi}{\partial \varphi^2} + \frac{2}{r^2 \sin \theta} \frac{\partial u_r}{\partial \varphi} + \frac{2 \cos \theta}{r^2 \sin^2 \theta} \frac{\partial u_\theta}{\partial \varphi}.$$

After some calculations, using these expressions, one finds that

$$u_r = -U_0 \cos \theta \left(1 - \frac{3a}{2r} + \frac{a^3}{2r^3} \right),$$
$$u_\theta = U_0 \sin \theta \left(1 - \frac{3a}{4r} - \frac{a^3}{4r^3} \right), \tag{6.91}$$
$$p = p_0 + \frac{3}{2} \frac{\mu U_0 a}{r^2} \cos \theta$$

is a solution of the Stokes equation that satisfies the boundary condition of zero velocity at the surface of the sphere and has the correct velocity at a large distance from the sphere. Here, p_0 is the ambient pressure.

The *stress* from the pressure on the sphere in the direction of flow is

$$(p|_{r=a} - p_0) \cos \theta = \frac{3\mu U_0}{2a} \cos^2 \theta. \tag{6.92}$$

The stress along the surface has only one component in the θ-direction having a magnitude of $(\partial u_\theta / \partial r)$. This stress has a component in the flow direction given by

$$\mu \left.\frac{\partial u_\theta}{\partial r}\right|_{r=a} \sin\theta = \frac{3\mu U_0}{2a}\sin^2\theta. \tag{6.93}$$

The total force per unit area in the flow direction is therefore

$$f_z = \frac{3\mu U_0}{2a}(\cos^2\theta + \sin^2\theta) = \frac{3\mu U_0}{2a}. \tag{6.94}$$

The combined effect of pressure and viscous stress on the surface of the sphere, in the direction of the flow, is independent of θ, and the integrated drag force is therefore given by $F_D = 4\pi a^2 f_z$, which may be written

$$\boxed{F_D = 6\pi \mu a U. \quad \text{(Stokes' law)}} \tag{6.95}$$

This is the famous *Stokes' law* for the drag force on a sphere of radius a moving with a velocity U.

With the Reynolds number defined by $\text{Re} = (\rho U a)/\mu$, we see that *Stokes'* law may be written as

$$F_D = 6\pi \cdot (\rho U^2 a^2) \cdot \frac{1}{\text{Re}}, \tag{6.96}$$

which has the similarity form given in Eq. (6.93). Einstein (1906, 1911) used this expression with his theory for the viscosity of sugar-solutions to estimate Avogadro's number $N_A \simeq 6.56 \cdot 10^{23}$ mol^{-1} (the correct value is $N_A = 6.022045 \cdot 10^{23}$ mol^{-1}). Eq. (6.95), which is a result of continuum hydrodynamics, has been reproduced using molecular-dynamics simulations of a "Lenard-Jones" fluid and shown to hold even for spheres comparable to molecular dimensions (Vergeles et al., 1995).

The expression for the drag force has been improved for $\text{Re} < 1$ using perturbation expansions (Proudman and Pearson, 1957):

$$F_D = 6\pi\mu a U \left[1 + \frac{3}{16}\text{Re} + \frac{9}{160}\text{Re}^2 \ln \text{Re} + \cdots\right]. \tag{6.97}$$

Experiments show that an eddy forms behind the sphere at $\text{Re} \simeq 24$. The eddy and wake start to oscillate at $\text{Re} \simeq 130$, but the flow remains *laminar* up to $\text{Re} \simeq 200$. With further increase in Re, turbulence gradually builds up (Taneda, 1956).

The Stokes equation neglects completely the convection of momentum in the original Navier–Stokes equation (the nonlinear term). C. W. Oseen (1910, 1945) pointed out that this neglect is not justified and proposed the equation (now called the *Oseen equation*) for stationary flow

$$\rho(\mathbf{U} \cdot \nabla)\mathbf{u} = -\nabla p + \mu\nabla^2\mathbf{u}, \quad \text{(Oseen equation)} \tag{6.98}$$

which replaces $(\mathbf{u} \cdot \nabla)$ in the Navier–Stokes Eq. (6.39) by $(\mathbf{U} \cdot \nabla)$, where $\mathbf{U}$ is the velocity of the sphere. The first order correction is then

$$F_D = 6\pi \mu a U \left[1 + \frac{3}{8} \text{Re} + \cdots \right], \qquad (6.99)$$

as shown already by Oseen (1913, 1945) and further improved upon later (see a discussion by Mazur and Weisenborn (1984)).

A small bubble of radius a rising in a fluid of density ρ is spherical when small. Stokes' law does not apply since the boundary condition is changed at the surface of the bubble. The result for the drag force is (see for instance Bird et al. (1987))

$$F_D = 4\pi \mu a U. \qquad \text{(For bubble)} \qquad (6.100)$$

Note that this result is *independent* of the surface tension. The drag force of a bubble is only $\frac{2}{3}$ of that experienced by a solid sphere. However, if the boundary conditions are modified, for example by proteins or surfactants that accumulate at the bubble–water interface, the boundary conditions are effectively changed to the non-slip boundary condition.

For small finite Re, inertial forces will deform the bubble, and the resulting shape is a balance among viscous, inertial and surface tension forces. Taylor and Acrivos (1964) have shown that the shape may be given by the radial function

$$R(\theta) = a \left[1 - \frac{5}{96} \text{Re} \, \text{Ca} \, (3 \cos^2 \theta - 1) \right], \qquad (6.101)$$

where Ca is the *capillary number* defined by $\text{Ca} = \mu U / \sigma$. The expression is valid for $\text{Re} \ll 1$ and $\text{Re} \, \text{Ca} \ll 1$ and shows that the rising bubble tends to become an ellipsoid flattened transverse to the flow.

Exercises

6.1 **Flow in a Rectangular Channel:**
   We shall look at the stationary flow along a *channel of rectangular cross section* with height $2H$ and width $2L$. Consider first the high Re-case where the Euler equation might be a relevant description. Find a stationary flow and pressure field that solves this equation and discuss the result.

6.2 Argue that for a time-independent flow along the channel, the Navier–Stokes equation reduces to the Stokes equation $-\nabla P + \mu \nabla^2 \mathbf{u} = 0$.

6.3 Assume that the solution of the Stokes equation with no-slip boundary condition may be written in the form

$$u_z(x, y) = \frac{\nabla P}{2\mu}(y^2 - H^2) + \sum_{n=0}^{\infty} a_n \cosh(k_n x) \cos(k_n y). \qquad (6.102)$$

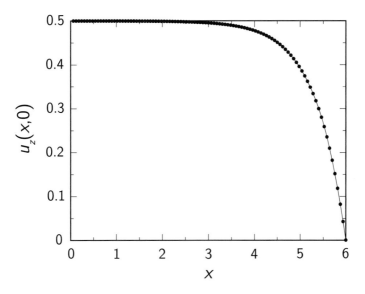

Figure 6.14 Flow field $u_z(x,0)$ when $H = 1$ and $L = 6$.

Show that this really is a solution to the Stokes equation. Require that the boundary conditions at $y = \pm H$ and at $x = \pm L$ are fullfilled, and show that

$$a_n = \frac{\nabla P}{\mu} \frac{2(-1)^n}{H k_n^3 \cosh(k_n L)} \quad \text{and} \quad k_n = \frac{(n + 1/2)}{H} \pi. \tag{6.103}$$

Place the origin in the middle of the channel.

6.4 Truncate the sum in Eq. (6.102) at $n = 5$ or 6, and plot the velocity field $u_z(x,0)$ for various values of H and L, setting $\nabla P/\mu$ to -1 (the pressure is falling in the positive z-direction). In Figure 6.14, the $L = 6$ and $H = 1$ case is shown.

6.5 Show that the averaged flow velocity $\bar{u} = \int dx\, dy\, u_z(x, y)/(4LH)$ takes the form

$$\bar{u} = -\kappa \frac{\nabla P}{\mu}, \tag{6.104}$$

where

$$\kappa = \frac{H^2}{3} \left(1 - \sum_{n=0}^{\infty} \frac{H}{L} \frac{6 \tanh(k_n L)}{((n + 1/2)\pi)^5} \right). \tag{6.105}$$

6.6 How is Eq. (6.105) related to the permeability in a two-dimensional channel? Take the $L \gg H$ limit of Eq. (6.105), and estimate the $n = 0$ correction term when $L = 6$ and $H = 1$.

6.7 Take $L = H = 1$ (a square channel), and calculate κ. How does this value of κ compare to that of a circular pipe with radius a and the same cross-sectional area (that is, $\pi a^2 = 4H^2$)? Discuss if this result is reasonable.

6.8 Explain why we must have $\kappa(H, L) = \kappa(L, H)$, and prove it numerically or analytically.

7

The Darcy Law

Permeability and the proportionality of flow with pressure gradient were first established by Henry Darcy (1856) in connection with his design and execution of the municipal water supply systems for the city of Dijon. His results are widely used for investigating all types of water flow through porous media, such as underground flow to wells, flow in soils being irrigated and the permeability of dam foundations. The flow of oil in reservoirs has been found to follow Darcy's law, and the unit of permeability is designated as the darcy. Darcy's experimental setup is shown in Figure 7.1.

A homogeneous filter bed of sand of height L is bounded by planes of area A. By varying the flow rate and measuring the pressure differences, Darcy found that the volume flow rate Q, with units m^3/s, was related to the head difference $h_1 - h_2$ by

$$Q = K'A \frac{h_1 - h_2}{L}, \tag{7.1}$$

where K' is a constant that depends on the type of sand used. The pressure at the bottom is $p_2 = \rho g h_2$, and the pressure at the inlet level is $p_1 = \rho g h_1 - \rho g L$. In the last expression, the term $\rho g L$ represents the hydrostatic pressure drop in going from (2) to (1). The difference in pressure at the two ends of the filter bed is

$$(p_1 - p_2) = \rho g(h_1 - h_2) - \rho g L. \tag{7.2}$$

Therefore, the observed relation may be rewritten (with $K = K'/\rho g$) as

$$Q = KA \left(-\frac{p_2 - p_1}{L} + \rho g \right). \tag{7.3}$$

This equation is one formulation of *Darcy's law*. The flow Q is zero if the pressure difference exactly equals the hydrostatic pressure difference so that $p_2 - p_1 = \rho g L$. In this case, $h_2 = h_1$ and the head difference vanishes. Note that the flow moves in

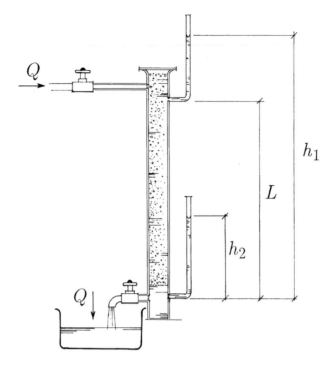

Figure 7.1 Darcy's experiment on water permeability. (From Darcy (1856).)

the direction of lower *piezometric head h*, as measured by water-filled manometers –
not in the direction of decreasing pressure (the pressure increases with depth).

(Nutting, 1930) introduced the *permeability k*, by the relation

$$K = k/\mu, \tag{7.4}$$

where μ is the fluid viscosity. The permeability is characteristic of the porous
medium, whereas the fluid viscosity enters separately in the constant K of Darcy's
law. The permeability k is dimensionally determined by Eq. (7.3):

$$[k] = \frac{[Q][L][\mu]}{[A][p]} = m^2. \tag{7.5}$$

Hence the dimension of k is area. The oil industry, however, uses the unit "darcy."[1]
For a substance with a permeability of 1 darcy, a pressure gradient of 1 atm/cm
produces a flow of $1\,cm^3/s$ through a cross section of $1\,cm^2$ for a fluid of viscosity
1 cp (as for water). With this definition, we find that

$$1\,\text{darcy} = 9.87 \cdot 10^{-9} cm^2 \simeq (1\,\mu m)^2. \tag{7.6}$$

---

[1] This name was first suggested by Wyckoff et al. (1933).

Table 7.1. *Permeabilities of various substances. (From Scheidegger (1974).)*

Substance	k in cm^2			Literature references
Berl saddles	$1.3 \cdot 10^{-3}$	to	$3.9 \cdot 10^{-3}$	Carman (1938)
Wire crimps	$3.8 \cdot 10^{-5}$	to	$1.0 \cdot 10^{-4}$	Carman (1938)
Black slate powder	$4.9 \cdot 10^{-10}$	to	$1.2 \cdot 10^{-9}$	Carman (1938)
Silica powder	$1.3 \cdot 10^{-10}$	to	$5.1 \cdot 10^{-10}$	Carman (1938)
Sand (loose beds)	$2.0 \cdot 10^{-7}$	to	$1.8 \cdot 10^{-8}$	Carman (1938)
Soils	$2.9 \cdot 10^{-9}$	to	$1.4 \cdot 10^{-7}$	Aronovici and Donnan (1946)
Sandstone ('oil sand')	$5.0 \cdot 10^{-12}$	to	$3.0 \cdot 10^{-8}$	Muskat (1937)
Limestone, dolomite	$2.0 \cdot 10^{-11}$	to	$4.5 \cdot 10^{-10}$	Locke and Bliss (1950)
Brick	$4.8 \cdot 10^{-11}$	to	$2.2 \cdot 10^{-9}$	Stull and Johnson (1940)
Bituminous concrete	$1.0 \cdot 10^{-9}$	to	$2.3 \cdot 10^{-7}$	McLaughlin and Goetz (1955)
Leather	$9.5 \cdot 10^{-10}$	to	$1.2 \cdot 10^{-9}$	Mitton (1945)
Cork board	$3.3 \cdot 10^{-6}$	to	$1.5 \cdot 10^{-5}$	Brown and Bolt (1942)
Hair felt	$8.3 \cdot 10^{-6}$	to	$1.2 \cdot 10^{-5}$	Brown and Bolt (1942)
Fiberglass	$2.4 \cdot 10^{-7}$	to	$5.1 \cdot 10^{-7}$	Wiggins et al. (1939)
Cigarette	$1.1 \cdot 10^{-5}$	to	$1.2 \cdot 10^{-5}$	Brown and Bolt (1942)
Agar-agar	$2.0 \cdot 10^{-10}$	to	$4.4 \cdot 10^{-9}$	Pallmann and Deuel (1945)

In Table 7.1, the permeabilities of various substances are given.

7.1 Derivation of Darcy's law

Darcy's law follows from the linearity of the Stokes equation – that is, the linear relation between ∇P and $\nabla^2 u$. In Chapter 5, we derived Darcy's law for the flow in channels and pipes: By inspection of Eqs (5.10) and (5.18), we see that $k = a^2/3$ for a channel (where a is the channel half-width), and $k = a^2/8$ for a pipe (where a is now the pipe radius). The prefactors $1/3$ and $1/8$ reflect the geometry of the conduit, which enter as boundary conditions in the Stokes equation.

The procedure of integrating the Stokes equation has been carried out more formally by Whitaker (1986) who wrote down expressions for the permeability in terms of integrals over an arbitrary pore geometry. In principle this procedure is the same as the one applied to get the channel and pipe permeabilities, in the sense that it delivers $k = Pa^2$ from the Stokes equations. Now a is a characteristic pore size, and the prefactor P reflects some average of the pore shapes.

7.2 Differential Form of Darcy's law

Darcy's law is easily generalized to infinitesimal layers (Muskat, 1937). With the volume flux (volume per area and time) defined by

$$U = Q/A, \tag{7.7}$$

we find that Eq. (7.3) may be written in the limit $L \to 0$ as

$$\boxed{\mathbf{U} = -\frac{k}{\mu}(\nabla p - \rho \mathbf{g}), \qquad \text{(Darcy's law)}} \tag{7.8}$$

where we have used $(p_2 - p_1)/L \to \nabla p$ for $L \to 0$, and $\mathbf{g} = (0, 0, -g)$. The volume flux is in the direction of the pressure gradient if the effect of gravity is neglected – as for horizontal flow. Note that $\mathbf{U}$, the *filtration velocity*, has the dimension m/s. This filtration velocity is also called the Darcy velocity, seepage velocity or specific discharge. Note again that with the z-axis vertical and up, we find for $\mathbf{U} = 0$ that $\frac{\partial p}{\partial z} + \rho g = 0$, which gives $p(z) = p(0) - \rho g z$ as expected.

The Darcy law is an example of a *constitutive equation*, that is, an equation that relates motion to forces.

The Darcy law alone does not specify the flow; it must be supplemented by the continuity equation. The continuity equation for a fluid moving in a porous medium is different from the continuity Eq. (6.1) for hydrodynamic flow. Instead, the continuity equation takes the form:

$$\frac{d}{dt}\int_V \phi \rho \, dV \qquad = \qquad -\int_S \rho \mathbf{U} \cdot d\mathbf{S}. \tag{7.9}$$

$$\text{Rate of increase of} \qquad\qquad \text{Rate of addition of mass}$$
$$\text{mass of fluid within } V \qquad\qquad \text{across the surface } S$$

The porosity ϕ on the left-hand side of Eq. (7.9) accounts for the fact that the fluid is excluded from the matrix. On the right-hand side, we have the Darcy velocity $\mathbf{U}$ instead of the fluid velocity $\mathbf{u}(\mathbf{r}, t)$, as it expresses the volume flow per unit area and time. In the limit $V \to 0$, we obtain the continuity equation for flow in porous media in the form

$$\boxed{\frac{\partial(\phi \rho)}{\partial t} + \nabla \cdot (\rho \mathbf{U}) = 0.} \tag{7.10}$$

Of course, this continuity equation is only valid on length scales much greater than the pore scale since both the porosity ϕ and the Darcy velocity $\mathbf{U}$ are defined as average quantities over some (macroscopic) region. Thus the limit $V \to 0$ is not to be taken too seriously. The averaging procedures required are in fact subtle and controversial; for a good discussion, see the classic book by Jacob Bear (1972), and Bear and Bachmat (1990).

The Darcy equation (7.8) and the continuity Eq. (7.10) may be combined to give a dynamic equation

$$\frac{\partial(\phi \rho)}{\partial t} = \nabla \cdot \left[\rho \frac{k}{\mu} (\nabla p - \rho \mathbf{g}) \right]. \tag{7.11}$$

Table 7.2. *Analogy between porous and electric stationary flow.*

	Porous flow	Electric flow
Potential	p	V
Current	U	j
Transport equation	$\mathbf{U} = -\frac{k}{\mu}\nabla p$	$\mathbf{j} = -\sigma \nabla V$
Transport coefficient	k/μ	σ

In order to solve this equation, even in the simplest situations, we need boundary conditions and an equation of state that relates the density to the pressure; that is, we need $\rho(p)$. Eq. (7.11) is non-linear unless simplifications apply.

For incompressible fluids in porous media with ϕ and k constant, the dynamical equation (7.11) simplifies for the stationary flow to the *Laplace equation* for the pressure:

$$\nabla^2 p = 0. \tag{7.12}$$

This equation is well known from many fields of physics, from electrostatics, elasticity theory, diffusion, currents in electrolytes etc. For electrical conduction, the analogous quantities are listed in Table 7.2. The electrical potential V is analogous to the pressure p, the conductivity σ is analogous to k/μ, the electric current is analogous to the Darcy flow and Ohm's law is analogous to Darcy's law. Note also the boundary conditions must also be taken into account when the analogy is discussed. We shall return to a more detailed discussion of this later.

It is possible to linearize Eq. (7.11) if the fluid has a constant compressibility, κ; that is, the density is given by an equation of state of the form (Scheidegger, 1974)

$$\rho = \rho_0 \exp(\kappa(p - p_0)), \qquad \kappa = \rho^{-1}\frac{\partial \rho}{\partial p}. \tag{7.13}$$

Using this expression in Eq. (7.11), we find

$$\frac{\partial \rho}{\partial t} = \nabla \cdot \left[D(\nabla \rho - \kappa \rho^2 \mathbf{g}) \right], \tag{7.14}$$

where the "diffusion constant" D for the density is given by

$$D = \frac{k}{\phi \mu \kappa}. \tag{7.15}$$

This type of equation enters the discussion when gas expands in a porous medium. If the gravity term in Eq. (7.14) can be ignored (as in horizontal flow), then the equation becomes a simple linear diffusion equation for the density, and the mathematical methods and intuition developed for diffusion equations and heat conduction can be used.

7.3 Model Calculations for the Permeability

In order to understand the physical basis of Darcy's law, many models have been suggested and discussed in the literature. The simplest of these models is the capillary model, which we discuss in the next section. For idealized models where spheres are placed in a regular lattice array, the hydrodynamic equations can be solved, and accurate numerical results have been obtained. Finally, the analogy to electric flow in inhomogeneous substances will be discussed.

7.4 The Capillary Model

The simplest model of a porous substance consists of a parallel array of capillaries of radius a and length L, as indicated in Figure 7.2. Such filters are available under the brand name NUCLEPORE. They are made by irradiating sheets of Lexan or other materials with heavy fission fragments. This irradiation produces nuclear tracks that may be etched preferentially by various chemicals to produce cylindrical pores typically 0.1 µm in diameter and 20 µm long. Filters with pore diameters in the range 1 nm to 10 µm are available commercially (see Figure 7.3 for an early example by the inventors Fleischer, Price and Walker (Fleischer, 1995; Fleischer and Price, 1963)). As discussed before (Eq. (5.17)), the volume flow rate [m^3/s] through a single capillary, ignoring end effects, is

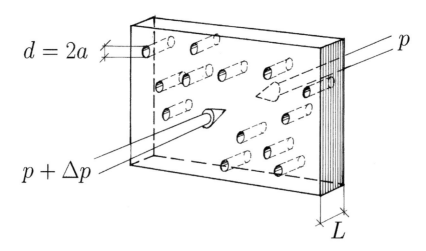

Figure 7.2 Filter consisting of capillaries.

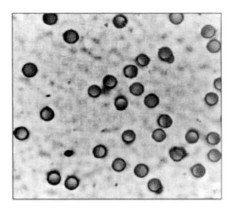

Figure 7.3 Nuclepore filter. (After Fleischer et al. (1964).)

$$Q = \frac{\pi a^4}{8\mu} \frac{\Delta p}{L},$$ (7.16)

where the pressure gradient is $\frac{\partial p}{\partial x} = -\Delta p/L$, and Δp is the magnitude of the pressure drop. The effective or average velocity in the pore is (see Eq. (5.21))

$$\langle u \rangle = \frac{Q}{\pi a^2} = \frac{a^2}{8\mu} \frac{\Delta p}{L}.$$ (7.17)

With n pores per unit area, the flow rate is

$$U = Qn = n \frac{\pi a^4}{8\mu} \frac{\Delta p}{L} = n\pi a^2 \langle u \rangle.$$ (7.18)

If we consider the membrane to be an infinitesimal cross section of a porous medium, as indicated in Figure 7.2, then we see that the flow per unit area and time is given by

$$U = \frac{k}{\mu} \frac{\Delta p}{L},$$ (7.19)

which in fact is Darcy's law with the permeability k given by

$$k = n \frac{\pi a^4}{8}.$$ (7.20)

Thus the capillary model leads directly to Darcy's law and gives an explicit expression for the permeability that depends only on the pore geometry. However, this model clearly is inadequate because it supports flow only in a single direction. There have been many attempts to generalize this model but without much success. The most important aspect of the model is that it shows that the *separation* of the fluid viscosity and the pore geometry in Darcy's law has the form $K = k/\mu$.

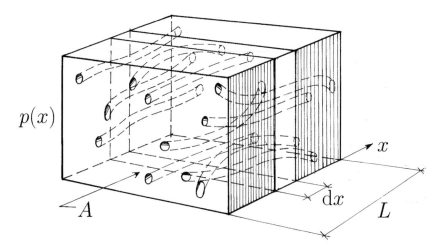

Figure 7.4 Model of a porous medium as a stack of membranes.

7.4.1 *k Expressed in Terms of Macroscopic Quantities*

For a porous medium consisting of an assembly of tortuous capillaries, the permeability in Eq. (7.20) may be expressed in terms of macroscopically measurable quantities such as the porosity ϕ and the specific surface S. The rewritten expression is in practice used to correlate observations of permeability with sample porosity.

The porosity of the sample illustrated in Figure 7.4 is

$$\phi = nA\frac{\pi a^2 L}{AL} = n\pi a^2. \tag{7.21}$$

Combining this result with Eq. (7.18), we find that

$$U = \phi \langle u \rangle. \tag{7.22}$$

Thus the volume flow velocity, or filtration velocity, U, is less than the average pore velocity, $\langle u \rangle$, by a factor ϕ. This result was first pointed out by Dupuit (1854).

Using the expression for ϕ, we may eliminate n in Eq. (7.20), and the permeability k may be written

$$k = \frac{\phi a^2}{8}. \tag{7.23}$$

The *specific surface area S* is the pore surface area per unit volume. This is a quantity that may be relatively easily measured by several methods. Clearly S is a quantity characteristic for a porous substance just as the porosity is. For the capillary model, S is

$$S = \frac{(nA)2\pi aL}{LA} = n2\pi a, \tag{7.24}$$

Table 7.3. *Values of K_0 for streaming flow in various cross sections. (After Carman (1937).)*

	Shape	K_0	Remarks
1.	Circle	2.0	Poiseuille's law
2.	Ellipses		
	Major axis = twice minor axis	2.13	—
	Major axis = 10× minor axis	2.45	—
3.	Rectangles		
	Length = breadth, i.e., square	1.78	—
	Length = 2× breadth	1.94	—
	Length = 10× breadth	2.65	—
	Length is infinite	3.0	—
4.	Equilateral triangle	1.67	—
5.	Pipes with cores		—
	Core set concentricaly	2.0–3.0	—
	Core set eccentricaly	1.7–3.0	Eccentricity <0.7
	Core set eccentricaly	1.2–2.0	Eccentricity >0.7

where A is the cross section and L the length of the sample. The expression for S may be combined with the expression (7.21) for the porosity to eliminate n, and we find

$$S = 2\frac{\phi}{a}. \tag{7.25}$$

The specific surface area has the dimension of inverse length, and we expect S^{-1} to be a typical pore size for any porous medium.

Using Eqs (7.23) and (7.25), the permeability may be written in terms of the macroscopically measurable quantities ϕ and S:

$$k = \frac{\phi^3}{2S^2} = \frac{\phi^3}{K_0 S^2}. \tag{7.26}$$

Clearly, the factor $K_0 = 2$ in the relation above comes from the geometry of the pores. The results for capillaries of various cross sections in Table 7.3 have been given by Carman (1937). This table shows that the shape of the cross section of the capillaries has very little effect on the permeability. K_0 ranges only between the limits 1.2 and 3. It is interesting to note that $K_0 = 2$ does not necessarily denote a circular cross section, nor even a shape resembling a circle.

Equation (7.26) is valid for a porous medium consisting of capillary pipes of equal radius, as in a Nuclepore filter. The permeability expression in Eq. (7.26) is expressed in a form involving only *porosity*, ϕ, and *specific surface*, S, which are both easily measured for rock samples. Thus it is interesting to compare Eq. (7.26) with experimental results.

7.5 Kozeny Expression for k

In order to extend Eq. (7.26) to a real porous substance with known porosity, ϕ, and specific surface area, S, one must in some way account for the fact that the path of fluid elements is not straight. This is done by introducing an effective length L_e. There are two effects to consider:

First, the pressure drop Δp acts over the effective capillary length L_e instead of the actual sample length, so that we may write the flow in any one of the *effective* capillaries as

$$Q = \frac{\pi a^4}{8\mu} \frac{L}{L_e} \frac{\Delta p}{L},$$

(7.27)

as in Eq. (7.16) with L_e replacing L (Kozeny, 1927).

Second, the relation between the flow per unit area, U, and Q is modified since the number of "capillaries" per unit area must satisfy the equation

$$n A \pi a^2 L_e = \phi A L,$$

(7.28)

which simply equates two expressions for the pore volume. Combining Eqs (7.26) and (7.28), we find that $U = nQ$ is given by Darcy's equation in the form

$$U = \frac{\phi a^2}{8\mathcal{T}^2 \mu} \frac{\Delta p}{L},$$

(7.29)

so that the permeability is given by

$$k = \frac{\phi a^2}{8\mathcal{T}^2}.$$

(7.30)

The ratio of effective length to sample length is called the *tortuosity*

$$\mathcal{T} = \frac{L_e}{L}.$$

(7.31)

The tortuosity also determines other properties of porous media, and it is commonly deduced from measurements of the electric resistivity of samples saturated with an. The specific surface area is given[2] by Eq. (7.24), as before, and we may therefore write the permeability in as

$$k = \frac{\phi^3}{K S^2},$$

(7.32)

---

[2] Two expressions for specific surface for the tortuous capillary model give $n A 2\pi a L_e = S L A$, so that $S = 2\phi/a$ as before.

where the *Kozeny constant* should be given by the *Carman relation*

$$K = K_0 \left(\frac{L_e}{L}\right)^2 = K_0 \mathcal{T}^2, \tag{7.33}$$

instead of Kozeny's result, Eq. (7.26). The expression for the specific permeability finally becomes

$$k = \frac{\phi^3}{K_0 \mathcal{T}^2 S^2}. \quad \text{(Carman–Kozeny)} \tag{7.34}$$

Equations similar to this one, containing S and ϕ, are commonly called Carman–Kozeny equations.

Carman (1937) observed that a dye streak injected in a packing of spheres typically made an angle of 45° to the flow direction; he concluded that $\mathcal{T} = L_e/L = \sqrt{2}$ is the proper value for the tortuosity, and therefore the Kozeny constant is estimated to be 4. Experimentally the Carman–Kozeny constant K is approximately 5, as shown in Figure 7.5, for random packings of spheres.

For beds of spheres packed to various porosities, the specific surface area is

$$S = 3(1 - \phi)/a. \tag{7.35}$$

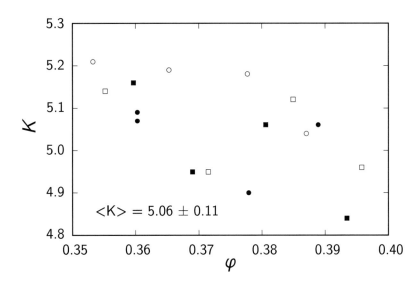

Figure 7.5 Carman–Kozeny constant K. Schriever's data (1930) for small glass spheres of diameter d. (○): $d = 1.025$ mm; (●): $d = 0.528$ mm; (□): $d = 0.443$ mm; (■): $d = 0.252$ mm. The average is $K = 5.06 \pm 0.11$. (After Carman (1937).)

With this result, the Carman–Kozeny expression (7.32) for the permeability takes the form[3]

$$k = \frac{a^2}{9K} \frac{\phi^3}{(1-\phi)^2}.$$

(7.36)

A most important aspect of this equation is the $\phi^3/(1-\phi)^2$ dependence. This factor is in agreement with experiments, as discussed by Carman (1937), and also with numerical solutions of the Stokes equations for lattice packings of spheres when the porosity is less than 0.5 (Zick and Homsy, 1982).

7.5.1 Flow at Finite Reynolds Numbers in Porous Media

Carman (1937) reviewed and discussed the earlier results in terms of dimension-less groups first introduced by Blake (1922), and Stanton and Pannell (1914). In a discussion of flow in porous media, the first dimensionless group to consider is the *Reynolds number*

$$\mathrm{Re} = \frac{\rho \langle u \rangle m}{\mu}.$$

(7.37)

Here the characteristic length scale was taken to be m, the so-called *hydraulic radius* defined by

$$m = \frac{\text{cross section normal to flow}}{\text{perimeter presented to flow}}.$$

(7.38)

For a circular pipe of radius a, one finds $m = a/2$. For pipes of uniform cross section, an alternate expression for m is

$$m = \frac{\text{volume of fluid}}{\text{surface presented to fluid}}.$$

(7.39)

If this expression is applied to a porous medium, the hydraulic radius is also given by

$$m = \frac{\phi}{S},$$

(7.40)

which is Kozeny's assumption. With the relation $U = \phi \langle u \rangle$, it follows that the Reynolds number for porous flow is given by

$$\mathrm{Re} = \frac{\rho U}{\mu S}.$$

(7.41)

---

[3] For Poisson porous media, we find $k = a^2 \phi^3 / 9 K_0 T^2 (\phi \ln \phi)^2$, where we have used that $S = 4\pi a^2 \phi n$, and $\ln \phi = -n(4\pi/3)a^3$ gives $S = 3a^{-1}\phi \ln \phi$.

A second dimensionless group is the *resistance coefficient* ψ:

$$\psi = \frac{F}{\rho \langle u \rangle^2}. \tag{7.42}$$

This is the ratio of the frictional force F, per unit area of the pore surface, and the kinetic energy of the flow per unit volume. The frictional force is obtained by noting that in the volume $V = AL$, the work done per unit time by the pressure drop, $U \Delta p A$, must equal the energy dissipated at the pore surface:

$$U \Delta p A = F \langle u \rangle S V, \tag{7.43}$$

where the fluid is taken to move at the velocity $\langle u \rangle$ in the pore. This gives the frictional force as

$$F = \frac{U}{\langle u \rangle S} \frac{\Delta p}{L} = \frac{\phi}{S} \frac{\Delta p}{L}. \tag{7.44}$$

The resistance coefficient can therefore be written in a form involving easily measured quantities, using $U = \phi \langle u \rangle$,

$$\psi = \frac{\phi^3}{\rho U^2 S} \frac{\Delta p}{L}. \tag{7.45}$$

In Figure 7.6, the resistance coefficient ψ is plotted as a function of the Reynolds number for many experiments on beds of spheres.

It should be noted that in Figure 7.6, the Reynolds number is twice Re defined in Eq. (7.41), whereas the resistance coefficient is four times the resistance coefficient in Eq. (7.45) as used by Carman. Such numerical differences in the definitions of dimensionless groups are common in hydrodynamics. As is seen from Figure 7.6, the plot of experimental results with the dimensional groups ψ and Re give a very nice data collapse, with all the experimental results on a single curve.

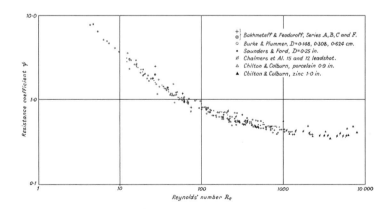

Figure 7.6 Resistance coefficient as a function of Reynolds number for beds of spheres. (After Mott (1951).)

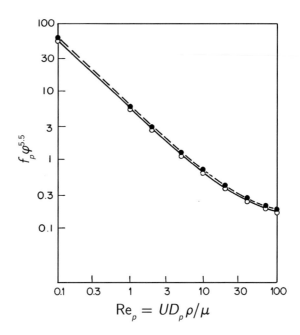

Figure 7.7 The particle scaled friction factor, $f_p\phi^{5.5}$, as a function of particle Reynolds number Re_p. (From Dullien (1992), after Rumpf and Gupte (1971).)

More recent correlations and dimensional arguments have been reviewed by Dullien (1992). He introduces the *friction factor*

$$f_p = \frac{D_p}{\rho U^2}\frac{\Delta p}{L} \tag{7.46}$$

and the particle Reynolds number

$$\mathrm{Re}_p = \frac{D_p U \rho}{\mu}. \tag{7.47}$$

Here D_p is the "surface average sphere diameter." In terms of these expressions, the permeability is given by $k = D_p^2/\mathrm{Re}_p f_p$. Rumpf and Gupte (1971) found for random packings of spheres with $0.35 < \phi < 0.7$, and $10^{-2} < \mathrm{Re}_p < 10^2$, (see Figure 7.7) that the experimental results are well fitted by the expression

$$f_p = c\frac{5.6}{\mathrm{Re}_p}\phi^{-5.5}, \tag{7.48}$$

with $c = 1.00$ and 1.05 for narrow and wide size distributions of the spheres, respectively.

Table 7.4. *Different porosity functions for low Reynolds number flow. (After Rumpf and Gupte (1971).)*

$1/f(\phi)$	Author
$(1-\phi)^2/\phi^3$	Blake (1922), Kozeny (1927), Carman (1937)
$(1-\phi)^2/\phi$	Zunker (1920)
$[(1-\phi)^{1.3}/(\phi-0.13)]^2$	Terzaghi (1925)
$[1.115(1-\phi)/\phi^{1.5}][(1-\phi)^2+0.018]$	Rapier (1949)
$69.43 - \phi$	Hulbert and Feben (1933)
$\phi^{-3.3}$	Slichter (1899)
$\phi^{-1.0}$	Krüger (1918)
$\phi^{-6.0}$	Hatch (1934), Mavies and Wilsey (1936)
$\phi^{-4.0}$	Fehling (1939)
$\phi^{-4.1}$	Rose (1945)
$\phi^{-5.5}$	Rumpf and Gupte (1971)

We note that Eq. (7.48), with the definitions (7.46) and (7.47), may be rewritten as

$$U = \frac{D_p^2 \phi^{5.5}}{5.6c\mu} \frac{\Delta p}{L},$$

(7.49)

which is Darcy's law. Therefore the permeability is given by

$$k_{RG} = \frac{D_p^2 \phi^{5.5}}{5.6c},$$

(7.50)

according to Rumpf and Gupte. They found that the porosity function $f(\phi) = \phi^{-5.5}$ fits the data best. A number of other porosity functions have been fitted to observations, see Table 7.4.

From Table 7.4, it is clear that a number of power-law-type relations fit observations equally well. It is well known that fits of power laws to experimental data in a limited range give unreliable results for the exponent. It is not to be expected that there is a general correlation between the permeability and porosity. A closed cell foam may have a large porosity and a negligible water permeability. Rocks with large, poorly connected pores will have an unusually low permeability for a given porosity. Thus we conclude that factors other than the porosity ϕ influence the permeability strongly. The comparison between the Carman–Kozeny result and the Rumpf and Gupte result is seen in Figure 7.7.

7.6 Katz–Thompson Model for Permeability[*]

(Katz and Thompson, 1986) proposed the following formula for the permeability of rocks:

$$K = c l_c^2 \frac{\sigma}{\sigma_0}, \tag{7.51}$$

where σ_0 is the conductivity of brine, σ the conductivity of the rock saturated with brine, c is a constant of the order of $1/226$ and l_c is a length scale obtained from mercury porosimetry. This celebrated formula works well and is much used. And it is based on percolation theory.

The conductivity ratio σ/σ_0 measures the connectivity of the pore space. The constant c is related to a correction between hydrodynamic flow and electrical flow; one has to remember that the fluid has zero velocity at the pore walls, whereas this is not true for the flow of electricity. Percolation theory enters in the determination of the length scale l_c.

We will now follow Katz and Thompson and work out what is the length scale l_c. They based their work on a very deep article by Ambegaokar, Halperin and Langer (1971), where they gave what would become the so-called AHL-argument. We now present this argument, not in the way AHL did it, but in a slightly modified way.

Suppose we wish to simulate a bond percolation problem on the computer. For now we wish to generate percolation clusters at the percolation threshold of some network. We set up the network and decide upon the two nodes between which we ask whether there is percolation or not – these are the boundary nodes. Following Section 4, we choose a probability p of whether or not bonds in the network are occupied. If $p > p_c$, the percolation threshold, there is at least one path connecting the two boundary nodes. But, how do we implement this in practice? In order to decide if a bond is "present" with probability p, we draw a random number r_i on the interval $[0, 1]$. If $r_i \leq p$, then bond i is present; otherwise it is absent.

So, we assign to each bond in the network a random number on the interval $[0, 1]$. We furthermore assign to each bond a status "present" or "absent."

We initiate the network setting all bonds to "absent." We then turn the status of the bond that has been assigned the smallest random number to "present." Then we do the same for the bond assigned the second smallest random number. We repeat this procedure until a first connected path of bonds in the "present" status appears between the two boundary nodes. The random number assigned to this last bond, r_c, is the percolation threshold for that particular network with those particular boundary nodes. Why is this? If we set $p < r_c$, this last bond will have status "absent," and there is no path. If $p > r_c$, this last bond has status "present," and there is at least one connected path. Hence, $p_c = r_c$.

Let us now determine p_c in a very different way. Drop the concept of "present" or "absent" bonds for now. Choose some path $\mathcal{P}$ between the two boundary nodes. Determine the largest random number assigned to the bonds along this path, that is, $\max_{i \in \mathcal{P}} r_i$. Then do this for every possible path between the two boundary nodes. The percolation threshold is given by

$$p_c = \min_{\mathcal{P}} \left(\max_{i \in \mathcal{P}} r_i \right), \tag{7.52}$$

that is, the percolation threshold is over all paths, the smallest of the largest random numbers along each path.

Let us now assign electrical resistances ρ_i to the links in the network we are considering. These resistances are drawn from a probability distribution $p(\rho)$ whose cumulative probability is $P(g) = \int_0^g dg' \, p(g)$. How do we do that? We have already assigned random numbers r_i to the links. Then, the resistances are given by

$$\rho_i = P^{-1}(r_i), \tag{7.53}$$

where P^{-1} is the inverse function to P.

Now, suppose that the distribution $p(\rho)$ is very broad. What does that mean? We use order statistics (see Section 3.5.2) to answer this question. Say the network contains N links. That means we have N resistances. We order them in an ascending sequence $\{\rho_1, \rho_2, \ldots, \rho_N\} \rightarrow \{\rho_{(1)} \leq \rho_{(2)} \leq \cdots \leq \rho_{(N)}\}$, and according to Eq. (3.30), we have

$$\frac{\langle \rho_{(n+1)} \rangle}{\langle \rho_{(n)} \rangle} = \frac{P^{-1}\left(\frac{n+1}{N+1}\right)}{P^{-1}\left(\frac{n}{N+1}\right)}, \tag{7.54}$$

where the averaging is over an ensemble of networks, all identical except for how the g_i's are distributed. When this ratio is large, the distribution $p(\rho)$ is broad. Consider for a moment a chain consisting of L resistors drawn from the distribution $p(\rho)$, connected in series. The total resistance R is then the sum

$$R = \sum_{n=1}^{L} \rho_{(n)} = \rho_{(L)} \left[1 + \sum_{n=1}^{L-1} \frac{\rho_{(n)}}{\rho_{(L)}} \right], \tag{7.55}$$

where we have ordered the sequence of resistors in an ascending series. In the limit of a very broad distribution, we have that $\rho_{(n)}/\rho_{(L)} \rightarrow 0$ for $n \neq L$, and the resistance is then given by

$$R = \rho_{(L)} = \max_i \rho_i. \tag{7.56}$$

We now return to our network. We start placing the resistances in ascending order one by on until the network percolates. What is the structure at that point? We have one blob of resistances on each side of the link that makes it percolate. Assuming a very broad distribution, the resistance of the two blobs connected in series with the bridging link will be dominated by the bridging link. This is precisely the link at the percolation threshold. When we continue the process of placing resistances after this point, those resistances will be so large that they do not change the total resistance between the boundary nodes. Hence, the total resistance of the network is given by

$$R = \min_{\mathcal{P}} \left(\max_{i \in \mathcal{P}} \rho_i \right). \tag{7.57}$$

This is the AHL-argument.

Katz and Thompson considered mercury injection into the porous medium, recording the increase in the injected volume of mercury as a function of the pressure across the sample, see Figure 7.10. The pressure necessary for the mercury to penetrate through a pore throat with diameter d is given by the Young–Laplace equation, to be discussed in Section 9.2:

$$\Delta p = \frac{4\gamma \cos \theta}{d}, \tag{7.58}$$

where γ is the surface tension and θ is the contact angle between the interface and the pore wall.

Katz and Thompson argue that the point with the largest slope in Figure 7.10 occurs when a connected mercury cluster forms across the sample. Here is the reasoning: Regard the pore throats as links in a network. Each link i in this network then has a threshold pressure associated with it, $\Delta p_i = 4\gamma \cos \theta / d_i$. Following the same logic as in the AHL-argument, the pressure associated with the link at the percolation threshold is given by

$$\Delta p_c = 4\gamma \cos \theta \, \min_{\mathcal{P}} \left(\max_{i \in \mathcal{P}} \frac{1}{d_i} \right). \tag{7.59}$$

When the mercury reaches this link, those that get invaded after are easier to invade, and the injected pore volume increases rapidly. Hence, we read off the pressure difference across the sample at this point. This gives us Δp_c and, through Eq. (7.58), the critical length $l_c = d$ for this pressure difference. This length scale then controls the entire permeability of the sample in line with the AHL-argument, and Eq. (7.51) for the permeability follows.

7.6.1 Reynolds Number Dependence of the Permeability*

From the curves in Figures 7.6–7.8, it is clear that there is a change in behavior near $Re = 1$, from the low velocity creeping flow to the non-linear regime, where the

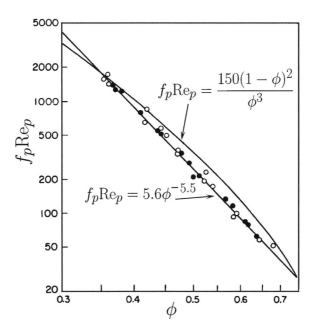

Figure 7.8 Comparison of the Rumpf and Gupte porosity function $\mathrm{Re}_p f_p = D^2/k = 5.5\phi^{-5.5}$ as a function of porosity, ϕ. The Carman–Kozeny equation (7.36) gives $\mathrm{Re}_p f_p = 36K(1-\phi)^2/\phi^3$. (From Dullien (1992), after Rumpf and Gupte (1971).)

inertial $(\mathbf{u} \cdot \nabla)\mathbf{u}$ term in the Navier–Stokes equation becomes important, and finally to the *turbulent* flow in the high velocity region. The phenomenological equation proposed by Forchheimer (1901) to describe this crossover is

$$\frac{\Delta p}{L} = \alpha\mu U + \beta\rho U^2. \tag{7.60}$$

The coefficient α must be identified with the inverse permeability $1/k$, in order to obtain Darcy's law at low Reynolds numbers. The so-called inertia parameter, β, has the dimension $[\beta] = \mathrm{m}^{-1}$, that is, an inverse length. The Forchheimer equation may be written in a dimensionless form by rearranging Eq. (7.60), to give the form suggested by Ahmed and Sunada (1969),

$$\frac{\Delta p}{L\beta\rho U^2} = 1 + \frac{\alpha\mu}{\beta\rho U}, \tag{7.61}$$

where the last term on the right-hand side is an inverse Reynolds number. With this form, a very large set of experimental results collapse onto a single curve, as shown in Figure 7.9.

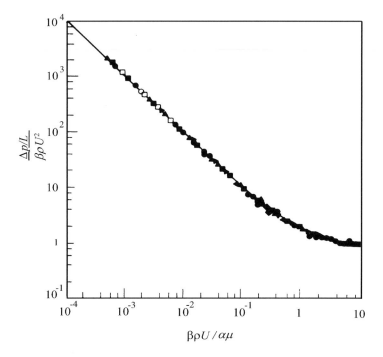

Figure 7.9 Dimensionless correlation based on the Forchheimer equation. (From Dullien (1992), after Ahmed and Sunada (1969).)

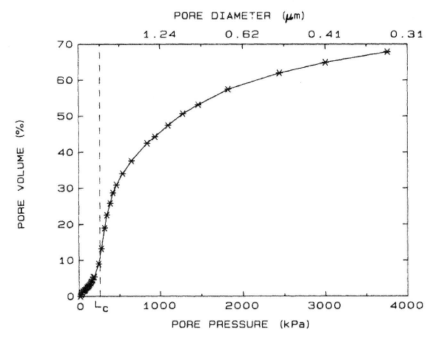

Figure 7.10 The injected pore volume vs. pressure during mercury injection in an eastern Utah sandstone. The point along the curve where the slope is largest corresponds to the percolation threshold. (From Katz and Thompson (1986).)

7.7 When the Porous Medium Is Rarified: The Brinkman Equation*

The Stokes equation, Eq. (6.56), is the low Reynolds number limit of the Navier–Stokes equation. Let us now supplement the pressure gradient with another force density (i.e. force per volume) affecting the flow, $\mathbf{f}$. The Stokes equation then becomes

$$\mu \nabla^2 \mathbf{U} = \nabla p + \mathbf{f}. \tag{7.62}$$

What if this additional force density is due to friction on the fluid from a porous medium? What if we regard the Darcy equation as manifestation of the friction force? That is, we rewrite the Darcy equation (7.8), while ignoring the gravity term for simplicity,

$$\mathbf{f} = -\frac{\mu}{k}\mathbf{U}. \tag{7.63}$$

Combining this equation with the Stokes equation (7.62), we get the Brinkman equation (Brinkman, 1949)

$$\mu \nabla^2 \mathbf{U} - \frac{\mu}{k}\mathbf{U} = \nabla p. \tag{7.64}$$

This equation was originally developed to describe flow past a sparse ensemble of fixed spherical particles. For a rigorous derivation, see Hinch (1977).

We may non-dimensionalize the Brinkman equation. If we set $\mathbf{x} = L\boldsymbol{\xi}$ and $t = T\tau$, where $\boldsymbol{\xi}$ and τ are dimensionless space and time variables respectively, it becomes

$$\nabla_\xi^2 \tilde{\mathbf{U}} - \frac{L^2}{k}\tilde{\mathbf{U}} = \nabla_\xi \tilde{p}, \tag{7.65}$$

where $\mathbf{U} = (L/T)\tilde{\mathbf{U}}$ and $p = (\mu/T)\tilde{p}$. The dimensionless parameter (L^2/k) determines the importance of the Darcy term compared to the Stokes term. This leads to the definition of a Brinkman screening length

$$L_B = \sqrt{k}. \tag{7.66}$$

When L_B is large, that is, the permeability is large, we will see Stokes flow as the porous medium will only slightly influence the flow. On the other hand, when L_B is small, the effect of the Stokes term vanishes, and we have flow that follows the Darcy equation.

If the porous medium consists of point-like particles, the screening decays exponentially (Capuani et al., 2003), whereas if the particles have a finite radius, the screening decays as $1/r^3$ where r is the distance (Koch and Brady, 1985).

We would use the Brinkman equation in cases where there are strong variations in permeability. A typical situation would be to study the flow above and in a sand packing. Above the sand, there is no Darcy term, and the flow will be purely governed by the Stokes equation. As we move into the sand, the permeability decreases, and hence we gradually will cross over to pure Darcy flow.

7.8 When the Flow Is Rarified: The Klinkenberg Correction*

One of the standard procedures to measure the permeability of rocks uses gases rather than liquids. Permeability is a property of the porous medium, not of the interstitial fluid, so it should not matter whether we use gases or liquids, right? No, wrong. The fallacy in this argument is that the Darcy equation (7.8) itself assumes liquids, and we use this equation to define the permeability. The permeability of a porous medium calculated from the Darcy equation will come out larger when gases are used compared to when liquids are used.

So what goes wrong in the Darcy equation when using gases? The difference lies in how gases behave near walls, compared to liquids. Near a wall, there is always a so-called Knudsen layer, where the molecules collide with the wall much more frequently than with each other. In liquids, this layer, whose thickness is given by the molecular mean free path, is very thin and can be ignored. However, in gases it can be considerable; for example, at pressures smaller than 0.1 atm, it is of the order of $100\,\mu$m. This layer causes slippage, that is, the velocity near the wall does not go to zero. Hence, the overall flow through the pores is larger for a given pressure difference in comparison to the case when there is no slippage.

It was Klinkenberg (1941) who first pointed this out. He proceeded to provide the following correction term to the permeability:

$$k = \frac{k_{\text{gas}}}{1 + b_K/\bar{p}}, \qquad (7.67)$$

where both k and k_{gas} are permeabilities measured using the definition of Darcy's law, but with k that would have been derived from measurements of liquid flow and k_{gas} from the actual gas flow. $\bar{p} = (p_{\text{inlet}} + p_{\text{outlet}})/2$ and b_K is the Klinkenberg constant. This approach works when the Knudsen number, that is, the ratio of the molecular mean free path to typical pore diameter, is in the range 0.01–0.1.

The Klinkenberg constant b_K depends on the porous medium and has to be determined for each. The way to determine it, as Klinkenberg describes, is to plot k_{gas} vs. the inverse of the mean pressure $\bar{p}$. From Eq. (7.67), we should find a straight line,

$$k_{\text{gas}} = k + k\frac{b_K}{\bar{p}}, \qquad (7.68)$$

which allows us to determine both the permeability k and the Klinkenberg constant b_K.

7.9 Non-Newtonian Flow

Non-Newtonian fluids are fluids that respond in a non-linear way to stress. There are many examples of such fluids. Ketchup is one. One can turn the bottle up side down and nothing happens. It is only when one jolts it that the ketchup flows. It has a *yield threshold* . It is also *shear thinning* (Koocheki et al., 2009). We wil get back to this later on. Another example is mayonnaise. This is also a yield threshold fluid, but the rheology is simpler here. We show another very nice example of a non-Newtonian fluid in Figure 7.11.

Figure 7.11 This bottle of French dressing contains a great example of a fluid possessing a yield threshold. The oil contains small pieces of carrot, onion, lettice, parsley, dill etc. These vegetables do not have the same density. Still, they stay exactly where they are in the fluid. If the oil had been Newtonian, only one of the vegatable species could have floated neutrally; the others would have ended up at the bottom or at the top. What is happening is that the buoyancy is not large enough to create shear forces high enough to exceed the yield threshold. So, the pieces remain trapped where they are, as in a solid. Untill we shake the bottle. When the vegetable pieces stop moving again after the shaking of the bottle, notice that they oscillate a few times before stopping completely. This signals that the fluid is acting as an elastic solid. (A black and white version of this figure will appear in some formats. For the colour version, refer to the plate section.)

Mayonnaise is an example of a *Bingham fluid*. Go back to Eq. (5.1), which defines the Newtonian fluid. We write it here as

$$\frac{\partial u_x}{\partial z} = \frac{1}{\mu} \sigma_{xz}. \tag{7.69}$$

The Bingham fluid introduces a *yield stress* σ_c, which modifies Eq. (7.69) to

$$\frac{\partial u_x}{\partial z} = \frac{1}{\mu} \begin{cases} \left[\sigma_{xz} - \text{sign}(\sigma_{xz})\sigma_c\right] & \text{if } |\sigma_{xz}| > \sigma_c, \\ 0 & \text{if } |\sigma_{xz}| \leq \sigma_c. \end{cases} \tag{7.70}$$

Hence, if the shear stress is larger than the yield threshold, the Bingham fluid flows essentially like a Newtonian fluid. On the other hand, for shear stresses less than the yield threshold, the fluid acts as a solid.

We derived the Hagen–Poiseuille law for a rectangular channel (5.9) in Chapter 5. We now derive the corresponding law for a Bingham fluid. The channel is the same as shown in Figure 5.1. We assume that the channel is so wide in the y-direction that we may disregard end effects. Our goal is to find the velocity profile $u = u(z)$, which we then integrate to find the volumetric flow rate Q, thus giving us the equivalent of Eq. (5.3).

As in the Newtonian case, we invoke no-slip boundary conditions along the walls of the channel at $z = -a$ and $z = +a$, that is, $u(-a) = u(+a) = 0$. Close to the walls, the shear σ_{xz} will be at the largest. Let us assume that it is larger that the yield threshold σ_c. At the center of the channel, $z = 0$, we must have that $\partial u/\partial z = 0$. From Eq. (7.70), we deduce that $|\sigma_{xz}| \leq \sigma_c$ here. Hence, there must be a length a_c, so that for $z = \pm a_c$, $|\sigma_{xz}| = \sigma_c$ and $|\sigma_{xz}| > \sigma_c$ for $-a < z < -a_c$ and $a_c < z < a$.

Let us now focus on the interval $a_c < z < a$. Here we have that

$$\sigma_{xz} = \sigma_c + \mu \frac{\partial u}{\partial z}. \tag{7.71}$$

By balancing the forces due to pressure gradient and shear acting on the volume element shown in Figures 5.2 and 5.3, we find that the Bingham fluid also obeys Eqs (5.3), (5.4) and (5.5). We integrate Eq. (5.5) twice to obtain

$$u(z) = -\frac{G}{2\mu} z^2 + C_1 z + C_2, \tag{7.72}$$

where C_1 and C_2 are integration constants. We have furthermore defined the pressure gradient G as in Eq. (5.6). Using the no-slip boundary condition at $z = a$ gives $C_2 = -G/2\mu \, a^2 - C_1 a$, so that we have

$$u(z) = \frac{G}{2\mu} (a^2 - z^2) - C_1(a - z). \tag{7.73}$$

At the inner boundary, $z = a_c$ where $\sigma_{xz} = \sigma_c$, we have from Eq. (7.71) that

$$\left.\frac{\partial u}{\partial z}\right|_{z=a_c} = 0. \tag{7.74}$$

Inserting this condition into Eq. (7.73) determines the second integration constant $C_1 = -G/\mu\, a_c$. Hence, we have

$$u(z) = \frac{G}{2\mu}(a + z - 2a_c)(a - z). \tag{7.75}$$

Working through the same steps for the region $-a < z < -a_c$ gives the velocity profile

$$u(z) = -\frac{G}{2\mu}(a - z - 2a_c)(a + z). \tag{7.76}$$

We now need to determine a_c. In the interval $-a_c \le z \le +a_c$ where $|\sigma_{xz} \le \sigma_c$, we have from Eq. (7.70) that

$$\frac{\partial u}{\partial z} = 0. \tag{7.77}$$

Hence, we have that

$$u(z) = u(0) \quad \text{for } -a_c \le z \le +a_c, \tag{7.78}$$

that is, the Bingham fluid moves as a solid block in this region. The force exerted on the block by the pressure drop across the channel $|\Delta p|$ is $2a_c\Delta p$. This force is counteracted by the frictional force on the block from the surrounding fluid in the liquid state. This force is $2|\sigma_{xz}|L = 2\sigma_c L$ – the factor 2 coming from there being two boundaries, and we find $\sigma_c L = |\Delta p|a_c$ or

$$a_c = \frac{\sigma_c}{|\Delta p|/L} = \frac{\sigma_c}{|G|}. \tag{7.79}$$

We now have the complete velocity profile,

$$u(z) = -\frac{G}{2\mu}\begin{cases} (a + z - 2a_c)(a - z) & \text{if } a_c < z \le a, \\ (a - a_c)^2 & \text{if if } -a_c \le z \le a_c, \\ (a - z - 2a_c)(a + z) & \text{if } -a \le z < -a_c, \end{cases} \tag{7.80}$$

where a_c is given by Eq. (7.79).

We may now integrate $u(z)$ over the interval $-a \le z \le +a$ to find the volumetric flow rate, Q, *under the assumption that there is flow*, that is, that $a_c < a$. If, on the

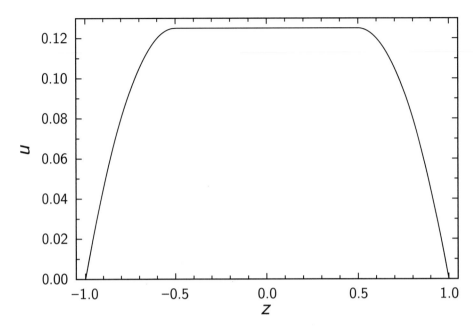

Figure 7.12 The velocity profile, Eq. (7.80), for a Bingham fluid where we have set $G = -1$, $\mu = 1$, $a = 1$ and $a_c = 1/2$ (in dimensionless units).

other hand, $a_c > a$, $|\sigma_{xz}|$ is everywhere less than the yield threshold σ_c and there is no flow. Hence, we have

$$Q = \frac{2Ga^3}{3\mu} \begin{cases} \left(1 - \frac{\sigma_c}{a|G|}\right)^2 \left(1 + \frac{1}{2}\frac{\sigma_c}{a|G|}\right) & \text{if } |\Delta p| > \frac{\sigma_c L}{a}, \\ 0 & \text{if } |\Delta p| \leq \frac{\sigma_c L}{a}. \end{cases} \qquad (7.81)$$

The average flow rate in the channel is given by $U = Q/2a$. This result should be compared to the Hagen–Poiseuille law, Eq. (5.9), valid for Newtonian fluids. We see that as $\sigma/|G| \to 0$, we recover the Newtonian case.

When $|\Delta p|$ is close to the threshold pressure $\sigma_c L/a$, this equation reduces to

$$Q = \frac{a^4}{\mu \sigma_c L^2} \operatorname{sign}(\Delta p) \begin{cases} \left(|\Delta p| - \frac{\sigma_c L}{a}\right)^2 & \text{if } |\Delta p| > \frac{\sigma_c L}{a}, \\ 0 & \text{if } |\Delta p| \leq \frac{\sigma_c L}{a}. \end{cases} \qquad (7.82)$$

Hence, we end up with a *quadratic* law rather the linear one we know from the Newtonian case.

There have been a number of numerical and theoretical studies of the behavior of Bingham fluids in porous media, see e.g., Talon and Bauer (2013); Chen et al. (2019). Given a porous medium, it turns out that there is a threshold pressure Δp_t.

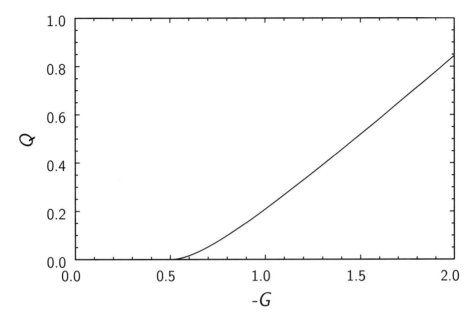

Figure 7.13 Volumetric flow rate Q as a function of the pressure gradient $G = -\partial p/\partial x$ for a Bingham fluid, Eq. (7.81). We have set $\sigma_c = 1/2$, $\mu = 1$ and $a = 1$ (in dimensionless units).

If $|\Delta p| \leq \Delta p_t$, where Δp is the pressure drop across the porous medium, there is no flow. If $|\Delta p| > \Delta p_t$, there are three flow regimes:

- When the pressure difference is very close to the threshold, there is flow in a single channel. The problem is then essentially the one we already considered when we generalized the Poiseuille flow in Eq. (7.81).
- As the pressure is increased, more and more channels open up, as shown in Figure 7.14. This leads to

$$Q \sim (|\Delta p| - \Delta p_t)^\beta, \qquad (7.83)$$

where $\beta = 2$ for a two-dimensional porous medium (Chen et al., 2019). This value is *universal*, that is, it does not depend on the details of the porous medium.
- When all channels have come into play, the flow is linear in Δp.

The Bingham fluid is the simplest of a family of non-Newtonian fluids, the Herschel–Bulkley fluids whose rheology is given by

$$\left(\frac{\partial u_x}{\partial z}\right)^n = \frac{1}{\mu}\begin{cases} \sigma_{xz} - \text{sign}(\sigma_{xz})\sigma_c & \text{if } |\sigma_{xz}| > \sigma_c, \\ 0 & \text{if } |\sigma_{xz}| \leq \sigma_c, \end{cases} \qquad (7.84)$$

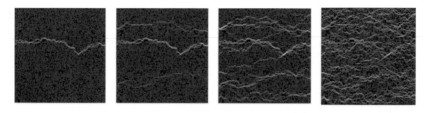

Figure 7.14 Flow regimes when a Bingham fluid moves in a porous medium with increasing pressure difference Δp (from Talon and Bauer (2013)). In the left-most picture, the pressure difference is low and there is only a single channel that is open, that is, the fluid in it moves. The next pictures show from left to right increasing pressure difference. The increasing pressure difference results in more and more channels opening up. (A black and white version of this figure will appear in some formats. For the colour version, refer to the plate section.)

where a new parameter n has entered. The Bingham fluid has $n = 1$. Ketchup, it turns out, is a Herschel–Bulkley fluid with n in the range 0.19 to 0.23. An example of a Herschel–Bulkley fluid with $n < 1$, hence being *shear thickening*, is corn starch mixed with water.

Bauer et al. (2019) analysed numerically and experimentally a three-dimensional porous medium filled with a Herschel–Bulkley fluid with $n = 0.5$ (numerically) and $n = 0.53$ (experimentally). They fitted their data to relation (7.83), finding three regimes. At pressure differences close to but larger than the threshold pressure Δ, there is essentially one open channel, and β is essentially $1/n$. For very high pressure differences when all channels are open, they find essentially the same result, $\beta = 1/n$. But, there is also a middle regime where they report a higher value for β. This is the regime where the number of open channels increases with increasing pressure difference. They reported the values 1.89, 2.45 and 1.89 in their experiments. Whether there is universality also in this case, as there is for the Bingham fluid, is unknown.

We will not pursue this field further here. It is under active study, and the dust has definitely not settled at the time of this writing.

Exercises

7.1 Permeability is governed by the narrowest part of the widest path:

In this problem, we will look at the permeability of a channel with a constriction and calculate its permeability as a function of the width of the opening and the width of its narrowest part. The channel has length L, height h and varying width $a(x) = b + 4(a0 - b)x^2/L^2$, where $-L/2 \le x \le L/2$ is the coordinate in the flow direction, so that $a = a_0$ at the ends, and $a = b < a_0$ in the center. We will assume $h \gg a_0$ and neglect the effect of the upper and lower walls.

Starting with the definitions in Eqs (7.3) and (7.4) using $g = 0$, we want to calculate k. The volume flux $Q = haU$, where U is the Darcy velocity, is fixed. Assuming that the velocity variation across the channel is parabolic, explain why the pressure gradient $\partial P / \partial x = -12\mu U / a^2(x)$.

7.2 Calculate the pressure drop along the whole channel

$$\Delta P = - \int_{-L/2}^{L/2} dx \, \partial P / \partial x.$$

Hint: You may want to solve the resulting integral using the substitution $x = (L/2)c \tan \phi$, where $c^2 = b/(a_0 - b)$.

7.3 Show that the you get the standard permeability $k = b^2/12$ when $b \to a_0$.

7.4 Take the $b \ll a_0$ limit and show that $k \to a_0^{(1/2)} b^{(3/2)}/(3\pi)$. Give an interpretation of the title of this problem based in this result.

7.5 **Pressure drop in a channel filled with Bingham fluid:**
It is easier to control the volumetric flow rate rather than the pressure drop in the laboratory. Hence, show by inverting Eq. (7.82) that the pressure drop $\Delta p = p(0) - p(L)$ in a channel with the geometry shown in Figure 5.1 of length L and height $2a$ is given by

$$\frac{\Delta p}{L} = \frac{\sigma_c}{2} \, \mathrm{sgn}(Q) \left(1 + 8|\tilde{Q}|\right) \left(1 + 2\cos\left[\frac{2}{3}\csc^{-1}\left(\left(1 + 8|\tilde{Q}|\right)^{3/2}\right)\right]\right),$$

where $\csc(\psi)$ is the cosecant $(= 1/\sin(\psi))$. We define

$$\tilde{Q} = \frac{\mu Q}{8\sigma_c a^2}.$$

7.6 Suppose now that the height of the channel a is not constant but depends on x, that is, $a = a(x)$. Suppose that the dependency is

$$a(x) = a_0 + \delta \, \cos\left(\frac{\pi x}{L}\right).$$

Caclulate to lowest order in δ Q as a function of Δp in this case. Hint: Use the lubrication limit.

7.7 **Velocity profile of the Herschel–Bulkley fluid:**
We consider here a Herschel–Bulkley fluid without a yield threshold. Its rheology is then given by

$$\left(\frac{\partial u_x}{\partial z}\right)^n = \frac{\sigma_{xz}}{\mu}.$$

Assuming a channel as in Figure 5.1 and no-slip boundary conditions for $z = \pm a$, find the velocity profile $u = u(z)$.

7.8 Calculate the average flow velocity in the channel U, and plot U as a function of the pressure gradient $G = -\partial p/\partial x$ for different values of n.

7.9 What are the qualitative differences between the velocity profiles and volumetric flow rates for $n > 1$ and $n < 1$?

8

Dispersion

Diffusive processes are abundant, giving rise to heat conduction in solids, the spreading of momentum in a viscous fluid and the statistical behavior of stock prices. To have a simple, mental picture of a diffusive process, consider a very small drop of colored dye in stationary water. After some time, we will see it spreading out over a larger area and thinning out in concentration.

When the diffusion happens on top of a background velocity field, the combined action of diffusion and advection with the flow is called *hydrodynamic dispersion*. Inside a porous medium, this process is particularely non-trivial. Dispersion in porous media is an important process in industry, pollution control and chromatography.

In this chapter we show how simple random walks give rise to diffusion and use the central limit theorem to derive the advection–diffusion equation. This equation, which describes things at the pore scale, may be upscaled, or *coarse grained*, to produce a description of dispersion on larger scales.

8.1 Random Walks

Probably the simplest and most important stochastic process is the *random walk*. It plays the same role in non-equilibrium statistical mechanics as the hydrogen atom does in atomic physics. Most of its basic properties are obtained by considering the walk in one dimension. We then have a system that can be a particle or a person who at given, separated times takes a step to the right with probability p or to the left with probability q. We then obviously have $p + q = 1$. After N steps, there will be R steps to the right and $L = N - R$ to the left. The net displacement to the right is $S = R - L$. If these walks are repeated a large number of times, the net displacement will vary from $S = +N$ to $S = -N$. There is only one walk that gives the maximal displacement $S = N$. It corresponds to having all the steps taken to the right. Since the probability for each such step is p, the probability for having this walk is $P_N(N) = p^N$. When there is one step to the left among these N steps, the net displacement is $S = N - 2$. It can occur at any of the N different times with probability q. Thus the overall probability of finding such a walk with $R = N - 1$ steps to the right is $P_N(N - 1) = Np^{N-1}q$.

In this way we see that the probability to find a walk with R steps to the right among all the N steps is given by the *Bernoulli probability distribution*

$$P_N(R) = \binom{N}{R} p^R q^{N-R}, \tag{8.1}$$

where $\binom{N}{R} = N!/R!(N-R)!$ is the number of such walks. The total probability for N steps irrespective of the number taken to the right is then

$$\sum_{R=0}^{N} P_N(R) = (p+q)^N = 1, \tag{8.2}$$

as it should be.

We can now use this normalized probability distribution to calculate different, average properties of a walk with N steps. For instance, the average number of steps to the right is

$$\langle R \rangle = \sum_{R=0}^{N} R P_N(R) = \sum_{R=0}^{N} \binom{N}{R} R p^R q^{N-R}. \tag{8.3}$$

The sum is easily done by writing $Rp^R = p(d/dp)p^R$. We then have

$$\langle R \rangle = \sum_{R=0}^{N} p\frac{d}{dp} p^R q^{N-R} \binom{N}{R} = p\frac{d}{dp} \sum_{R=0}^{N} \binom{N}{R} p^R q^{N-R}$$
$$= p\frac{d}{dp}(p+q)^N = Np(p+q)^{N-1} = Np. \tag{8.4}$$

This result is to be expected since of N steps, a fraction p is taken to the right. Since the displacement in one walk is $S = 2R - N$, we find for the average over a great many walks

$$\langle S \rangle = 2\langle R \rangle - N = 2Np - N(p+q) = N(p-q). \tag{8.5}$$

The symmetric walk has the same probability of going to the left as to the right. Then $p = q = 1/2$, and the average displacement is zero as expected.

In many cases we are interested in just the magnitude of the displacement irrespective of which direction it is. This can be obtained from the average $\langle R^2 \rangle$. Using the same trick as above for $\langle R \rangle$, we find

$$\langle R^2 \rangle = \sum_{R=0}^{N} R^2 P_N(R) = \left(p\frac{d}{dp}\right)^2 \sum_{R=0}^{N} \binom{N}{R} p^R q^{N-R}$$

$$= \left(p\frac{d}{dp}\right)^2 (p+q)^N = pN(p+q)^{N-1} + p^2 N(N-1)(p+q)^{N-2} \tag{8.6}$$

$$= pN + p^2 N(N-1) = (Np)^2 + Npq.$$

Since the absolute displacement is given by $S^2 = (2R - N)^2$, we obtain for the average

$$\langle S^2 \rangle = 4\langle R^2 \rangle - 4N\langle R \rangle + N^2$$

$$= N^2(4p^2 - 4p + 1) + 4Npq \tag{8.7}$$

$$= N^2(p-q)^2 + 4Npq.$$

The first term on the last line is just the square of the average displacement. Hence we can write the result as

$$\Delta S^2 \equiv \langle S^2 \rangle - \langle S \rangle^2 = 4Npq, \tag{8.8}$$

which gives the fluctuation around the average final position of the walk. For the symmetric walk, we have $\langle S \rangle = 0$ and thus

$$\langle S^2 \rangle = N. \tag{8.9}$$

We see that displacement increases with time as $|S| = N^{1/2}$, where the exponent is characteristic for random walks. For an ordinary, directed walk, we know that it is one. The random walk is, for obvious reasons, much more inefficient in covering a given distance.[1]

When the number of steps N is very large, we can approximate the Bernoulli distribution (8.1) by a Gaussian distribution. This is most easily shown in the symmetric case when $p = q = 1/2$. The probability for a net displacement of S steps to the right is then

$$P(S, N) = \left(\frac{1}{2}\right)^N \frac{N!}{R!\, L!}. \tag{8.10}$$

---

[1] It is easy to derive the central result Eq. (8.9) more directly. If the position of the walker after N steps is S_N, it will be at $S_{N+1} = S_N + 1$ with probability p at the next moment of time or at $S_{N+1} = S_N - 1$ with probability q. Squaring these two equations and taking the average, we obtain the recursion relation

$$\langle S^2 \rangle_{N+1} = \langle S^2 \rangle_N + 1$$

in the symmetric case $p = q = 1/2$. Starting at the origin with $\langle S^2 \rangle_0 = 0$, we then obtain $\langle S^2 \rangle_1 = 1$, $\langle S^2 \rangle_2 = 2$ and so on.

Table 8.1. *Probabilities for a random walk with ten steps.*

S	0	2	4	6	8	10
$P_{10}(S)$	0.246	0.205	0.117	0.044	0.010	0.001
Gauss	0.252	0.207	0.113	0.042	0.010	0.002

Using Stirling's formula, $n! = \sqrt{2\pi n}\, n^n e^{-n}$, we then get

$$P(S, N) = \sqrt{\frac{N}{2\pi RL}}\, e^{N \log N - R \log R - L \log L - N \log 2}$$

$$= \sqrt{\frac{N}{2\pi RL}}\, e^{-R \log (2R/N) - L \log (2L/N)}, \tag{8.11}$$

where $2R/N = 1 + S/N$, $2L/N = 1 - S/N$ and $RL = (1/4)(N^2 - S^2)$. Now expanding the logarithms to second order in S/N, we find

$$P(S, N) = \sqrt{\frac{2}{\pi N}} \exp\left(\frac{-S^2}{2N}\right). \tag{8.12}$$

In the prefactor, we have made the approximation $1 - S^2/N^2 \simeq 1$. This is due to the exponent which forces S^2 to be of order N. Thus $N^2/S^2 \propto 1/N$ and can be neglected when N is very large. The result is seen to be an ordinary Gaussian distribution with average value $\langle S \rangle = 0$ and width $\langle S^2 \rangle = N$, as we already have found.

The approximate Gaussian formula is quite accurate also when S is of the same order as N and N is rather small. This is illustrated in Table 8.1, where we have compared the approximate probability (8.12) with the exact Bernoulli distribution (8.1) for a walk with $N = 10$ steps.

When the number of steps N in the walk gets very large, we can assume that each step is very small of length a. If the walk takes place along the x-axis, the final position of the walker will then be $x = S\delta x$, which we now can assume is a continuous variable. Similarly, if the time interval τ between each consecutive step is also very small, the walk takes a time $t = N\tau$, which will also be a continuous variable. The probability for the walker to be at position x at time t is then, from Eq. (8.12),

$$p(x, t) = \frac{1}{2\delta x}\sqrt{\frac{2\tau}{\pi t}} \exp\left(-\frac{x^2 \tau}{2(\delta x)^2 t}\right). \tag{8.13}$$

The factor $1/2\delta x$ in front is needed to normalize this continuous probability distribution since the separation between each possible final position in walks with the same number of steps is $\Delta x = 2\delta x$. Introducing the diffusion constant

$$D_m = \frac{\delta x^2}{2\tau},$$ (8.14)

we can write the result as

$$p(x,t) = \frac{1}{\sqrt{4\pi D_m t}} \exp\left(\frac{-x^2}{4D_m t}\right).$$ (8.15)

At any given time, it is a Gaussian distribution, which is very peaked around $x = 0$ at early times and gets flatter at later times. Physically, this just reflects the fact that the probability to find the particle away from the origin increases with time. The area under the curve, however, remains constant. This is due to the normalization of the probability distribution, which in the discrete case was given by Eq. (8.2). Now it becomes

$$\int_{-\infty}^{\infty} dx\, p(x,t) = 1,$$ (8.16)

as is easy to check by direct integration.

Equipped with this normalized probability distribution, we can now calculate different properties of the continuous random walk. The mean position of the walker at time t is

$$\langle x(t) \rangle = \int_{-\infty}^{\infty} dx\, x p(x,t) = 0,$$ (8.17)

as we already know. Again it must be stressed that this average of the final position x is taken over a large number of symmetric walks, all starting at position $x = 0$ at time $t = 0$ and lasting a time t. The absolute displacement is similarly given by the average

$$\langle x^2(t) \rangle = \int_{-\infty}^{\infty} dx\, x^2 p(x,t) = 2D_m t,$$ (8.18)

which is just the continuous version of the discrete result, Eq. (8.9).

Random walks can easily be generalized to higher dimensions. For example, in two dimensions the walker can move in four directions at every step in time: up or down, right or left. In the symmetric case the probabilities for these four possibilities are each equal to $1/4$. If we let a random walk go on indefinitely, we will discover that it eventually comes back to the initial point one or more times and that the curve completely covers the two-dimensional plane. Random walks in three dimensions have neither of these properties.

The continuous probability distribution in three dimensions can be obtained from the product of three probabilities of Eq. (8.12), one for each direction. On the average, every time step will involve a step in each direction. The probability density for finding the walker at the position $\mathbf{x} = (x, y, z)$ is thus

$$p(\mathbf{x},t) = \left(\frac{1}{2a}\right)^3 \left(\frac{2}{\pi N/3}\right)^{3/2} \exp\left(\frac{-\mathbf{x}^2}{2Na^2}\right)$$

$$= \frac{1}{(4\pi D_m t)^{3/2}} \exp\left(\frac{-\mathbf{x}^2}{4D_m t}\right), \tag{8.19}$$

where $D_m = a^2/2\tau$ is the diffusion constant in three dimensions. With the unit normalization

$$\int d^3x\, p(\mathbf{x},t) = 1, \tag{8.20}$$

we now find the mean position of the walker at time t as

$$\langle x_i(t)\rangle = \int d^3x\, x_i\, p(\mathbf{x},t) = 0, \tag{8.21}$$

while the squared displacement follows from the average

$$\langle x_i(t)x_j(t)\rangle = \int d^3x\, x_i x_j\, p(\mathbf{x},t) = 2\delta_{ij} D_m t. \tag{8.22}$$

It gives as expected $\langle \mathbf{x}^2(t)\rangle = \langle x^2(t)\rangle + \langle y^2(t)\rangle + \langle z^2(t)\rangle = 6D_m t$, which also follows directly from the one-dimensional result, Eq. (8.18). Here the factor 6 comes from the fact that a timestep includes a step in the x, y or z direction. If, on the other hand, each timestep includes *simultaneous* steps in all directions, we must replace $t \rightarrow t/3$, and

$$\langle \mathbf{x}^2(t)\rangle = \langle x^2(t)\rangle + \langle y^2(t)\rangle + \langle z^2(t)\rangle = 2D_m t \tag{8.23}$$

becomes independent of dimension. This is the standard expression.

8.2 The Central Limit Theorem

In the previous chapter, a random walk was shown to have a Gaussian distribution. The central limit theorem generalizes this result to any sum of steps of random length, but finite variance. The last condition may be broken if the distribution of step lengths x_i converges more slowly than $1/x^3$ for large x_i. We will use this result to describe diffusion in terms of the diffusion equation.[2]

Gaussian distributions are highly common and may be found everywhere in nature, politics and the economy. For that reason, we should look for an explanation that does not rely on physical law.

---

[2] The central limit theorem is also a useful tool to understand many of the random processes we are surrounded by. Fluctuations are everywhere, and the theorem will sometimes let you distinguish between a fluctuation and a trend. If, for instance, a newspaper reports that the total tax income in the city of Oslo is dramatically decreasing, it may be a simple calculation to find that the decrease is within the expected year-to-year variation.

Very often a variable X of interest is the sum of a large number of increments, so that

$$X = \sum_{i=1}^{N} x_i, \tag{8.24}$$

where the x_i are random variables with a distribution $p_0(x)$ and zero mean $\langle x \rangle = 0$. This distribution may have any form, but the variance of x must be finite, so that we may define

$$\sigma^2 = \int dx \, x^2 p_0(x). \tag{8.25}$$

The question now is what the distribution for X is. The probability of finding a given X is the sum of the probabilities of all the sets $\{x_i\}$ that sum up to X. This sum may be written

$$p(X) = \int dx_1 \, dx_2 \ldots dx_N \, p_0(x_1) p_0(x_2) \ldots p_0(x_N) \delta \left(X - \sum_{i=1}^{N} x_i \right). \tag{8.26}$$

To progress, we take the Fourier transform of $p(x)$:

$$\hat{p}(k) = \frac{1}{2\pi} \int dX \, e^{-ikX} p(X)$$

$$= \frac{1}{2\pi} \int dX \, e^{-ikX} \int dx_1 \ldots dx_N \, \delta \left(X - \sum_{i=1}^{N} x_i \right) p_0(x_1) \ldots p_0(x_N)$$

$$= \frac{1}{2\pi} \int dx_1 \ldots dx_N \, \exp \left(-ik \sum_{i=1}^{N} x_i \right) p_0(x_1) \ldots p_0(x_N) \tag{8.27}$$

$$= \frac{1}{2\pi} \prod_{i=1}^{N} \int dx_i \, e^{-ikx_i} p_0(x_i)$$

$$= (2\pi)^{N-1} \prod_{i=1}^{N} \hat{p}_0(k) = \frac{1}{2\pi} (2\pi \, \hat{p}_0(k))^N,$$

where $\hat{p}_0(k)$ is the Fourier transform of $p_0(x)$ The above result is a generalized convolution theorem. We then invert the transform to get back to $p(X)$:

$$p(X) = \int dk \, e^{ikX} \hat{p}_0(k) = \frac{1}{2\pi} \int dk \, e^{ik'X/N} (2\pi \, \hat{p}_0(k'/N))^N, \tag{8.28}$$

where we have made the substitution $k' = Nk$. We may write $\hat{p}_0(k)$ in terms of $p_0(x)$ as

$$\hat{p}_0(k) = \frac{1}{2\pi} \int dx \, e^{-ikx} p_0(x) = \frac{1}{2\pi} \sum_{n=0}^{\infty} \frac{(-ik)^n \langle x^n \rangle}{n!}, \tag{8.29}$$

where we have used the Taylor series for the exponential. Since $\langle x \rangle = 0$, the linear term in $\hat{p}_0(k)$ vanishes. The zeroth order term is $1/(2\pi)$, and the second order term is $-(k\sigma)^2/(4\pi)$. For that reason we may make the following approximations:

$$(2\pi \hat{p}_0(k/N))^N \approx \left(1 - \frac{(k\sigma)^2}{2N^2}\right)^N \rightarrow \exp\left(-\frac{k^2\sigma^2}{2N}\right) \tag{8.30}$$

when $N \rightarrow \infty$. In the above we have used the formula $(1 + x/N)^N \rightarrow e^x$ with $x = -(k\sigma)^2/2N$. Using the above approximation in Eq. (8.28), we get

$$
\begin{aligned}
p(X) &= \int dk \, e^{ikX} \hat{p}_0(k) \\
&= \frac{1}{2\pi} \int dk \, \exp\left(ikX - \frac{1}{2}k^2 N\sigma^2\right) \\
&= \frac{1}{\sqrt{2\pi N\sigma^2}} \exp\left(-\frac{X^2}{2N\sigma^2}\right),
\end{aligned}
\tag{8.31}
$$

where, at the end, we have performed the Gaussian integral. This result is the central limit theorem. It shows that X has a Gaussian distribution with variance $\langle X^2 \rangle = N\sigma^2$.

8.3 Advection–Diffusion Equation

We start in one dimension and consider the particle, or molecule, as it performs a random walk due to molecular collisions. If the molecule starts out at $x = 0$ at $t = 0$, we may write for its position

$$x(t) = \sum_{i=1}^{N_0} \Delta y(t_i), \tag{8.32}$$

where the step-length $\Delta y(t_i)$ has a finite variance $\sigma^2 = \langle \Delta y^2 \rangle$. For this reason we may apply the central limit theorem and immediately write down the probability of finding the particle at some other position at a later time:

$$p(x,t) \propto \exp\left(-\frac{x^2}{2N_0\sigma^2}\right). \tag{8.33}$$

We will not need the microscopic information of σ^2, and we observe that $N_0 \propto t$, so we can make the replacement $2N_0\sigma^2 \rightarrow 4D_m t$, where D_m is the diffusion constant. Writing things like this, Eq. (8.33) may immediately be used to show that $\langle x^2 \rangle = 2D_m t$. By normalization we then get

$$p(x,t) = \frac{1}{\sqrt{4\pi D_m t}} \exp\left(\frac{-x^2}{4D_m t}\right), \tag{8.34}$$

which is just Eq. (8.15) again. Note that

$$p(x,t) = \frac{1}{\sqrt{4\pi D_m t}} \exp\left(\frac{-x^2}{4D_m t}\right) \rightarrow \delta(x) \tag{8.35}$$

when $t \rightarrow 0$. This is just the mathematical statement that we initially know where the particle is located.

8.3.1 Green's Function

If we have $N \gg 1$ particles, initially all at the position x_0, their concentration

$$C(x,t) = \frac{\Delta N}{\Delta x} = Np(x,t), \tag{8.36}$$

and we note that their number is conserved as

$$\int dx \, C(x,t) = N. \tag{8.37}$$

It is always possible to consider an arbitrary concentration profile at some time $t = 0$ as a sum of delta-functions through the identity

$$C(x,0) = \int dx_0 \, C(x_0,0)\delta(x - x_0). \tag{8.38}$$

If the particles do not interact with each other, their evolution is independent, so that each delta-function evolves independently according to Eq. (8.34). This means that the concentration at some later time t is

$$C(x,t) = \int dx_0 \, C(x_0,0)\frac{1}{\sqrt{4\pi D_m t}} \exp\left(-\frac{(x - x_0)^2}{4D_m t}\right), \tag{8.39}$$

where the probability function

$$\frac{1}{\sqrt{4\pi D_m t}} \exp\left(-\frac{(x - x_0)^2}{4D_m t}\right) \tag{8.40}$$

is usually referred to as the Green's function for diffusion.

8.3.2 Conservation Law

Figure 8.1 illustrates a one-dimensional system populated with particles and the net current J_x (number of particles per unit time) passing a given point $x > 0$. Since particles are neither created nor destroyed, this current equals the rate of change of

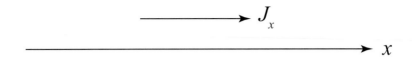

Figure 8.1 Diffusive current J_x along the x-axis.

the particle number to the right of x. Taking all particles to start out at $x = 0$, this statement of particle conservation may be written

$$
\begin{aligned}
J_x &= \frac{\mathrm{d}}{\mathrm{d}t} \int_x^\infty \mathrm{d}x'\, N p(x',t) \\
&= \frac{\mathrm{d}}{\mathrm{d}t} \int_u^\infty \frac{\mathrm{d}u'}{\sqrt{4\pi}} N \mathrm{e}^{-u'^2/4} \\
&= \frac{x}{2t} \frac{N}{\sqrt{4\pi D_m t}} \exp\left(-\frac{x^2}{4 D_m t}\right) \\
&= \frac{x}{2t} N p(x,t) = -D_m \frac{\partial}{\partial x} N p(x,t),
\end{aligned}
\tag{8.41}
$$

where we have made the substitution $u = x/\sqrt{D_m t}$. The relation

$$
J_x = -D_m \frac{\partial C(x,t)}{\partial x}
\tag{8.42}
$$

is known as Fick's law, and since it is linear in C, it must hold for any sum of initial delta-function-shaped initial concentrations and, therefore, for any initial concentration profile.

In three dimensions, we can locally align the x-axis with ∇C. Then there are locally no variations in the transverse directions, and by symmetry, the diffusive flow must be in the direction of the gradient. This implies that Eq. (8.42) generalizes to

$$
\mathbf{J} = -D_m \nabla C.
\tag{8.43}
$$

In three dimensions, $C(\mathbf{x},t)$ is a particle number per volume, rather than length, and $\mathbf{J}$ is the number of particles per unit time and area that flows through space. The diffusion constant has, as always, dimensions of length^2/time.

Fick's law holds, as we have seen, for particles that do not interact with each other. For small gradients it may be generalized to the case of interacting particles by the introduction of a C-dependent diffusion coefficient $D_m(C)$.

Fick's law describes the situation when individual particles have equal probability of moving in any direction. The net flow that results only happens because of concentration gradients. If there is a background flow velocity $\mathbf{u}$ of some carrier fluid, the current must be modified to

$$\mathbf{J} = C\mathbf{u} - D_m \nabla C, \qquad (8.44)$$

just as the case was when $p \neq q$ for the random walkers.

Now, the fact that particles are not created or annihilated may be stated by saying that inside a given volume V in space, the particle number change is entirely due to flow across the surface of that volume, or in mathematical terms:

$$\frac{d}{dt} \int_V dV \, C(\mathbf{x}, t) = -\int dS \cdot \mathbf{J}. \qquad (8.45)$$

Using Eq. (8.44) this becomes

$$\frac{d}{dt} \int_V dV \, C(\mathbf{x}, t) = \int dS \cdot (D_m \nabla C - C\mathbf{u})$$
$$= \int dV \, \nabla \cdot (D_m \nabla C - C\mathbf{u}), \qquad (8.46)$$

where Fick's law and Gauss' theorem have been used to re-write the surface integral. By simply moving the right-hand side to the left and taking the time-derivative inside the integral, this equation becomes

$$\int_V dV \left(\frac{dC}{dt} + \nabla \cdot (C\mathbf{u}) - D_m \nabla^2 C \right) = 0, \qquad (8.47)$$

which holds for any integration volume V. Therefore, the integrand must be zero, and

$$\boxed{\frac{dC}{dt} = D_m \nabla^2 C - C\nabla \cdot \mathbf{u},} \qquad (8.48)$$

where the substantive derivative is given by Eq. (6.15). This equation is known as the *advection–diffusion equation* and results from the combination of Fick's law, flow and local mass conservation. The last $(C\nabla \cdot \mathbf{u})$-term describes changes in C due to density changes of the carrier fluid and vanishes if this is incompressible $(\nabla \cdot \mathbf{u} = 0)$.

For a spatial delta-pulse of concentration in three-dimensional Eq. (8.48) has the solution

$$p(\mathbf{x}, t) = p(x, t)p(y, t)p(z, t), \qquad (8.49)$$

which is easily checked by writing the equation in the form

$$\left(\frac{\partial}{\partial t} - \frac{\partial^2}{\partial x^2} + \frac{\partial^2}{\partial y^2} + \frac{\partial^2}{\partial z^2} \right) p(\mathbf{x}, t) = 0. \qquad (8.50)$$

By the same arguments of linearity as above, the concentration in three-dimensional space may be written

$$C(\mathbf{x},t) = \int d^3x_0 \frac{C(\mathbf{x}_0,0)}{(4\pi D_m t)^{3/2}} \exp\left(-\frac{(\mathbf{x}-\mathbf{x}_0)^2}{4D_m t}\right). \tag{8.51}$$

This is formally the general solution to the initial value problem given by the diffusion equation and a given initial condition $C(\mathbf{x},0)$.

8.3.3 Diffusion of Temperature

While Eq. (8.48) is taken to reflect the conservation of a concentration or mass, diffusion is a mechanism that governs other processes too, like the spreading of conserved energy. In particular, if the energy is just the internal energy given by the local heat capacity and temperature, this is the case.

Let the energy inside a small volume element $dE = \epsilon dV$, where ϵ is the energy density, which we will take only to depend on the temperature T. This means that a change $d\epsilon = c_v dT$, where c_v is the heat capacity per unit volume. If dV is kept fixed and the temperature changes inside it over time, $\partial dE/\partial t = dV c_v \partial T/\partial t$, so that the total rate of energy change

$$\frac{d}{dt}\int dV\epsilon = \int dV c_v \frac{\partial T}{\partial t}. \tag{8.52}$$

Let the energy inside a small volume element $dE = \epsilon dV = \rho c_V dV$, where ϵ is the energy density. Then the local conservation of energy is expressed as the statement that the only energy change inside any given volume is that caused by the current across the surface of that volume:

$$\int dV \frac{\partial \epsilon}{\partial t} = -\int d\mathbf{S}\cdot\mathbf{j}_\epsilon, \tag{8.53}$$

where the energy current may, under rather general conditions, be written as

$$\boxed{\mathbf{j}_\epsilon = -\lambda\nabla T,} \tag{8.54}$$

where λ is the thermal conductivity. Eq. (8.54) is known as Fourier's law and plays the same role for the flow of energy as Fick's law does for the flow of mass. Using Gauss' law on Eq. (8.53) and inserting Eq. (8.54) immediately gives the diffusion equation for the flow of heat as

$$\boxed{\frac{\partial T}{\partial t} = D_T\nabla^2 T,} \tag{8.55}$$

where the thermal diffusivity $D_T = \lambda/c_V$.

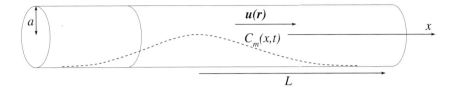

Figure 8.2 Capillary pipe of radius a with a concentration profile that spreads out over a distance L.

8.4 Taylor Dispersion

Taylor (1953) studied the dispersion of tracers in a capillary pipe (radius a) at low flow velocities. He found that the tracers dispersed relative to a plane that moves with the average flow velocity U "*exactly as though it were being diffused by a process which obeys the same law as molecular diffusion.*" However, the diffusion coefficient is quite different from the molecular one. The cross-sectional average of the tracer concentration is taken, and it is the spreading of this that is measured by the new diffusion-, or rather dispersion coefficient $D_{\parallel}$. We will follow Taylor's original derivation of this coefficient.

The low Reynolds number flow through a pipe such as that in Figure 8.2 is given in Eq. (5.22) and may be written

$$u(r) = u_0 \left(1 - \frac{r^2}{a^2}\right), \tag{8.56}$$

where r is the distance from the center of the pipe and u_0 the maximum velocity. The cross-sectional average

$$U = \frac{2}{a^2} \int_0^a dr\, r u(r) = \frac{u_0}{2}. \tag{8.57}$$

The corresponding average of the concentration is defined as

$$C_m = \frac{2}{a^2} \int_0^a dr\, r C(r,x). \tag{8.58}$$

If there were no molecular diffusion at all, this flow profile would cause an endless stretching of any initial concentration profile, with a steady increase of the concentration gradients. In particular, there would be a stationary layer near the wall, were $u \approx 0$, and a central part that would move at the maximum speed u_0. The result of such a deformation is in fact that, eventually, any amount of molecular diffusion would be significant over the length scales over which C varies.

For non-zero D_m, the effects of molecular diffusion are indeed the key to the whole phenomenon at hand. As long as the pipe is sufficiently narrow and long, we will always arrive in the regime where diffusion has had time to even out the concentration variations across the channel, while there are still large scale variations along it. The main physical effect is that molecular diffusion causes tracer molecules to traverse streamlines and therefore move with the flow at variable speeds. The fact that their average speed converges to U explains Taylor's main finding. The time needed for transverse C-variations to relax is

$$t_\perp = \frac{a^2}{D_m}. \tag{8.59}$$

Using this, it is in fact straighforward to get an estimate of the *longitudinal dispersion coefficient* $D_\parallel$ by noting that the squared longitudinal displacement during this time is roughly

$$(u_0 t_\perp)^2 \sim D_\parallel t_\perp, \tag{8.60}$$

which by insertion of Eq. (8.59) gives

$$D_\parallel \sim \frac{u_0^2 a^2}{D_m}. \tag{8.61}$$

It is rather surprising perhaps that $D_\parallel$ is *inversely* proportional to D_m. This is best understood by noting that the longer two molecules stay on, or close to their respective streamlines, the further they move apart. It is only the molecular diffusion that may cause a change of streamlines, and for this reason a larger D_m should be expected to cause a smaller $D_\parallel$.

We shall take the observation time $t \gg t_\perp$ as well as the time needed for a longitudinal variation in C to pass, $t_\parallel = L/u_0 \gg t_\perp$. We are therefore in the regime defined by the condition

$$\frac{a^2}{D_m} \ll \frac{L}{u_0}. \tag{8.62}$$

However, even though the transverse variations have essentially relaxed, we shall assume that the second derivative

$$\frac{\partial^2 C}{\partial x^2} \ll \frac{\partial^2 C}{\partial r^2}, \tag{8.63}$$

which is consistent with the fact that transverse gradients in C keep building up due to the velocity gradients.[3]

---

[3] It is possible in the end to verify the self-consistency of this assumption by using the final evolution equation for C_m that we are after.

Writing the Laplacian in cylindrical coordinates, the advection–diffusion Eq. (8.48) takes the form

$$D_m \left(\frac{\partial^2 C}{\partial r^2} + \frac{1}{r} \frac{\partial C}{\partial r} + \frac{\partial^2 C}{\partial x^2} \right) = \frac{\partial C}{\partial t} + u(r) \frac{\partial C}{\partial x}, \tag{8.64}$$

where we have used the assumption that there are no angular variations, that is, $\partial C/\partial\phi = 0$, and also that the flow is incompressible, $\nabla \cdot \mathbf{u} = 0$. Eq. (8.63) suggests dropping the $(\partial^2 C/\partial x^2)$-term. Doing this and introducing the radial coordinate $z = r/a$ yields

$$\frac{D_m}{a^2} \left(\frac{\partial^2 C}{\partial z^2} + \frac{1}{z} \frac{\partial C}{\partial z} \right) = \frac{\partial C}{\partial t} + u(z) \frac{\partial C}{\partial x}. \tag{8.65}$$

Now, we will transform this equation to the frame of reference that moves with U. This is acheived by the coordinate transform $x_1 = x - Ut$, or

$$\frac{\partial C(x,t)}{\partial t} = \frac{\partial C(x_1,t)}{\partial t} - U \frac{\partial C(x_1,t)}{\partial x_1}, \tag{8.66}$$

$$\frac{\partial C(x,t)}{\partial x} = \frac{\partial C(x_1,t)}{\partial x_1}. \tag{8.67}$$

These results may also be obtained from Eq. (6.15) by noting that the substantive derivative equals $\partial/\partial t$ in the co-moving frame of reference where the flow velocity (in this case, the average flow) vanishes.

We are interested in obtaining a description of the dispersion in this co-moving frame. In these coordinates, Eq. (8.65) takes the form

$$\frac{D_m}{a^2} \left(\frac{\partial^2 C}{\partial z^2} + \frac{1}{z} \frac{\partial C}{\partial z} \right) = (u(z) - U) \frac{\partial C}{\partial x_1} + \frac{\partial C}{\partial t}. \tag{8.68}$$

To see what this equation describes, we integrate it over a volume slice that is located between x_1 and $x_1 + \Delta x$ and apply Gauss' theorem to make the volume integral a surface integral. This yields

$$\frac{D_m}{a^2} \int_{z=1} d\mathbf{S} \cdot \nabla C = \int_{x_1, \, x_1+\Delta x} dS \, (u(z) - U)C + \frac{d}{dt} \int dV \, C, \tag{8.69}$$

where the subscripts x_1, $x_1 + \Delta x$ denote that the surface integral is to be taken over the two surfaces of these constant x_1-values. The left side vanishes because the boundary condition that the capillary walls are impenetrable implies that

$\mathbf{dS} \cdot \nabla C = 0$ at $z = 1$. Note that here $\mathbf{dS}$ points along the surface normal of the pipe. The first term on the right may be written $\Delta Q = Q(x_1 + \Delta x, t) - Q(x_1, t)$, where the current

$$Q(x_1, t) = 2\pi a^2 \int_0^1 dz\, z(u(z) - U)C(x_1, z, t). \tag{8.70}$$

When we take Δx to be small, the last term may be written $\pi a^2 \Delta x\, \partial C_m / \partial t$, so that we are left with

$$\frac{\partial C_m}{\partial t} = -\frac{1}{\pi a^2} \frac{\partial Q}{\partial x_1}, \tag{8.71}$$

which expresses the conservation of C_m.

It can be seen from Eqs (8.70) and (8.57) that $Q = 0$ unless C varies with z. Since we have discarded the $(\partial^2 C / \partial x^2)$-term, there is no diffusive term in the x_1-direction. In order to obtain these z-variations, we use Eq. (8.65) and make the approximation of being in a steady state, so that $\partial C / \partial t \approx 0$, and we take $\partial C / \partial x_1$ to be z-independent, so that we may replace it by $\partial C_m / \partial x_1$. These two approximations keep the leading order terms and yield the ordinary differential equation

$$\frac{D_m}{a^2} \left(\frac{\partial^2 C(x_1, z)}{\partial z^2} + \frac{1}{z} \frac{\partial C(x_1, z)}{\partial z} \right) = (u(z) - U) \frac{\partial C_m(x_1)}{\partial x_1}. \tag{8.72}$$

It has the simple polynomial solution

$$C(x_1, z) = C(x_1, 0) + \frac{a^2 u_0}{8 D_m} \left(z^2 - \frac{z^4}{2} \right) \frac{\partial C_m(x_1)}{\partial x_1}, \tag{8.73}$$

which is easily verified by substitution. Inserting this result in Eq. (8.70), the tracer flux across the constant x_1-surface becomes

$$Q = -\frac{\pi U^2 a^4}{48 D_m} \frac{\partial C_m(x_1)}{\partial x_1}, \tag{8.74}$$

where the factor of 48 comes from working out the integral over the polynomial in z, and we have substituted $u_0 = 2U$.

Eq. (8.71), which describes the dispersive behavior in the co-moving frame, now takes the form

$$\frac{\partial C_m(x_1, t)}{\partial t} = D_\| \frac{\partial^2 C_m}{\partial x_1^2}, \tag{8.75}$$

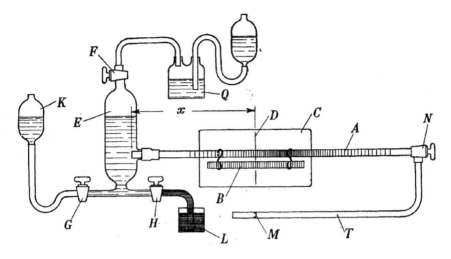

Figure 8.3 The experimental setup used by Taylor (1953).

where the dispersion cofficient

$$D_{\parallel} = \frac{a^2 U^2}{48 D_m}.$$ (8.76)

This is the famous result by Taylor. It is quite remarkable: If a small pulse of tracer is injected at $x = 0$, say, in the lab frame of reference, the tracer region will be displaced at a velocity U and spread out symmetrically as it moves along. This means that a central part of the fluid, which moves at twice the speed u_0, will absorb tracer as it catches up with this region, and then, as it passes through, it will give up the tracer again and eventually move on as pure solvent at speed u_0.

Taylor decided to verify this prediction experimentally, using the setup in Figure 8.3. The QFE-part of this system is designed to control the pressure, or suction. The container labeled K delivers distilled water and L a mixture of water and $KMnO_4$ from which a tracer pulse may be delivered. The capillary pipe leading up to the outlet valve has an internal diameter of 0.5 mm. The pipe B contains a series of mixtures with known reference concentrations, so that by comparing the color in the main pipe A, different concentrations at different positions x were measured.

The resulting measurements are shown in Figure 8.4, which indeed shows how the concentration moving at velocity U spreads out symmetrically. Fitting these curves to Gaussians also allows the determination of $D_{\parallel}$ and by Eq. (8.76), the determination of the molecular diffusion constant D_m. These measurements were found to be consistent with independent results, thus supporting the theory further.

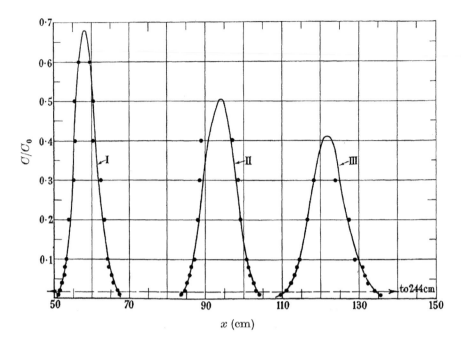

Figure 8.4 The concentration profiles measured at late times. (After Taylor (1953).)

8.5 Geometric Dispersion*

Having considered the case of dispersion in an open pipe, we now add a porous medium to it.

Consider a column packed with a fine powder, with a typical particle diameter d, and saturated with a fluid flowing at a constant average velocity, U, as indicated in Figure 8.5. Since we are primarily interested in longitudinal dispersion, we consider the one-dimensional version of the problem, *plug flow*, where a thin layer of marker molecules is found as a sheet across the sample at $x = 0$ for $t = 0$.

At a later time t, the average position of the marker molecules will be at $x = Ut$. However, the tracer molecules will be smeared around the average position. As we have seen, when the fluid moves, mixing occurs because of velocity gradients that arise in even the simplest flow geometries. In a porous medium, there is an additional mixing or *dispersion* due to the stagnation points that separate flow lines and bring the random nature of the medium into play. Again the spread of a tracer obeys the law of diffusion but with a (longitudinal) dispersion coefficient $D_{\parallel}$ replacing the ordinary molecular diffusion coefficient.

In order to get the behavior of $D_{\parallel}$ with U, we consider a tracer molecule that is moving along with the flow at an average velocity U. Moreover, we shall assume that there is a correlation time $\tau = d/U$ such that for $t > \tau$, the fluctuations in

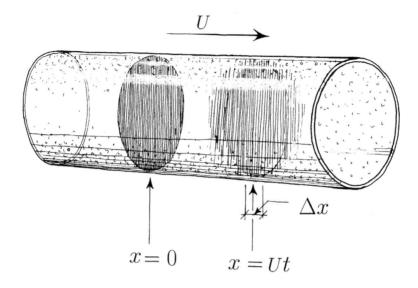

Figure 8.5 Dispersion of tracers convected by flow through a porous medium. The flow is one-dimensional since the average velocity, U, is constant over the cross-section. A sheet of tracers starts at $x = 0$. After a time t, the average position of the tracers is $x = Ut$ with a characteristic spread of Δx.

velocity become uncorrelated. This means that we may decompose the displacement of the tracer particle into uncorrelated steps of typical length d so that

$$X(t) = \sum_{i=1}^{N}(U\tau + \Delta x_i),\tag{8.77}$$

where the random part of the step length also has a typical magnitude $U\tau$, so that $\langle \Delta x_i \Delta x_j \rangle = (U\tau)^2 \delta_{ij}$. Therefore

$$\langle \Delta X^2 \rangle = \sum_{i=1}^{N}\langle \Delta x_i^2 \rangle = N\langle \Delta x_i^2 \rangle = \frac{t}{\tau}(U\tau)^2 = (Ud)t.\tag{8.78}$$

We may then write

$$\langle \Delta X^2 \rangle = 2D_\parallel t,\tag{8.79}$$

with

$$D_\parallel = \frac{Ud}{2}.\tag{8.80}$$

This result should be compared to that of Taylor dispersion, where the correlation time $\tau \sim t_\perp = a^2/D_m$ given in Eq. (8.59). In this case τ is U-independent, and so $D_\parallel \sim U^2$ instead of the above $D_\parallel \sim U$ dependence.

To characterize the competition between molecular diffusion and convective dispersion, one defines the *Peclet number*

$$\text{Pe} = \frac{Ud}{D_m}. \tag{8.81}$$

This dimensionless number may be viewed as the ratio of two time scales. The time it takes to diffuse a typical pore size, d, is $t_{\text{diff}} \simeq d^2/D_m$, whereas the advection time for tracers is estimated by $t_{\text{flow}} = d/U$. With these expressions we find that Pe may be expressed as $\text{Pe} = t_{\text{diff}}/t_{\text{flow}} = (Ud)/D_m$.

At very low flow velocities, $\text{Pe} \ll 1$, dispersion is controlled by molecular diffusion, and we expect the form

$$D_\parallel = \gamma D_m, \quad \text{for Pe} \ll 1. \tag{8.82}$$

Here the dispersion coefficient is equal to the diffusion coefficient reduced by a factor γ since diffusion confined to the pore space is less efficient than unconstrained diffusion. The coefficient γ depends on the pore structure. For packings of beads or sands, $\gamma \simeq 0.7$ (de Ligny, 1970).

At high velocities, dispersion should no longer depend on molecular diffusion but rather on the (average) flow velocity U and the geometric mixing length, which is of the same order of magnitude as the typical pore size d. We expect that

$$D_\parallel \sim Ud = D_m \text{Pe}, \quad \text{for Pe} \gg 1. \tag{8.83}$$

These limiting forms may be obtained from the *empirical* expression (de Ligny, 1970; Dullien, 1992)

$$D_\parallel = D_m \left(\gamma + \lambda_\parallel \frac{\text{Pe}}{1 + C_\parallel/\text{Pe}} \right). \tag{8.84}$$

Experimental results on fluid flow in sand packs give $\lambda_\parallel \simeq 2.5$ and $C_\parallel \simeq 8.8$ (de Ligny, 1970). Taylor dispersion corresponds to moderate to small Pe. Note that Eq. (8.62) may be written $\text{Pe} \ll L/a$ with a Peclet number defined as $\text{Pe} = u_0 a/D_m$. In this regime, Eq. (8.84) will take the approximate form

$$D_\parallel \approx D_m \left(\gamma + \lambda_\parallel \frac{\text{Pe}^2}{C_\parallel} \right), \tag{8.85}$$

which is consistent with the U^2 behavior of Eq. (8.76) provided the γ-term may be considered much smaller than the other term, which is usually the case for Taylor dispersion.

If, in the one-dimensional flow geometry shown in Figure 8.5, the tracer starts in a small volume instead of a sheet across the sample, one finds that the tracer also disperses in a direction perpendicular to the flow direction. This dispersion is characterized by a transverse dispersion coefficient $D_\perp$. For very low flow velocities, transverse dispersion is also controlled by diffusion so that $D_\perp = D_\parallel = \gamma D_m$ for $U \to 0$. However, in the regime of convective dispersion, that is, for Pe $\gg 1$, one finds that $D_\perp \ll D_\parallel$. Again an empirical relation connects low and high Peclet number regimes (Dullien, 1992):

$$D_\perp = D_m \left(\gamma + \lambda_\perp \frac{\text{Pe}}{1 + C_\perp/\text{Pe}} \right), \tag{8.86}$$

where empirical values are $\lambda_\perp \simeq 0.08$ and $C_\perp \simeq 78$. For large Peclet numbers (Pe $\gg 78$), Eqs (8.84) and (8.86) lead to the asymptotic expression $D_\perp \simeq D_\parallel/31$. This relatively simple picture gets modified in the presence of stagnation effects, whereby the fluid can get trapped in porous grains where the flow velocity is zero. Porous rocks are not all that homogeneous, and further dispersion is caused by variations in permeability and porosity.

8.5.1 An Experiment on Geometric Dispersion*

The effects of diffusion and mixing due to flow through a random structure are easily demonstrated. A channel $1 \times 5 \times 30$ cm was filled with glycerol that was pumped through the channel by a peristaltic pump. The channel was blocked by a random array of cylinders as seen in Figure 8.6.

The fluid was glycerol with a viscosity $\mu = 900$ cp $= 0.9$ Ns/m^2. In front of the random array of cylinders that traverse the 1 cm gap between the Perspex windows, we injected a line of colored glycerol. We used Nigrosin as a dye, which was also used by Hiby (1962) in a similar experiment. For the success of the experiment, it is essential that the dye solution has the same density as the fluid in the chamber and that there are no temperature gradients in the system. In the sequence of photographs, Figure 8.6, we see that an initially straight streak of dye became deformed and snaked through the array of cylinders as the clear glycerol was pumped from the left to the right in Figure 8.6. The dye string got stuck at the stagnation point of each cylinder, where it became thinner and thinner as the flow proceeded. When the flow direction was reversed, the dye string, marking the flow, retraced its steps – and finally the original straight dye streak was recovered! If, however, the flow proceeded longer before it was reversed, we observed that the 'time reversed' flow did not reproduce the original dye streak exactly. We observed a spread at positions that corresponded to the stagnation points of the random array (see Figure 8.7(b)). The physical basis of this effect is simple. Since the fluid cannot flow past the stagnation points in the array, fluid elements containing dye virtually stop at these points, get stretched and must thin down, eventually to molecular dimensions. A detailed discussion of this phenomenon can be found in a paper by Oxaal et al. (1994).

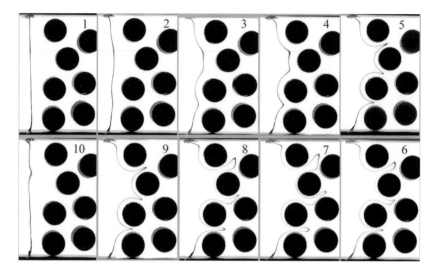

Figure 8.6 The reversible flow through a random array of cylinders. The flow of a glycerol solution was from left to right in parts 1–7. At part 7 the flow was reversed. The line of dye (nigrosine) tracer molecules were convected with the flow and could not pass the stagnation points defined by the cylinders that traverse the channel. As the flow was reversed (part 7), the flow and tracer molecules retraced their paths, and finally (in part 10) the initial straight line of dye was recovered. The dye line 'echo' was not completely straight in its original place. This error was due to imperfections in the experiment, such as temperature gradients and the fact that the original dye streak was not in the symmetry plane of the channel.

As an example of the qualitative arguments needed in the discussion of complex phenomena, such as flow in porous media, let us get some feeling for the parameters involved by a discussion of the simplified problem in Figure 8.7(a). Here a single cylinder was placed in a duct. In Figure 8.7(a), the white line marks the position of the initial dye streak, that is, at $t = 0$. After a time t, the dye streak was deformed as seen in Figure 8.7(a). The two lobes of the dye streak reached a distance $x = ut$, where u is the maximum velocity in the channel. Let a be the initial dye streak radius. The dye in a segment of length equal to the cylinder diameter, d, must be distributed over a string of length $L \simeq 2ut$, therefore it is thinned down to an average radius b given by the relation

$$\pi b^2 L = \pi a^2 d. \tag{8.87}$$

In the time $t = L/2u$, the molecular diffusion gave rise to a mean square spread in b of $\Delta b^2 = 2D_m t = D_m L/u$. At the return of the dye streak to its original position, the returned streak will now have a spread Δa, due to molecular diffusion, which dominated when the streak was thinned down at the stagnation point. In the

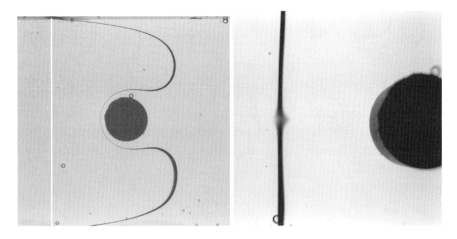

Figure 8.7 The deformation of a dye streak around an obstacle. Glycerol flows from left to right in a channel where the flow is blocked by a single cylinder traversing the channel. (a) The nigrosine colored glycerol dye streak started at the white line. The streak became very thin at the stagnation point since it was stretched there. When the flow was reversed and the dye streak returned to its original position, a broadening of the original line at a position that corresponded to the stagnation point was observed. The broadening is due to diffusion. (After Flekkøy et al. (1996).)

reversed flow, instead of thinning down, the streak in the stagnation region gets pushed back together and thickened to its original thickness, but with an increased spread. The relative spreads $\Delta a/a$ and $\Delta b/b$ must be equal; therefore, we find using Eq. (8.87) that

$$(\Delta a/a)^2 = \frac{D_m L}{b^2 u} = \frac{1}{Pe}\left(\frac{L}{a}\right)^2, \tag{8.88}$$

where again Pe is the Peclet number. At high Pe, the relative spread is small because diffusion is negligible compared with advection. The critical value of the Peclet number is obtained by setting $\Delta a/a = 1$, with the result

$$Pe^* = (L/a)^2. \tag{8.89}$$

The condition that the stagnation point dispersion is negligible, and therefore the flow appears to be reversible, is given by the condition $Pe > Pe^*$, which can always be achieved by increasing the velocity. However, as we will discuss later, to obtain the simple reversible creeping flow as observed in our experiments, we must also satisfy the condition of low Reynolds number flow,

$$Re = \frac{\rho u d}{\mu} \ll 1, \tag{8.90}$$

where the density is ρ, and μ is the fluid viscosity. Combining these two conditions, we find that

$$D_m(L/a)^2 \ll ud \ll \mu/\rho. \tag{8.91}$$

As is seen from this condition, the experiment is always possible for small displacements L. But in order to have a reasonable value of L, without too much smearing by diffusion, a high viscosity fluid is preferred. The relation between the viscosity and the molecular diffusion is given approximately by the *Einstein–Stokes relation* (Einstein, 1905)

$$D_m = \frac{k_B T}{6\pi \mu R_m}, \tag{8.92}$$

where R_m is the molecular radius, k_B is Boltzmann's constant and T is the absolute temperature of the fluid. With this expression for D_m, we rewrite the combined condition (8.91) of reversible flow as

$$\left(\frac{L}{a}\right)^2 \ll \frac{6\pi \mu^2 R_m}{\rho k_B T}. \tag{8.93}$$

Therefore, a high viscosity is necessary, and a large molecular radius is favorable for reversible flow.

8.6 First Arrival Times*

We will now pose a question that needs a very different approach to answer than those we have posed so far: When do we see the first trace of a tracer on the other side? To be more precise, we inject a tracer in the porous medium at the inlet point. When do we *first* detect it at the outlet?

Following Devillard (1993), we will assume that we are at the high Peclet number limit where the dispersion mechanism is geometric. Furthermore, we see the porous medium as a network where each link (=pore) i has a length l_i and a maximum velocity v_i associated with it; the maximum velocity here refers to the maximum of the transversal velocity profile in the link. The shortest time a tracer will take in crossing the link is then given by

$$t_i = \begin{cases} \frac{l_i}{v_i}, & \text{with the flow direction in the link,} \\ \infty, & \text{against the flow direction in the link,} \end{cases} \tag{8.94}$$

where the two possibilities distinguish between the tracer attempting to cross the link in the direction of the flow in the link or against the direction of the flow in the link – in which case it would take "forever." Suppose now $\mathcal{P}$ is some path from the inlet to the outlet of the porous medium. What would be the first arrival time *along this particular path*, $T_\mathcal{P}$? It would be

$$T_\mathcal{P} = \sum_{i \in \mathcal{P}} t_i. \tag{8.95}$$

The path $\mathcal{P}$ that has the smallest first arrival time is the one that causes the tracer to be seen first, and we have

$$T_{\text{first}} = \min_\mathcal{P} T_\mathcal{P} = \min_\mathcal{P} \sum_{i \in \mathcal{P}} t_i. \tag{8.96}$$

This expression has a strong resemblance to the central expression in another problem that received enormous attention in the nineties: The directed polymer problem (Barabási and Stanley, 1995). Here is a very short summary of that problem: Suppose we have a polymer that interacts strongly with its frozen but disordered environment: Each part of the polymer adds an interaction energy to the total energy depending on where in this environment the part is placed. What is then the minimum total interaction energy, and what would the corresponding configuration be when back turns are not allowed (thus accounting for the term "directed") and there are no spatial correlations in the environment? If the local interaction energy is ϵ_i, then the mininum energy of the polymer given by

$$E_{\text{min}} = \min_\mathcal{P} \sum_{i \in \mathcal{P}} t\epsilon_i. \tag{8.97}$$

This expression should be compared to Eq. (8.96).

In the first arrival time problem, the link traversal times are correlated since the velocities are correlated. Devillard (1993) showed numerically that these correlations are not strong enough to change the first arrival time problem away from the directed polymer problem. More precisely, when considering the scaling properties of the two problems, the corresponding scaling exponents are the same; they are in the same universality class.

Let us therefore describe some of these scaling properties. We base this discussion on the results of Talon et al. (2013). We show in Figure 8.8 a square network oriented at 45° compared to the coordinate system shown. Each link has been given a random number drawn from the spatially uncorrelated distribution $p(t) \sim t^{-2}$ with $t \geq 1$. Each link will have many paths that pass through it – paths that start at the lower edge and end at the upper edge. For each link, only one path will be the one with the smallest $T_\mathcal{P}$. The links have then been colored according to the value of $T_\mathcal{P}$ for that minimal path. No assumption has been made that the paths are directed. The ensuing colored landscape has been named a *pathscape*. As is apparent, there is a lot of structure in this pathscape.

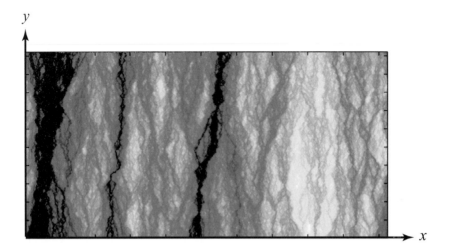

Figure 8.8 The minimal path passing through every link in this square lattice placed at 45° compared the x-axis. All paths start somewhere along the lower edge and end somewhere along the upper edge. The distance between the lower and upper edge is L. Darker color means smaller $T_\mathcal{P}$. (Adapted from Talon et al. (2013).) (A black and white version of this figure will appear in some formats. For the colour version, refer to the plate section.)

By examining the pathscape in Figure 8.8 closely, it will be clear that paths with a given color normally do *not* start at the lower edge and end at the upper edge. This is due to branching since a path with a higher $T_\mathcal{P}$ may overlap with one that has smaller $T_\mathcal{P}$, in which case only the one with the smaller $T_\mathcal{P}$ will be visible. This makes it possible to readily identify those paths that are local minima: those will be spanning paths. We show in Figure 8.9 the spanning paths in Figure 8.8. According to the theory of directed polymers, the density of spanning minimal paths, N_{path}, scales as

$$N_{\text{path}} \sim L^{-2/3} \tag{8.98}$$

in two dimensions. This means in terms of the first arrival problem that the distance between the points where the tracer locally first appears scales as $L^{2/3}$.

The spanning paths themselves are self-affine, having a Hurst exponent $H = 2/3$. This means that the roughness of the paths scales in the same manner as the distance between them.

We show in Figure 8.10 the ordered sequence of first arrival times, where n is the ordering. The curve to the left shows all paths, whereas the curve to the right is a zoom-in on a portion of it. Figure 8.11 shows a histogram N over the step sizes Δ

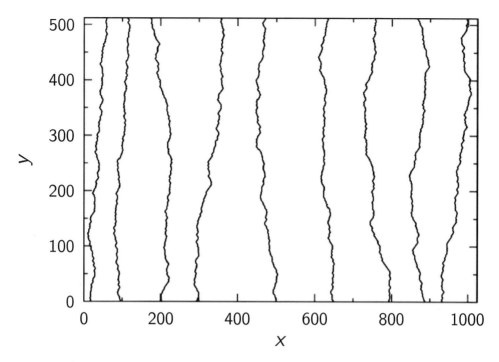

Figure 8.9 The spanning paths of Figure 8.8. (From Talon et al. (2013).)

in Figure 8.10. The histogram shows a power law, $N \sim \Delta^{-2.5}$. The exponent 2.5 has not been explained so far.

Just one more remark before we leave this theme. Let us return to Eq. (8.96). Suppose now that the disorder is very strong. That is, the distribution of times t_i is very broad. In this case, the sum over times in a given path $\mathcal{P}$ will be dominated by the largest value, that is,

$$\sum_{i \in \mathcal{P}} t_i \rightarrow \max_{i \in \mathcal{P}} t_i, \tag{8.99}$$

so that we have

$$T_{\text{first}} = \min_{\mathcal{P}} \left(\max_{i \in \mathcal{P}} t_i \right). \tag{8.100}$$

We have seen such an expression before: See Eq. (7.57) for the resistance of a highly disordered conductor. As Ambegaokar et al. (1971) showed, this is the percolation problem in disguise. In this limit, the first arrival time problem becomes a percolation problem, and the first arrival time will be given by the inverse of the maximum velocity in the link that is identified through the AHL argument, see the discussion in Section 7.6.

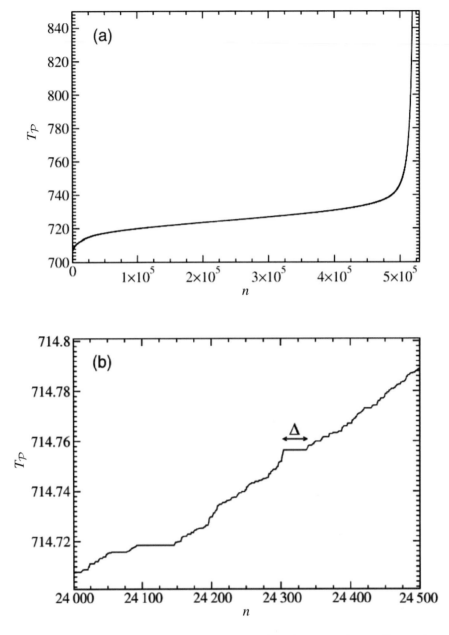

Figure 8.10 (a) The ordered sequence of first arrival times T_P, the smallest to the left, the largest to the right. The parameter n is the order of a given path: the path with the smallest T_P is assigned $n = 1$ and so on. (b) A zoom-in on a portion of the figure to the left. We see a *devil's staircase*. We define a step size Δ in the figure. (From Talon et al. (2013).)

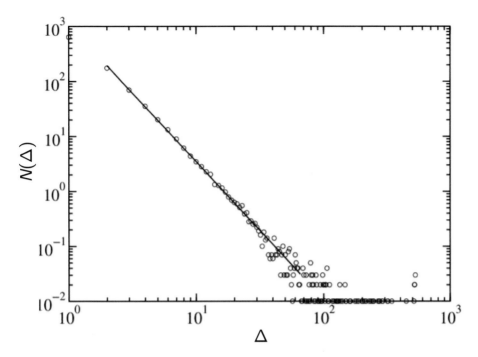

Figure 8.11 Histogram over the step sizes Δ, defined in Figure 8.10. (From Talon et al. (2013).)

Exercises

8.1 **Sinusoidal concentration oscillations:**
    We shall study the relaxation of an intial oscillation concentration profile $C(x,0) = \sin kx$ in one dimension. Use Eq. (8.39) to find $C(x,t)$.

8.2 Take the spatial Fourier transform of Eq. (8.48) with $u = 0$, and derive a k-dependent equation for $C_k(t)$. Use the initial condition $C(x,0) = \sin kx$ in to derive $C_k(t)$ and from this $C(x,t)$.

8.3 **Temperature in a wine cellar:**
    We shall take the temperature on the surface of the ground to vary in a sinusoidal way over the year and obtain the temperature variations in a cellar $\Delta T(z,t)$ at a depth of $z = 4\,\mathrm{m}$. Use the one-dimensional version of Eq. (8.55), and look for a solution of the form $\Delta T(z,t) = \Delta T_0 e^{i(\omega t - kz)}$, where $\omega = 2\pi/\mathrm{year}$ and $\Delta T_0 = 15°\mathrm{C}$.

8.4 Determine k and the amplitude of $\Delta T(z,t)$. Take the thermal diffusivity of the ground soil to be that of water: $D_T = 1.4 \cdot 10^{-7}\,\mathrm{m}^2/\mathrm{s}$.

8.5   How would this result change if the cellar started at $z = 2\,\mathrm{m}$?

8.6   **The advection–diffusion equation and relativity:**

In the advection–diffusion equation, we shall not consider velocities near the speed of light. However, we can still apply the principle of relativity that it must be possible to write the laws of nature in a form that does not change upon the change of frame of reference. Consider the Galilei transformations between two inertial frames S and S' that move with relative velocity $\mathbf{u}_0$:

$$t = t' \tag{8.101}$$

and

$$\mathbf{x} = \mathbf{x}' + \mathbf{u}_0 t' \tag{8.102}$$

and show that dC/dt keeps its form under this transformation. Show also that, as a result, Eq. (8.48) is invariant.

9

Capillary Action

A tube, the bore of which is so small that it will only admit a hair (*capilla*), is called a capillary tube. When such a tube of glass, open at both ends, is placed vertically with is lower end immersed in water, the water is observed to rise in the tube, and to stand within the tube at a higher level than the water outside. The action between the capillary tube and the water has been called Capillary Action, and the name has been extended to many other phenomena which have been found to depend on properties of liquids and solids similar to those which cause water to rise in capillary tubes.

Thus begins J. C. Maxwell's famous article in Encyclopedia Britannica (1876). The understanding of the surface phenomena involved in capillary rise has contributions from many of the great names in science starting with Leonardo da Vinci, considered to be the discoverer of capillary action. Maxwell goes on as follows:

The forces which are concerned in these phenomena are those which act between neighboring parts of the same substance, and which are called forces of cohesion, and those which act between portions of matter of different kinds, which are called forces of adhesion. These forces are quite insensible between two portions of matter separated by any distance which we can directly measure.

Here we have a macroscopic phenomenon, easy to observe as in capillary rise, that depends on the interactions on the atomic, or molecular, level. Maxwell's article was written before science had clear concepts concerning atoms, molecules, and heat – concepts we now take for granted. Capillary phenomena were investigated by Hooke (1661), Newton, Young, Laplace, Gay-Lussac, Gauss, Poisson, Faraday, Rayleigh, Kelvin, van der Waals, Boltzmann, Gibbs, and many others. The development of the early ideas and experiments are very nicely described in Maxwell's article. A readable introduction in Rowlinson and Widom's book *Molecular Theory of Capillarity* (1982) also describes the evolution of the ideas of how capillary phenomena can be understood in terms of molecular interactions.

Capillary phenomena, such as *surface tension*, adhesion, adsorption, surface films, wetting, flotation, detergency, emulsions and foams, are important in many practical situations in daily life, in industry, and in oil recovery and production. The use of soap and detergents in washing depends on capillary phenomena. Capillary phenomena represent a very large field of science and engineering. Excellent books on the subject are Adamson's standard reference: *Physical Chemistry of Surfaces* (1982), the book by Rowlinson and Widom already mentioned, and Israelachivili's book *Intermolecular and Surface Forces* (1992). de Gennes (1985) has written a clear review of many aspects of surface tension, wetting and spreading.

In the following sections, we discuss fundamental aspects of interface dynamics, surface tension and *contact angle* required for a discussion of the simultaneous flow of several fluids in porous media, so called multiphase flow. The formation of bubbles in fluids and droplets in vapors are important in oil production; therefore, we discuss nucleation theory in an introductory way.

9.1 Surface Tension Thermodynamics

The understanding of surface tension and processes that change the surface, such as capillary waves, detergents, condensation, evaporation and nucleation phenomena, require thermodynamic concepts and relations not ordinarily covered in introductory courses on thermodynamics and statistical physics. However, some excellent texts that introduce the relevant thermodynamic arguments exist, such as Landau and Lifshitz' book *Statistical Physics* (1987b), and the review of *wetting* by M. Schick (1990).

To start, consider a drop of water surrounded by water vapor. The drop will evaporate, stay in (metastable) equilibrium or grow, depending on the thermodynamic conditions given by temperature, pressure, solutes in the drop and chemical potentials. It is simplest to work in an ensemble that has a variable number of particles.

We will consider the interface between a fluid and its vapor, such as a drop of water in a cloud. We will also discuss the probability for the creation of a drop in a super-saturated vapor. Then we discuss the solid–liquid–vapor system, which leads to a discussion of wetting, contact angles and other properties characteristic of three-phase systems.

9.1.1 Reversible Work and Surface Tension

Let us discuss the interface between two "bodies," that is, two fluids, a fluid and a gas, a fluid and a solid, or two solids that are in contact over an interface area A. Let the interface increase by a small amount δA, for instance by changing the shape of a drop in another fluid.

The *reversible work*[1] in such a process is proportional to δA:

$$\delta W_{\min} = \sigma\, \delta A. \tag{9.1}$$

Here the *surface tension* is an energy per area. This energy reflects the fact that, in at least one of the phases, attractive inter-molecular forces keep the molecules together. When new surface is created, work must be done to overcome these forces.

The reversible work depends on the thermodynamic conditions for the system. For processes at constant pressure, temperature and number of molecules, $\delta W_{\min} = \delta G$, where G is the Gibbs free-energy of the system as a whole. In a system where the chemical potential μ, volume V and temperature T are fixed, we have $\delta W_{\min} = \delta \Omega$, where Ω is the thermodynamic "Landau Potential." If $\sigma < 0$, Eq. (9.1) states that negative reversible work has to be supplied in order to increase the interface by δA. But this means that the system would be able to do work on the surroundings by increasing its surface; the surface would grow indefinitely – the two phases would mix. Hence, for the interface to exist at all we must have $\sigma > 0$. Eq. (9.1) corresponds to the expression $\delta W_{\min} = -P\delta V$ for the reversible work done in a change of volume of a system.

For a system of two phases of the same substance in a given volume, V, separated by an interface of area A, containing N molecules, the first law of thermodynamics takes the form

$$dE = T dS + \mu dN + \sigma dA. \tag{9.2}$$

Here the internal energy is E, the entropy is S, the temperature is T, the chemical potential is μ and σ is the surface tension. The surface term enters simply as an additional work term. At equilibrium, the temperature and the chemical potential are uniform in the system and therefore equal for the two phases.

However, since the pressure is *not* uniform in a system with curved interfaces between the two phases of the same substance, it is more convenient to work with the Landau potential $\Omega = E - TS - \mu N = -PV$ relating the independent variables T, μ (and the volume V) instead of the energy E with the independent variables S, N and volume. The first law may be written in terms of Ω as

$$d\Omega = -S dT - N d\mu + \sigma dA. \tag{9.3}$$

For a system in an *external bath* that fixes the temperature $T = T_0$ and the chemical potential $\mu = \mu_0$, the reversible work is the change in Ω. Note that for a fluid in equilibrium with its own vapor, the chemical potential is the same in the two phases, and the chemical potential is not independent of temperature but given by an *equation of state* $\mu = \mu_{\mathrm{coex}}(T)$. The surface tension for such a system is therefore a function of only one independent variable, either the temperature or the chemical potential.

---

[1] That is, the minimum work required to perform the change, or the negative of maximum work the system can perform on its surroundings.

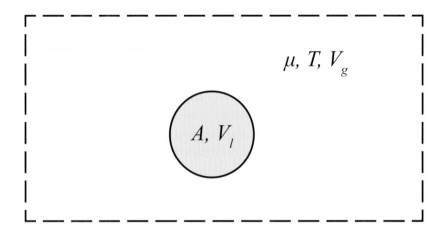

Figure 9.1 A liquid drop, volume V_ℓ and area A, in its vapor at a fixed chemical potential μ and temperature T.

9.1.2 Thermodynamic Surface Quantities

The thermodynamic potentials may be separated into "volume" and surface terms. At fixed T and μ, Eq. (9.3) gives $d\Omega = \sigma dA$, which may be integrated to give $\Omega_s = \sigma A$, and for the complete system we have[2]

$$\Omega = \Omega_0 + \Omega_s = \Omega_0 + \sigma A, \tag{9.4}$$

$$\Omega_0 = -P_\ell V_\ell - P_g V_g. \tag{9.5}$$

Here we use subscript s to indicate surface quantities and 0 for bulk quantities. For the system illustrated in Figure 9.1 we have

$$V_\ell + V_g = V, \quad \text{where } V \text{ is the total volume,} \tag{9.6}$$

$$n_\ell V_\ell + n_g V_g = N, \quad \text{where } N \text{ is the number of molecules.} \tag{9.7}$$

Here $n_\ell(\mu, T)$ and $n_g(\mu, T)$ are the number densities of molecules in the liquid and in the gas phase, respectively. In writing Eqs (9.6) and (9.7), we have assumed that there is no extra volume associated with the surface, that is, $V_s = 0$; and there is no excess number of particles in the surface, that is, $N_s = 0$.

The surface entropy S_s, the Helmholtz and Gibbs surface free-energy F_s and surface energy E_s, are given by

$$S_s = -\left(\frac{\partial \Omega_s}{\partial T}\right)_{\mu, A} = -\frac{d\sigma}{dT} A, \tag{9.8}$$

---

[2] This argument ignores the possibility that the interface moves. A small drop in a vapor has translational, rotational and vibrational degrees of freedom that contribute to the free energy as a whole but are not directly related to the surface free energy σ.

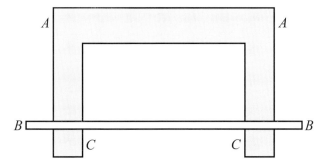

Figure 9.2 A piece of metal (AA) with a small metal bar (BB) is dipped into a soap solution. When taken out, the rectangle (AACC) is covered by a soap film. The strip (BB) will move as the soap film contracts (after Dupre).

$$F_s = \Omega_s + \mu N_s = \sigma A, \quad \text{since } N_s = 0, \tag{9.9}$$

$$G_s = F_s + P V_s = \sigma A, \quad \text{since } V_s = 0, \tag{9.10}$$

$$E_s = F_s + T S_s = \left(\sigma - T\frac{d\sigma}{dT}\right) A. \tag{9.11}$$

Thus σ is the Helmholtz or Gibbs surface free-energy per unit surface area.

Dupre described an experiment in which the surface energy σ can be measured. If the metal piece (BB) in Figure 9.2 is moved so as to increase the area $a \times b$ by increasing the distance $b = AB$, work $F db = a\sigma db$ is done. Here F is the force on the metal piece (BB) of length a. We may consider σ to be either the surface free-energy per unit area or the surface tension per unit contour. Maxwell (1876) defined surface tension as follows:

Definition 9.1 The tension of a liquid surface across any line on the surface is normal to the line, is the same for all directions of the line and is measured by the force across an element of the line divided by the length of that element.

9.2 The Young–Laplace Law

For a drop of fluid in its own vapor at given temperature T and chemical potential μ, the total Landau potential of the system is given by Eqs (9.4) and (9.5), that is,

$$\Omega = -P_\ell V_\ell - P_g V_g + \sigma A. \tag{9.12}$$

Since $V = V_\ell + V_g$, and because in equilibrium the two phases must have the same chemical potential $\mu = \mu_\ell(P_\ell, T) = \mu_g(P_g, T)$, it follows that the pressures are uniform in each phase (but $P_\ell \neq P_g$), and Eq. (9.12) becomes

$$\Omega = -P_\ell V_\ell - P_g(V - V_\ell) + \sigma A. \tag{9.13}$$

Now, let the droplet have a volume $V_\ell = 4\pi R^3/3$ and area $A = 4\pi R^2$, and consider a small change in radius δR, because molecules condense on the droplet, then we find that

$$\delta\Omega = \left(\frac{\partial\Omega}{\partial R}\right)_{T,\mu,V}\delta R = \left(-(P_\ell - P_g)4\pi R^2 + \sigma 8\pi R\right)\delta R. \tag{9.14}$$

Note that P_ℓ should not be considered a function of V_ℓ, since P_ℓ is given as the solution of $\mu_\ell(P_\ell, T) = \mu$, for processes at fixed μ and T. Similarly P_g is fixed. The Landau potential will decrease as the system approaches thermal equilibrium at fixed μ, T and V, and $\delta\Omega = 0$ at equilibrium. Solving Eq. (9.14) with $\delta\Omega = 0$ gives the Young–Laplace law

$$\boxed{P_\ell - P_g = \frac{2\sigma}{R}. \qquad \text{(Young–Laplace)}} \tag{9.15}$$

Here, σ is a surface free-energy.

9.2.1 The Vapor Pressure of a Drop

The chemical potential in the liquid and in the vapor must be equal at equilibrium; therefore, the pressures satisfy the equations

$$\mu_\ell(P_0, T) = \mu_g(P_0, T), \quad R \to \infty \tag{9.16}$$
$$\mu_\ell(P_\ell, T) = \mu_g(P_g, T), \quad R \text{ finite.} \tag{9.17}$$

If the pressures do not deviate much from the equilibrium vapor pressure P_0, then we may expand the chemical potentials as follows (with $i = \ell$ or g):

$$\mu_i(P_i, T) = \mu_i(P_0, T) + \left(\frac{\mu_i}{P_i}\right)_T \delta P_i + \cdots . \tag{9.18}$$

We note that the chemical potential is the Gibbs free-energy per particle $\mu = G/N$. Since

$$dG = -S\,dT + V\,dP + \mu\,dN, \tag{9.19}$$

we get

$$V_i = \left(\frac{G_i}{P_i}\right)_{TN}, \tag{9.20}$$

and the *molecular volume* is

$$v_i = \left(\frac{\mu_i}{P_i}\right)_T, \tag{9.21}$$

and it follows that

$$(P_\ell - P_0)v_\ell = (P_g - P_0)v_g. \tag{9.22}$$

This equation may be combined with the Young–Laplace equation (9.15) to give expressions for the pressures independently:

$$P_\ell - P_0 = \frac{2\sigma}{R}\frac{v_g}{v_g - v_\ell} \simeq \frac{2\sigma}{R}, \tag{9.23}$$

$$P_g - P_0 = \frac{2\sigma}{R}\frac{v_\ell}{v_g - v_\ell} \simeq \frac{2\sigma}{R}\frac{v_\ell}{kT}P_0, \tag{9.24}$$

where the approximations use that typically $v_\ell \ll v_g$ and that the vapor has a low pressure, so that the ideal gas law may be used to give $v_g = kT/P_g \simeq kT/P_0$. The pressure inside the drop, P_ℓ, is slightly larger than the pressure difference given by the Young–Laplace equation (9.15) since the vapor pressure of the drop is larger than the vapor pressure of a flat interface, which is P_0. This is expressed in Eq. (9.24).

9.3 Young's Law

Drops of water on solid surfaces are a familiar phenomenon. Depending on the properties of the solid surface, the drop may take a variety of shapes, as illustrated in Figure 9.3, that depend on the contact angle.

The *equilibrium angle of contact* θ_0 is a thermodynamic variable that depends on three surface free-energies:

$\sigma_{\ell,g}$ the free energy per unit area of the liquid–gas interface, that is, the liquid–vapour interface.

$\sigma_{s,\ell}$ the solid–liquid free energy per unit area.

$\sigma_{s,g}$ the solid–gas (or vapor) free energy per unit area. In general, molecules from the vapor adsorb onto the solid surface, and $\sigma_{s,g}$ refers to that equilibrium situation.

(a)   (b)   (c)

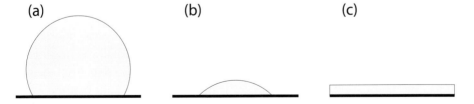

Figure 9.3 A small drop in equilibrium on a horizontal surface. (a) and (b) correspond to partial wetting with an equilibrium contact angle in the range $0 < \theta < \pi$. (c) corresponds to complete wetting, and the drop spreads to a film in between the liquid and the vapor.

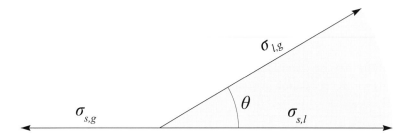

Figure 9.4 The balance of forces at the vapor–liquid–solid intersection is consistent with Young's law.

The equilibrium contact angle is given by Young's law:

$$\sigma_{s,g} = \sigma_{s,\ell} + \sigma_{\ell,g} \cos \theta_0, \qquad \text{(Young)} \qquad (9.25)$$

Young (1805) did not formulate his "law" in mathematical terms. In the beginning of his paper he states:

But it is necessary to premise one observation, which appears to be new, and which is equally consistent with theory and with experiment; that is, that for each combination of a solid and a fluid, there is an appropriate angle of contact between the surfaces of the fluid exposed to air and to the solid. This angle, for glass and water, and in all cases where a solid is perfectly wetted by a fluid, is evanescent: for glass and mercury, it is about 140°, in common temperatures, and when the mercury is moderately clean.

Here he claims priority to the concept of an angle of contact. He clearly states that the angle of contact is characteristic of the fluid (water or mercury), "air" and the solid (glass). He also implies that purity is relevant.

There are several ways to derive Young's law. The most common method is to interpret the surface free-energy as a tension.

Using Maxwells definition of surface tension, that is, that σ is a *force* across a line per unit length, mechanical equilibrium then requires that the forces along the solid surface balance, and Eq. (9.25) follows (see Figure 9.4).

The forces should also balance in the vertical direction. This is, however, not possible, since with the given geometry one needs to compensate the upward component $\sigma_{\ell,g} \sin \theta_0$ by a downward component that must finally come from the solid, and the problem with the force argument is that it really assumes that the solid is non-deformable. Thus we are led to the more general case of a drop resting on a deformable medium – such as another fluid. The conclusion from such a discussion is that Eq. (9.25) is indeed correct in the limit of a non-deformable solid (see Rowlinson and Widom (1982) for a discussion).

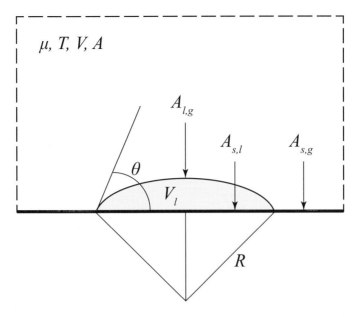

Figure 9.5 A drop resting on a solid surface at given chemical potential μ and temperature T. The open system has volume V and wall area A accessible to the vapor. The drop has radius of curvature R and angle of contact θ.

9.3.1 Angle of Contact Thermodynamics

The angle of contact can be understood as a thermodynamic variable that reaches the value θ_0 in thermal equilibrium. For the system illustrated in Figure 9.5, the total Landau potential is

$$
\begin{aligned}
\Omega &= -P_\ell V_\ell - P_g(V - V_\ell) + \sigma_{\ell,g} A_{\ell,g} + \sigma_{s,\ell} A_{s,\ell} + \sigma_{s,g}(A - A_{s,\ell}) \\
&= \Omega_v - (P_\ell - P_g)V_\ell + \sigma_{\ell,g} A_{\ell,g} + (\sigma_{s,\ell} - \sigma_{s,g})A_{s,\ell},
\end{aligned}
\tag{9.26}
$$

where $\Omega_v = -P_g V + \sigma_{s,g} A$ is the Landau potential in the absence of the drop. This expression for Ω differs from Eq. (9.12) by the additional surface terms. Here the volumes and areas of the drop resting on the surface may be expressed in terms of the radius R of the spherical cap and the angle of contact θ, or the variable $u = \cos\theta$:

$$
A_{s,\ell} = \pi (R \sin\theta)^2 = \pi R^2(1 - u^2),
\tag{9.27}
$$

$$
A_{\ell,g} = \int_0^\theta d\tilde{\theta}\, 2\pi R^2 \sin\tilde{\theta} = 2\pi R^2 \int_{\cos\theta}^1 du = 2\pi R^2(1 - u),
\tag{9.28}
$$

$$
V_\ell = \pi R^3 \int_0^\theta d\tilde{\theta}\, \sin^3\tilde{\theta} = \pi R^3 \int_{\cos\theta}^1 du\, (1 - u^2)
$$

$$
= \pi R^3 \left(\frac{2}{3} - u + \frac{1}{3}u^3 \right).
\tag{9.29}
$$

We have already shown that the pressure difference is given by the Young–Laplace law (9.15): $P_\ell - P_g = 2\sigma_{\ell,g}/R$, which when used with the expressions above leads to the following expression for Ω:

$$\Omega = \Omega_v + \frac{2}{3}\pi R^2 \sigma_{\ell,g}(1 - u^3) + \pi R^2 (\sigma_{s,\ell} - \sigma_{s,g})(1 - u^2). \tag{9.30}$$

In thermal equilibrium at fixed μ and T, we must have $(\partial\Omega/\partial u)_{\mu,T} = 0$, a condition that with Eq. (9.30) leads directly to Young's law (9.25). Thus we may consider the contact angle θ to take on a value θ_0 in thermal equilibrium.

9.4 Capillary Rise

A fluid rises to a height h in a capillary if the contact angle is less than $\pi/2$. The Young–Laplace pressure across the meniscus is

$$P_g - P_\ell = \frac{2\sigma_{l,g}}{R} = 2\sigma_{\ell,g}\frac{\cos\theta}{a}, \tag{9.31}$$

with the highest pressure being on the vapor side. This pressure difference must be equal to the difference in hydrostatic pressure in the two phases, given by

$$P_\ell(h) = p_0 - \rho_\ell g h, \qquad P_g(h) = p_0 - \rho_g g h. \tag{9.32}$$

Here g is the acceleration due to gravity, and p_0 is the reference pressure. Inserting these results in Eq. (9.31), we find that the fluid will rise to a height given by

$$h = \frac{2\sigma_{\ell,g}\cos\theta}{(\rho_\ell - \rho_g)ga}. \tag{9.33}$$

The capillary rise $h < 0$ if the fluid is non-wetting, that is, $\theta > \pi/2$, which is the case for mercury in glass capillaries. Often the notion of a *capillary length* λ (or capillary constant) is introduced. The rise of a fluid in capillary wetted by the fluid is

$$\boxed{\lambda^2 = \frac{2\sigma_{\ell,g}}{g\Delta\rho} = ha. \qquad \text{(Capillary constant)}} \tag{9.34}$$

Here $\Delta\rho = \rho_\ell - \rho_g$. The capillary length λ determines the length scale of most capillary phenomena. For water, $a = 3.93$ mm at $0°$C and decreases with increasing temperature, becoming zero at the critical point.

9.4.1 Capillary Rise Height

We have capillary pipe of three sizes. The volume V marked on the pipe is $V = \pi a^2 L$, where L is the distance of the fill mark from the end of the pipe, and a is the inside radius of the capillary, which is estimated to be $a = \sqrt{V/\pi L}$.

Table 9.1. *Capillary constant for water. Acceleration of gravity:*
$g = 980.665 \, cm/s^2$.

t	(°C)	0	20	25
σ	(N/m)	0.07.6	0.073	0.072
ρ	(g/cm^3)	0.998647	0.9982041	0.9970449
μ	(poise = g/cm s)		0.010019	0.008904
α	(cm)	0.3927	0.3855	0.3837

The capillary rise is

$$h = \frac{2\sigma_{l,g}\cos\theta}{(\rho_l - \rho_g)ga}. \tag{9.35}$$

Here the acceleration of gravity is $g = 980 \, cm/s^2$, and the density difference between water and air is approximately $\Delta\rho \simeq 1.0 \, g/cm^3$. The angle of contact is taken to be $0°$ for the *receding* meniscus. For a fluid, the capillary constant is defined as

$$\alpha^2 = ha = \frac{2\sigma_{l,g}}{g\Delta\rho}. \tag{9.36}$$

Experimental values for water are found in Table 9.1.

9.4.2 The Surface Tension of Water

We may measure the surface tension of water by observing the capillary rise, h, in capillaries, and using the expression

$$\sigma_{l,g} = h\frac{ag\Delta\rho}{2\cos\theta} \simeq \frac{1}{2}hag\Delta\rho, \tag{9.37}$$

where we assume that water wets the glass capillaries, that is, $\theta \simeq 0$. Dip the tip of a capillary into water, and it will rise. However, in practice the water does not rise to the expected height, due to *contact angle hysteresis*. The point is that although the contact angle is an equilibrium thermodynamic quantity, which in principle only depends on the fluid, the vapor (gas) and the solid wall, this equilibrium is in practice not reached. The contact line remains in a non-equilibrium position, in a thermodynamic meta-stable state, due to surface irregularities and impurities that pin the surface locally. Therefore, one obtains better results by dipping the capillary a bit too deep, withdrawing it slowly, until a lower meniscus is formed as the capillary leaves the surface of the water. The height of capillary rise, h, can be approximately measured by a ruler to, say, 5%. In Table 9.3, the results of such

Table 9.2. *Dimensions of VWR Scientific glass capillaries*

V (μl)	L (cm)	a (cm)	Cat. No.
5	5.5	0.0170	53432-706
20	7.62	0.0289	53432-740
50	7.30	0.0467	53432-783

Table 9.3. *The capillary rise of water at room temperature in glass capillaries of different diameter, and the estimated surface tension.*

V (μl)	5	20	50
a (cm)	0.0170	0.0289	0.0467
h (cm)	9.65	5.25	3.20
	9.72	5.98	3.10
	9.82	5.22	3.15
	9.60	5.20	3.12
⟨h⟩ (cm)	9.70	5.41	3.16
σ (dyn/cm)	80.8	76.7	72.5

measurements for VWR glass tubes whose properties are listed in Table 9.2 are shown, together with the estimated surface tension.

9.4.3 Imbibition in Blotting Paper

Blotting paper is designed to absorb fluids quickly. We use strips of blotting paper that have been laminated between two sheets of plastic in order to prevent evaporation during the wicking process.

When one end of the laminated paper is dipped into water, a front is seen to rise in the paper. For the blotting paper used here, the capillary height, and therefore the capillary constant for the paper, is not known. However, the capillary height, h, is well above the length, L, of the paper, and therefore it does not matter much whether the paper is vertical or horizontal during the paper *imbibition* process.

The process of imbibition is due to the water wetting the paper, which can be thought off as a porous medium. Thus there is a capillary pressure

$$\Delta p_c = \frac{2\sigma}{a}, \tag{9.38}$$

Table 9.4. *The effective pore radius K_0a, for chalk and two types of blotting paper.*

Sample	D (cm²/s)	K_0a (μm)
238	0.0491 ± 0.0002	0.271
bb-18	0.0263 ± 0.0004	0.145
chalk	0.0278 ± 0.0004	0.143

where a is a characteristic pore *radius* for the paper. The paper has a *permeability k*, and the flow velocity is therefore given by *Darcy's law*:

$$U = -\frac{k}{\mu}\frac{dp}{dx}. \tag{9.39}$$

The front velocity is simply $(dx/dt) = U$, the pressure gradient is $dp/dx = -\Delta p_c/x$, and we find

$$\frac{dx}{dt} = \frac{2k\sigma}{a\mu}\frac{1}{x}. \tag{9.40}$$

We find that a solution of this equation has the same form as Eq. (9.47):

$$x^2 = 2Dt, \quad \text{with } D = \frac{2\sigma k}{a\mu}. \tag{9.41}$$

The constant D characterizes the imbibition process of a given paper with a given fluid. D has the dimension as a diffusion constant, that is, cm²/s. The permeability of the paper is not known, and we write is as $k = K_0a^2/8$, where K_0 is a factor that measures the deviation from what would be expected in a capillary pipe. We may determine D from simple experiments, and for known σ and μ, we get an expression for the effective paper pore radius:

$$\boxed{K_0a = 4\frac{\mu D}{\sigma}.} \tag{9.42}$$

Figure 9.6 shows that Eq. (9.41) accurately describes the imbibition front. The slope is $2D = 0.0526 \pm 0.0007 \, \text{cm}^2/\text{s}$.

We found the results shown in Table 9.4. Thus, we find that the typical pore radius a is a fraction of a micron, unless the numerical constant K_0 happens to be large. We expect $K_0 \simeq 10-100$, and therefore the pore size to be of the order of a few microns. We need to determine k independently if we are to estimate a more precisely. To do that we would need to measure the flow through the paper as a function of the

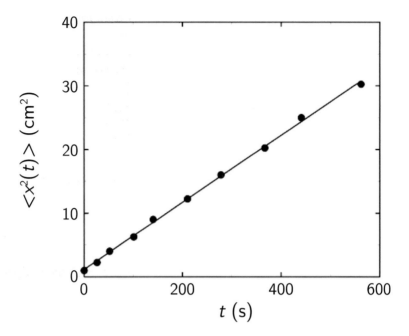

Figure 9.6 Imbibition front position as a function of time for bb18 blotting paper.

pressure gradient, and then find k from the Darcy law. Such an experiment requires more sophisticated equipment.

The results for chalk and for type 238 blotting paper are shown in Figures 9.8 and 9.7.

The Porosity of Blotting Paper

The Ahlstrom Electrophoresis and Blotting paper (Grade 238, size 10×15 cm) is a paper that will imbibe water spontaneously. With a micro pipette, filled to the mark, touch the paper and an ellipsoidal wet spot will grow until the capillary is empty. Thus the blotting paper (thickness of $d \simeq 0.035$ cm) is porous. The pores are smaller than the diameter of the capillaries, and therefore capillary action will draw the water from the capillary to the paper. Table 9.5 shows the results of the average dimensions in two experiments for each capillary size. Table 9.6 shows results for ordinary blotting paper. The results in the tables show that the paper is anisotropic, with the ellipticity of the spot being approximately $e \simeq 0.9$. This anisotropy is a result of the paper making process. The results in Tables 9.5 and 9.6 allow us to estimate the porosity of the paper to be $\phi \simeq 0.45$ for both types of blotting paper.

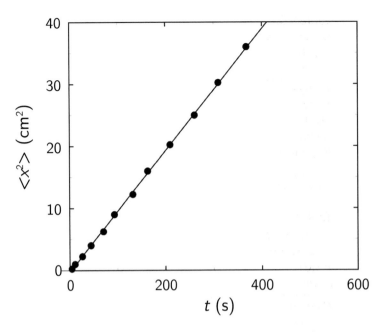

Figure 9.7 Imbibition front position as a function of time for 238 blotting paper.

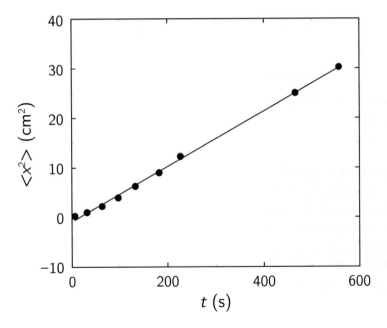

Figure 9.8 Imbibition front position as a function of time for chalk.

Table 9.5. Porosity of Grade 238 blotting paper.

V (µl)	5	20	50
a (cm)	0.0170	0.0289	0.0467
D_a (cm)	0.7	1.40	2.10
D_b (cm)	0.63	1.18	1.85
D_b/D_a	0.90	0.84	0.88
A (cm²)	0.346	1.30	3.05
V_{tot} (cm³)	0.0118	0.0441	0.1037
$\phi = V/V_{tot}$	0.42	0.45	0.48

Table 9.6. *Porosity of ordinary blotting paper*
(DZ/165 # BBL-18); d = 0.058 cm.

V (µl)	5	20	50
a (cm)	0.0170	0.0289	0.0467
D_a (cm)	0.6	1.08	1.60
D_b (cm)	0.52	0.90	1.42
D_b/D_a	0.87	0.83	0.89
A (cm²)	0.25	0.76	1.78
V_{tot} (cm³)	0.014	0.044	0.104
$\phi = V/V_{tot}$	0.35	0.45	0.48

9.4.4 The Dynamics of Capillary Rise

Fluid rises in a capillary due to surface tension. At what rate does water in the capillary rise?

A Horizontal Pipe

If the capillary pipe is horizontal, the meniscus will move with a velocity controlled only by the surface tension σ, the angle of contact θ, the fluid viscosity μ and the capillary radius a. Use a coordinate system where x is along the capillary pipe, with $x = 0$ at the entrance of the pipe where it is immersed in water.

The pressure in the air is p_0. The pressure in the fluid, just below the flat interface between the water and the air, is is also p_0. Therefore, the pressure in the fluid at the entry of the capillary is

$$p(x = 0) = p_0. \tag{9.43}$$

Let the radius of the capillary be a, as before. The radius of curvature of the meniscus depends on the angle of contact, θ, and is given by $R_c = a/\cos\theta$. The pressure inside the fluid, just below the surface of the meniscus, is therefore

$$p(x) = p_0 - \frac{2\sigma}{R_c},\tag{9.44}$$

where the Young–Laplace pressure difference is positive if the angle of contact is less than $\pi/2$. The pressure drop in the water inside the capillary, $\Delta p = p(0) - p(x)$, is distributed over the length x, which is the position of the meniscus relative to the inlet. There is therefore a pressure gradient in the fluid given by

$$\frac{dp}{dx} = -\frac{2\sigma}{R_c x}.\tag{9.45}$$

For a capillary with a circular cross-section, as we discuss here, the meniscus will move with a velocity given by:

$$\frac{dx}{dt} = U = \frac{Q}{\pi a^2} = -\frac{a^2}{8\mu}\left(\frac{dp}{dx}\right).\tag{9.46}$$

It follows that the position of the meniscus is given by

$$\boxed{x = a\sqrt{t/\tau}, \quad \text{where } \tau = 2\mu R_c/\sigma = \frac{2\mu a}{\sigma \cos\theta}.}\tag{9.47}$$

The characteristic time, τ, depends on surface tension, angle of contact and the capillary radius. A large viscosity and a low surface tension leads to slow imbibition. Note that the characteristic time τ is proportional to the radius of curvature of the meniscus. For the situation where the angle of contact is zero, one may find the characteristic pore size by measuring the position of the meniscus as a function of time.

A Vertical Pipe

Taking gravity into account, one finds that flow occurs only if there is a gradient in hydraulic potential

$$\Phi = p + \rho g z.\tag{9.48}$$

Thus, for $\Phi = Cte$, or $dp/dz = -\rho g$, we find the well known result from hydrostatics that the pressure increases with depth. The gradient in Φ is then

$$\frac{d\Phi}{dz} = \frac{[(p_0 - 2\sigma/R_c) + \rho g z] - p_0}{z} = \rho g - \frac{2\sigma}{z R_c}.\tag{9.49}$$

The equilibrium state is reached, as in Eq. (9.35), for

$$z = h = \frac{2\sigma}{R_c \rho g}.\tag{9.50}$$

If the meniscus is not at $z = h$, then it will move up for $z < h$ and move down for $z > h$. A pressure gradient in the capillary will drive a flow as given by Darcy's law for the capillary. The average flow velocity is also the rate of advancement of the meniscus; therefore, we have

$$\frac{dz}{dt} = U = \frac{Q}{\pi a^2} = -\frac{a^2}{8\mu}\left(\frac{d\Phi}{dz}\right).$$
(9.51)

Here μ is the viscosity of the fluid. Therefore, we find that the velocity of the meniscus is given by:

$$\frac{d(z/h)}{dt} = \frac{1}{4\tau}\mathrm{Bo}\left(1 - \frac{h}{z}\right) = \frac{2}{\tau}\left(\frac{a}{z} - \frac{a}{h}\right),$$
(9.52)

where h is the capillary rise, Eq. (9.35), and we have introduced the dimensionless *Bond number*, Bo, which measures the relative importance of gravitational to capillary forces:

$$\mathrm{Bo} = \frac{g\rho a^2}{\sigma\cos\theta} = 2\frac{a}{h}.$$
(9.53)

The characteristic time is given as in Eq. (9.47). Eq. (9.52) shows that the meniscus will rise for $z < h$ and fall for $z > h$, as expected. In the limit $z \ll h$, we find the same result as in Eq. (9.47). If we start *near $z = h$*, we may solve Eq. (9.52) explicitly:

$$h - z(t) = (h - z(t_0))\exp\left(-\frac{t - t_0}{\tau_h}\right),$$
(9.54)

where the relaxation time to reach height $z = h$ is

$$\tau_h = \frac{8}{(\mathrm{Bo})^2}\tau = 2\tau\left(\frac{h}{a}\right)^2.$$
(9.55)

We find that this relaxation time for a given capillary has the form $\tau_h \sim \sigma\mu$, and therefore it increases with viscosity, as expected; but τ_h also increases with surface tension, since σ also determines the capillary rise h that set the relevant length scale.

9.5 Bubble Flow in a Capillary

We start by presenting a simple model for bubble flow in capillaries when neither fluid wets the surface of the pipe, that is, there are no fluid films. We will in Section 9.5.1 consider the case when there is wetting and thus films.

Hence, we consider a capillary pipe, with length L and average radius r_0. We orient the pipe along the x axis so that it starts at $x = 0$ and ends at $x = L$. The radius of the pipe varies as

$$r = r_0 \left[1 - a \cos \left(\frac{2\pi x}{l} \right) \right]^{-1}, \tag{9.56}$$

where $l = L/N$ where N is an integer. The dimensionless parameter $0 \le a < 1$ determines the amplitude of the radius variation.

Let us place the midpoint of a bubble at position x_b. We assume there is an interface between the two fluids at $x = x_b - \Delta x/2$. The pressure drop across the interface Δp is given by the Young–Laplace expression (9.31), which in this case takes the form where we assume that the more wetting fluid is to the left of the interface:

$$\begin{aligned} p_c(x_b - \Delta x/2) &= -\frac{2\sigma}{r(x_b - \Delta x/2)} \\ &= \frac{2\sigma}{r_0} \left[1 - a \cos \left(\frac{2\pi (x_b - \Delta x)}{l} \right) \right], \end{aligned} \tag{9.57}$$

where σ is the surface tension.

Let us now assume that there is another interface at $x_b + \Delta x/2$, this time with the more wetting fluid to the right of the interface. The pressure drop across this interface is

$$p_c(x_b + \Delta x/2) = -\frac{2\sigma}{r_0} \left[1 - a \cos \left(\frac{2\pi (x_b + \Delta x/2)}{l} \right) \right]. \tag{9.58}$$

These two interfaces define a *bubble* of less wetting fluid inside the more wetting fluid. The total capillary force on the bubble of length Δx is

$$\begin{aligned} \Delta p_c(x_b) &= p_c(x_b - \Delta x/2) + p_c(x_b + \Delta x/2) \\ &= -\frac{4\sigma a}{r_0} \sin \left(\frac{2\pi \Delta x}{l} \right) \sin \left(\frac{2\pi x_b}{l} \right). \end{aligned} \tag{9.59}$$

We define $\gamma = (4\sigma a/r_0) \sin(2\pi \Delta x_b/l)$ for simplicity. With a pressure difference Δp between the outlet and inlet of the capillary pipe, the volumetric flow rate q is given by the Washburn equation (Washburn, 1921)

$$q = -\frac{\pi r_0^4}{8\mu_{\text{eff}} L} (\Delta p - \Delta p_c(x_b)). \tag{9.60}$$

The effective viscosity is given by

$$\mu_{\text{eff}} = \left(1 - \frac{\Delta x}{L} \right) \mu_w + \left(\frac{\Delta x}{L} \right) \mu_n, \tag{9.61}$$

where μ_w and μ_n are the viscosities of the more wetting and the less wetting fluids, respectively. The mobility in the Washburn equation, $(\pi r_0^4)/(8\mu_{\text{eff}} L)$, has been approximated by using the average radius r_0.

We recall that x_b is the coordinate of the center of the bubble. By using that $\pi r_0^2 dx_b/dt = q$, Eq. (9.60) gives the equation of motion for the bubble,

$$\frac{dx_b}{dt} = -\frac{r_0^2}{8L\mu_{\text{eff}}}\left[\Delta p + \gamma \sin\left(\frac{2\pi x_b}{l}\right)\right]. \tag{9.62}$$

Let us now introduce a dimensionless position variable variable $\Theta = 2\pi x_b/L$ and a dimensionless time variable $\tau = \gamma t \pi r_0^2/(4L^2 \mu_{\text{eff}})$. Furthermore, we assume the pressure drop Δp to be negative, $\Delta p = -|\Delta p|$. The equation of motion may then be written

$$\frac{d\Theta}{d\tau} = \frac{|\Delta p|}{\gamma} - \sin(\Theta). \tag{9.63}$$

This is a well known equation from dynamical systems theory: it is the equation for the overdamped driven pendulum (Strogatz, 1994).

The period T_τ (measured in units of τ) it takes the bubble to move from a position Θ to a position $\Theta + 2\pi$ is

$$T_\tau = \int_0^{T_\tau} d\tau = \int_0^{2\pi} \frac{d\Theta}{d\Theta/d\tau} = \frac{2\pi\gamma}{\sqrt{\Delta p^2 - \gamma^2}}, \tag{9.64}$$

for $|\Delta p| > \gamma$. For $|\Delta p| \leq \gamma$, the period diverges.

We may now calculate the average speed when keeping the pressure difference Δp constant and assuming $|\Delta p| > \gamma$:

$$\left\langle \frac{d\Theta}{d\tau} \right\rangle = \frac{1}{T_\tau}\int_0^{T_\tau} d\tau \frac{d\Theta}{d\tau} = \frac{1}{\gamma}\sqrt{\Delta p^2 - \gamma^2}. \tag{9.65}$$

Now, transforming back to the center-of-mass position x_b and then to the volumetric flow rate q, we find that the average flow rate is given by

$$\langle q \rangle = -\frac{\pi r_0^4}{8\mu_{\text{eff}}L}\,\text{sign}(\Delta p) \begin{cases} \sqrt{\Delta p^2 - \gamma^2} & \text{if } |\Delta p| > \gamma, \\ 0 & \text{if } |\Delta p| \leq \gamma. \end{cases} \tag{9.66}$$

For pressure differences $|\Delta p|$ close to but larger than γ, we may approximate this equation by

$$\langle q \rangle = -\frac{\pi r_0^4}{8\mu_{\text{eff}}L}\sqrt{2\gamma}\,\text{sign}(\Delta p)\sqrt{|\Delta p| - \gamma}. \tag{9.67}$$

For large pressure differences where $|\Delta p| \gg \gamma$, we recover Darcy's law,

$$\langle q \rangle = -\frac{\pi r_0^4}{8\mu_{\text{eff}}L}\Delta p. \tag{9.68}$$

We will expand on this result considerably in Section 12.1.

9.5.1 Bretherton Bubbles

What if the more wetting fluid wets the surface of the capillary so that there is a film between the bubble and the walls? We are then dealing with Bretherton bubbles, which are long compared to the diameter of the pipe.

Fairbrother and Stubbs (1935) observed the curious phenomenon that bubbles of non-wetting fluid move *faster* than the average flow rate in the capillary when there are films present. Bretherton (1961) solved the problem analytically by the method of asymptotic matching. He considered a straight capillary pipe, that is, no sinusoidal variation in pipe radius.

Bretherton argued that as the wetting fluid velocity in front of and behind the bubble are the same and equal to the seeping velocity of the wetting fluid, v_w, the bubble, moving at a velocity dx_b/dt, sweeps out the same volume per time unit as the wetting fluid in front and behind it does. If the radius of the circular pipe is r and the film thickness is δ, we then get

$$\frac{dx_b}{dt}\pi(r-\delta)^2 = v_w \pi r^2. \tag{9.69}$$

Bretherton then went on to define a relative excess velocity W as $W = 1 - (1 - \delta/r)^2$ so that $(dx_b/dt)(1 - W) = v_w$. Under the assumption that the viscosity of the non-wetting fluid is $\mu_n = 0$ and the wetting fluid viscosity is μ_w, he determined δ and thereby found that

$$W = C_B \left(\frac{\mu_w v_n}{\sigma}\right)^{2/3} \tag{9.70}$$

or

$$\frac{dx_b}{dt}\left[1 - C_B \left(\frac{3\mu_w v_n}{\sigma}\right)^{2/3}\right] = v_w, \tag{9.71}$$

where σ is the interfacial tension between the two fluids. We recognize the capillary number

$$Ca = \frac{\mu_w v_n}{\sigma}. \tag{9.72}$$

The constant $C_B \approx 1.29$. Eq. (9.71) is exact in the limit $Ca \to 0$. The variable W does not depend on how long the bubble is.

9.6 Funicular Flow in a Capillary

If we stretch the length of the Bretherton bubble – the theme of the previous section – to the length of the capillary itself, we are dealing with funicular flow. This geometry represents an unstable situation since surface tension will tend to break

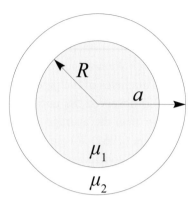

Figure 9.9 Two-fluid flow in a capillary. The fluid (1) occupies a cylindrical region of radius $r \leq a$.

up the cylindrical fluid (1) into drops.[3] However, the simple cylindrical geometry permits us to solve the Stokes equations for the flow exactly,[4] and the simple solution allows us to illustrate some general aspects of two-phase flow. We find that the equations for this simple two-phase flow are *not* consistent with the generalized Darcy equations (12.34) and (12.35).

Consider two-phase fluid flow through a pipe of radius a, as illustrated in Figure 9.9. Fluid (1), with viscosity μ_1, is assumed to occupy the region $0 < r < R$, and the other fluid (2), with viscosity μ_2, occupies the region $R < r < a$. Find the velocities and flow rates of fluid (1) and (2). We will write the results in terms of the *saturation* of phase (1) $S = S_1$, where

$$S = (R/a)^2. \tag{9.73}$$

The pressure in the two fluids may be different, and the capillary pressure, that is, the difference in pressure between the concave and the convex sides of the interface, is given by

$$p_c = (p_1 - p_2) = \sigma/R = \frac{\sigma}{a\sqrt{S}}. \tag{9.74}$$

---

[3] The surface free-energy per unit length of the cylindrical interface between the two fluids is $A_{cyl} = 2\pi R$.
The same volume converted into drops, of maximal radius a, has an area $A_{drop} = 2\pi (R/a)^2/a$. Therefore we have $\sigma A_{cyl}/\sigma A_{drop} = (3/2)(R/a) < 1$ if $R < 2a/3$. Thus the cylindrical arrangement is unstable with respect to the formation of droplets. This instability is driven by the surface tension. Even in the case $2/3 < R < 1$, we find that the cylindrical geometry may reduce its surface free-energy, by creating droplets that have a cylindrical shape of radius a, with spherical end-caps, that will conserve volume (per unit length) and decrease the interface area per unit length.

[4] In fact the stationary solution of the Stokes equation is also a solution of the Navier–Stokes equation for flow that is cylindrically symmetric, because $(\mathbf{u} \cdot \nabla)\mathbf{u} = 0$ in that case.

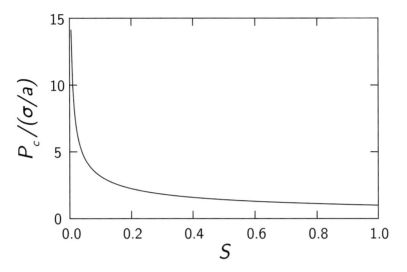

Figure 9.10 The capillary pressure, P_c, as a function of saturation, S, for two-phase flow in a capillary.

Here the radius of curvature along the capillary is infinite and does not contribute to the capillary pressure. The capillary pressure for this system is shown in Figure 9.10.

Each fluid must satisfy Eq. (5.13) for each fluid independently:

$$\mu_1 \frac{\partial}{\partial r}\left(r \frac{\partial u_1}{\partial r} \right) = r \frac{\partial p_1}{\partial r}, \quad 0 < r < R, \tag{9.75}$$

$$\mu_2 \frac{\partial}{\partial r}\left(r \frac{\partial u_2}{\partial r} \right) = r \frac{\partial p_2}{\partial r}, \quad R < r < a. \tag{9.76}$$

The solutions have the form (see Eq. (5.14)):

$$u_1(r) = A_1(a^2 - r^2) + C_1, \qquad\qquad 0 < r \le R, \tag{9.77}$$
$$u_2(r) = A_2(a^2 - r^2) + B_2 \ln r + C_2, \qquad R < r \le a. \tag{9.78}$$

Substitution of Eqs (9.77) and (9.78) in Eqs (9.75) and (9.76) determines the amplitudes,

$$A_1 = \frac{1}{4\mu_1}G_1 \quad \text{and} \quad A_2 = \frac{1}{4\mu_2}G_2, \tag{9.79}$$

where we have introduced the gradients $G_i = -(\partial p_i/\partial x)$.

The boundary conditions determine the other constants:

1. *The velocity must vanish at the pipe wall:*

$$u_2(a) = B_2 \ln a + C_2 = 0 \implies C_2 = -B_2 \ln a. \tag{9.80}$$

2. *The velocity of the two fluids, u_1 and u_2, must be equal at $r = R$:*

$$A_1(a^2 - R^2) + C_1 = A_2(a^2 - R^2) + B_2 \ln R + C_2$$
$$= A_2(a^2 - R^2) + B_2 \ln(R/a), \tag{9.81}$$

and we obtain an expression for C_1:

$$C_1 = -(A_1 - A_2)(a^2 - R^2) + B_2 \ln(R/a). \tag{9.82}$$

3. *The shear stress is continuous across the interface:*

$$\mu_1 \left(\frac{\partial u_1}{\partial r} \right) \bigg|_R = \mu_2 \left(\frac{\partial u_2}{\partial r} \right) \bigg|_R, \tag{9.83}$$

otherwise a volume element that contains the interface would rotate. Using Eqs (9.75) and (9.76), we find

$$B_2 = -2 \left(\frac{\mu_1}{\mu_2} A_1 - A_2 \right) R^2. \tag{9.84}$$

This result may be used with Eq. (9.82) to find

$$C_1 = -(A_1 - A_2)(a^2 - R^2) - 2 \left(\frac{\mu_1}{\mu_2} A_1 - A_2 \right) R^2 \ln(R/a). \tag{9.85}$$

Collecting the results for the constants, we may write the solution

$$u_1 = A_1(a^2 - r^2) - (A_1 - A_2)(a^2 - R^2)$$
$$- 2 \left(\frac{\mu_1}{\mu_2} A_1 - A_2 \right) R^2 \ln(R/a), \qquad\qquad 0 < r \le R, \tag{9.86}$$

$$u_2 = A_2(a^2 - r^2) - 2 \left(\frac{\mu_1}{\mu_2} A_1 - A_2 \right) R^2 \ln(R/a), \qquad R < r \le a. \tag{9.87}$$

The flow of each fluid may be found by integrating the expression

$$Q_i = \int_0^{2\pi} \int_0^a dr \, d\varphi \, r u_i(r), \tag{9.88}$$

to give

$$Q_1 = \frac{1}{2} \pi a^4 A_1 \left(S^2 - 2(\mu_1/\mu_2)^2 S^2 \ln S \right)$$
$$+ \frac{1}{2} \pi a^4 A_2 \left(2S - 2S^2 + 2S^2 \ln S \right), \tag{9.89}$$

$$Q_2 = \frac{1}{2} \pi a^4 A_1 \frac{\mu_1}{\mu_2} \left(2S + 2S^2 \ln S - 2S^2 \right)$$
$$+ \frac{1}{2} \pi a^4 A_2 \left(1 - 4S + 3S^2 - 2S^2 \ln S \right). \tag{9.90}$$

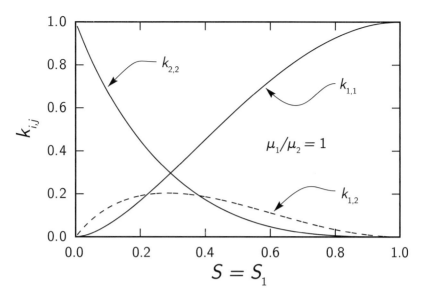

Figure 9.11 The relative permeabilities $k_{1,1}$, $k_{2,2}$ and the cross-term permeability $k_{1,2}$ as a function of saturation $S = S_1$, for $\mu_1/\mu_2 = 1$.

In terms of the average flow velocities (Darcy velocity): $U_i = Q_i/\pi a^2$, and with the *permeability* $k = a^2/8$ we find that the generalization of the Darcy equation for two fluid flow for this case has the form:

$$U_1 = -k\frac{k_{1,1}}{\mu_1}\frac{\partial p_1}{\partial x} - k\frac{k_{1,2}}{\mu_2}\frac{\partial p_2}{\partial x}, \tag{9.91}$$

$$U_2 = -k\frac{k_{2,1}}{\mu_2}\frac{\partial p_1}{\partial x} - k\frac{k_{2,2}}{\mu_2}\frac{\partial p_2}{\partial x}, \tag{9.92}$$

where we have introduced the symmetric *relative permeability matrix*:

$$
\begin{aligned}
k_{1,1} &= S^2 - (\mu_1/\mu_2)S^2 \ln S^2, \\
k_{1,2} &= 2S - 2S^2 + S^2 \ln S^2, \\
k_{2,1} &= 2S - 2S^2 + S^2 \ln S^2, \\
k_{2,2} &= 1 - 4S + 3S^2 - S^2 \ln S^2.
\end{aligned}
\tag{9.93}
$$

The cross-terms $k_{1,2} = k_{2,1}$ represent the viscous coupling between the fluids. Such cross-terms have been observed in experiments on porous rocks (Kalaydjian, 1990) and in lattice gas simulations for two phase flow in porous media (Gunstensen and Rothman, 1993; van Genabeek and Rothman, 1996). We see that Eqs (9.89) and (9.90) with our expression for k_{ij} give $U = U_1 + U_2 = -(k/\mu)(\partial p/\partial x)$ for $\mu_1 = \mu_2 = \mu$; that is, we recover Darcy's law for the total flow when the two fluids have the same viscosity. Note that the cross terms both have a $1/\mu_2$ factor in Eqs (9.91) and (9.92). We chose this form because then $k_{1,2} = k_{2,1}$ do not contain any viscosities and exhibit the general symmetry required from the theory of irreversible processes.

The general form of transport equations in non-equilibrium statistical physics is (Landau and Lifshitz, 1987b)

$$J_i = \sum_j L_{ij} X_j. \tag{9.94}$$

Here X_i are the *generalized force* (given by $X_i = -(\partial p_i / \partial x)$ in the present case), J_i are the *currents* ($J_i = U_i$ here) and L_{ij} are the transport coefficients given by

$$L_{11} = k_{1,1}/\mu_1, \qquad L_{12} = L_{21} = k_{1,2}/\mu_2, \qquad L_{22} = k_{2,2}/\mu_2. \tag{9.95}$$

The general *Onsager relations* for the transport coefficients

$$\boxed{L_{ij} = L_{ji} \qquad \text{(Onsager)}} \tag{9.96}$$

are satisfied for the flow Eqs (9.91) and (9.92) in the example discussed here.

We find that the cross-term permeability for $\mu_1/\mu_2 = 1$ is of the same order of magnitude as the relative permeabilities themselves (see Figure 9.11). In this case we have a maximum $k_{1,2} = 0.204$ for $S = 0.285$; at this saturation we have $k_{1,1} = 0.285$ and $k_{2,2} = 0.308$.

In the situation that $\mu_1/\mu_2 < 1$, that is, the viscosity of the fluid in the core is less than that of the fluid near the wall, we find that the relative permeabilities do not change much from the situation discussed above. In Figure 9.12, we have

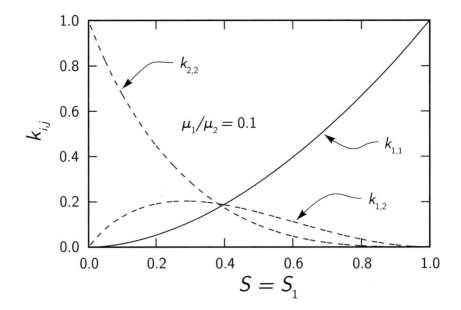

Figure 9.12 The relative permeabilities $k_{1,1}$, $k_{2,2}$ and the cross-term permeability $k_{1,2}$ as a function of saturation $S = S_1$, for $\mu_1/\mu_2 = 0.1$.

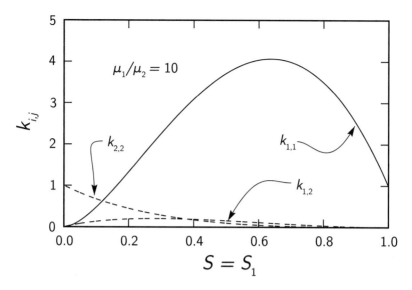

Figure 9.13 The relative permeabilities $k_{1,1}$, $k_{2,2}$ and the cross-term permeability $k_{1,2}$ as a function of saturation $S = S_1$, for $\mu_1/\mu_2 = 10$.

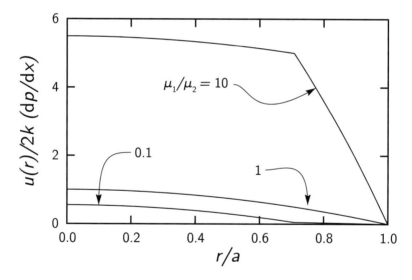

Figure 9.14 The velocity as a function of radius at a saturation of $S = 0.5$ for several values of the viscosity ratio for the situation where $dp_1/dx = dp_2/dx$.

illustrated the case $\mu_1/\mu_2 = 0.1$, which shows that the cross-term permeability is more important in this situation. Another feature of this situation is that $k_{1,1} \sim (1 - S)$ for $S \to 1$. The $S^2 \ln S^2$ term in $k_{1,1}$ disappears in the limit $\mu_1/\mu_2 \to 0$, and we find the behavior similar to what is shown in Figure 9.12.

The relative permeability curves change drastically when $\mu_1 \gg \mu_2$ (we illustrate the case $\mu_1/\mu_2 = 10$ in Figure 9.13). In this limit, the low-viscosity fluid near the

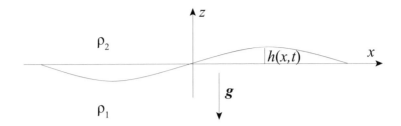

Figure 9.15 The height of an interface between two fluids of densities ρ_1 and ρ_2.

wall *"lubricates"* the flow of the high-viscosity fluid in the center of the capillary. The effect is that $k_{1,1}$ becomes much larger than one, that is, a small gradient in p_1 will give a much larger current than would be expected without lubrication.

In Figure 9.14, we show the velocity as a function of radial position for $S = 0.5$ at various viscosity ratios, in the situation where the gradients dp_i/dx in both phases are identical. The effect of lubrication is evident for $\mu_1/\mu_2 = 10$.

Exercises

9.1 **Capillary waves:**

We shall study surface waves that result from the combined action of gravity and capillary forces, as is illustrated in Figure 9.15, where the fluids are labeled $i = 1, 2$. We shall assume that the wave may be described by an interface height $h(x,t) = h_0 e^{i(kx-\omega t)}$ where the amplitude-to-wavelength ratio $hk \ll 1$, and use the Navier–Stokes equation to obtain the dispersion relation $\omega = \omega(k)$.

We will neglect viscous effects and take the flow field around the interface to start from $\mathbf{u} = 0$ everywhere and then result from turning on some distant pressure perturbation. Explain on the basis of Eq. (6.64) why there must be a field $\psi_i(\mathbf{x})$ such that $\mathbf{u}_i = \nabla \psi_i$, where $\psi_i = \psi_{0i}\, e^{i(kx-\omega t)}$ at $z = 0$. Explain why $\nabla^2 \psi_i = 0$, and obtain $\psi_i(\mathbf{x},t)$ from this.

9.2 Also, explain why the $hk \ll 1$ condition allows us to drop the $(\mathbf{u} \cdot \nabla)\mathbf{u}$-term in Eq. (6.36), and show that this equation may be written

$$\nabla(-i\omega\rho_i\psi_i + P_i) = \rho_i\mathbf{g}. \tag{9.97}$$

9.3 To linear order in h, the curvature of the interface is $h''(x)$. Explain why the pressure drop $\Delta p(z = h)$ across it is $-\sigma k^2 h(x,t)$.

9.4 Since $\dot{h}(h) \approx \dot{h}(0)$, show that $\psi_{oi} = \frac{i\omega}{k} \text{sign}(z)h_0$.

9.5 By taking the z-component of Eq. (9.97) and then integrating it in the z-direction, find an expression for the pressure p_i. Take the difference $\Delta p = p_2(z = h) - P_1(z = h)$ and show that it gives

Figure 9.16 Two swimmers, a snake and a duck.

$$\frac{\omega^2}{k}(\rho_1 + \rho_2)h = \gamma k^2 h - \Delta\rho g h + (C_2 - C_1), \qquad (9.98)$$

where C_i are the integration constants that come from the integration. Argue
that $(C_2 - C_1) = 0$.

9.6 Show the dispersion relation

$$\boxed{\omega^2 = \frac{\sigma}{\rho_1 + \rho_2}k^3 + \frac{(\rho_1 - \rho_2)g}{\rho_1 + \rho_2}k.} \qquad (9.99)$$

9.7 The surface tension of the air–water interface is approximatively
$\sigma = 0.07\,\text{N/m}$. At which wavelength λ_c will the two terms on the right hand
side of Eq. (9.98) be equal? Discuss the types of waves you will get when λ is
larger or smaller than λ_c. Derive the group velocity $v_g = d\omega/dk$ for $\lambda \ll \lambda_c$
and for $\lambda \gg \lambda_c$. How does v_g change with λ in these two cases?

9.8 At which wavelength is the phase velocity $v_f = v_g$? Use this result to discuss
the different wake patterns trailing the two swimmers in Figure 9.16.

10

The Hele–Shaw Cell and Linear Stability Analysis

In the Hele-Shaw cell, the medium is just described by two parallel plates, and we will study the interesting case where a less visous fluid displaces a more viscous one. In this case the characteristic wavelength of the displacment pattern may obtained analytically via so-called linear stability analysis. The term "linear" refers to the fact that the theory is linearized in the (small) front perturbations. In general, however, in more complex porous media, these perturbations are not small, and the theory becomes non-linear.

10.1 Viscous Fingering and Linear Stability

A Hele-Shaw cell consists of two transparent plates separated a distance b. Hele-Shaw (1898) studied the flow of water around various objects placed in the cell. He visualized the flow *streamlines* by injecting a dye to produce colored streamlines. These experiments verify directly that the fluid flow in a Hele-Shaw cell with a small b is the "creeping flow" characteristic of low Reynolds numbers. If the plate separation is increased, turbulent flow with confused streamlines arise at moderate flow velocities.

The velocity in an infinitely wide channel $\mathbf{U} = \langle \mathbf{u} \rangle$ was discussed in Section 5.1 with the main result in Eq. (5.8). This equation may be written in the form of the Darcy equation (7.8) for flow in the Hele-Shaw cell illustrated in Figure 10.1:

$$\mathbf{U} = -\frac{k}{\mu}\nabla(p + \rho\hat{g}z), \tag{10.1}$$

component of the acceleration of gravity along the z-coordinate of the cell. For a cell placed in the horizontal position, we therefore have $\hat{g} = 0$. The viscosity of the fluid is μ, and the permeability of the Hele-Shaw cell is

$$k = \frac{b^2}{12}. \tag{10.2}$$

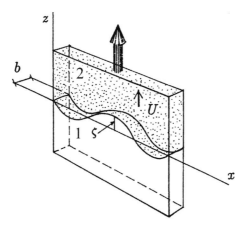

Figure 10.1 Geometry of the Hele-Shaw channel.

Note that the velocity in Eq. (10.1) is the *average* velocity over the thickness of the cell. For incompressible fluids, the equation of continuity gives

$$\nabla \cdot \mathbf{U} = -\frac{k}{\mu} \nabla^2 (p + \rho \hat{g} z) = 0. \tag{10.3}$$

The Laplace equation (10.3) is characteristic of potential problems encountered in electrostatics, in diffusion problems and in many other fields, and consequently we call flows controlled by Eq. (10.3) potential flow. A solution of the flow problem requires the boundary conditions to be specified in addition.

We will discuss the situation illustrated in Figure 10.1, where a fluid (1) displaces another fluid (2). The interface between the two fluids is controlled by capillary forces when the fluids are at rest and there is a pressure difference between the two fluids at the interface,

$$(p_1 - p_2) = \sigma \left(\frac{1}{R_x} + \frac{1}{R_y} \right), \tag{10.4}$$

where σ is the interfacial tension between the two fluids. The two radii of curvature R_x and R_y, describe the interface locally, as indicated in Figure 10.2. We define the radii of curvature to be positive if they have their center in fluid (1). We find $R_y = b/2 \cos \theta$, where θ is the contact angle between fluid (2) and the plates of the Hele-Shaw cell. We will assume that $R_x \gg R_y$. We have that $p_1 > p_2$ when the fluids are at rest and (2) is the wetting fluid.

Now let us inject the fluid (1) at a constant rate $\bar{U}$ at $z = -\infty$ and withdraw fluid (2) at the same rate at $z = \infty$. The interface between the two fluids will then move with a velocity $\bar{\mathbf{U}} = (0, 0, \bar{U})$ along the z-axis. However, it turns out that the interface is *unstable* if the viscosity of the driving fluid is smaller than the viscosity

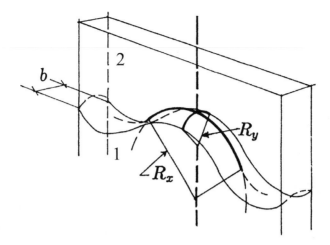

Figure 10.2 Geometry of the fluid–fluid interface.

of the fluid being driven. Engelberts and Klinkenberg (1951) coined the term *viscous fingering* in relation to their observation of such instabilities when water drives oil out of a porous medium. Flow in porous media also follow Eqs (10.1) and (10.3), and therefore the flow in Hele-Shaw cells is often used to model the flow in porous media. However, as we shall see, there are important differences, and the validity of using the Hele-Shaw cell as a model of flow in porous media remains an open question.

The theory of viscous fingering was developed and compared to experiments independently by Saffman and Taylor (1958) and Chuoke et al. (1959). Recently there has been a growing interest in the field, and many new theoretical and experimental results have been published; see Bensimon et al. (1986) for a review and the book *Fractals* (Feder, 1988) for a discussion of *fractal viscous fingering*.

The physics of viscous fingering lies in the geometry of the moving boundary. Assume that a pressure difference is maintained over the length of a finite Hele-Shaw cell. The pressure in the air is constant since we ignore its viscosity, and the largest pressure gradient appears at the longest finger. Thus a finger that gets ahead of the rest of the interface will move faster than the rest and grow further – this is clearly an unstable situation.

10.2 Linear Stability Analysis

To analyze the stability of the moving interface, we first note that in a frame of reference moving with the interface at a velocity $\bar{U}\mathbf{e}_z$ in the z-direction, a small *perturbation*, $\mathbf{u}$, in the velocity satisfies a modified version of Eq. (10.1) for each of the fluids:

$$\mathbf{u}_i = -\nabla \chi_i. \tag{10.5}$$

Here we have introduced the *velocity potential*

$$\chi_i = \frac{k}{\mu_i} \left(p_i + P_i(t) + \frac{\mu_i}{k} \bar{U} z + \rho_i \hat{g} z \right), \tag{10.6}$$

which then satisfies the Laplace equations

$$\nabla^2 \chi_i = 0. \tag{10.7}$$

The velocity potential χ_i gives the fluid velocity directly by Eq. (10.5). The velocity potential is linear in pressure, in the driving velocity $\bar{U}$ and in the gravity term. In addition, the velocity potential is linear in an arbitrary time dependent pressure term $P_i(t)$ that must be determined from the boundary condition that at the interface we must have $\mathbf{u}_1 = \mathbf{u}_2 = (\partial \zeta / \partial t)$, where $\zeta(t)$ is the interface position.

In order to test the stability of the advancing interface, we follow standard practice (Chuoke et al., 1959; Saffman and Taylor, 1958) and assume that the straight interface is perturbed by a sinusoidal displacement so that in the frame of reference moving with the average velocity $\bar{U}$, the position of the interface is given by the real part of

$$\zeta = \epsilon \exp(\gamma t + iqx), \tag{10.8}$$

as illustrated in Figure 10.1. The wavelength of the perturbation is $\lambda = 2\pi/q$. The growth rate of the perturbation is γ. For a stable interface, the perturbation ζ will decay in time, and $\gamma < 0$. If the growth rate is positive $\gamma > 0$, a perturbation of infinitesimal amplitude ϵ will grow exponentially. We shall limit ourselves to a discussion of the onset of instability, assume the perturbation to be *infinitesimal* and neglect terms of order ϵ^2 and higher.

The velocity of the interface in the moving frame must equal the fluid velocities, so that we have the boundary conditions

$$\frac{\partial \zeta}{\partial t} = -\left(\frac{\partial \chi_1}{\partial z} \right)_{z=\zeta} = -\left(\frac{\partial \chi_2}{\partial z} \right)_{z=\zeta}. \tag{10.9}$$

As is easily seen, the following solutions of the Laplace equations (10.7)

$$\chi_1 = -\frac{\gamma}{q} \epsilon \exp(qz + iqx + \gamma t),$$
$$\chi_2 = \frac{\gamma}{q} \epsilon \exp(-qz + iqx + \gamma t), \tag{10.10}$$

satisfy the boundary conditions given in Eq. (10.9) to first order in ϵ. The velocity potential of the perturbation $\chi_1 \to 0$ as $z \to -\infty$ and $\chi_2 \to 0$ as $z \to \infty$, so that the perturbation is localized at the interface.

We now insert Eq. (10.6) with $z = \zeta$ into the pressure Eq. (10.4) using the expressions for ζ and X_i to get

$$
\begin{aligned}
\sigma \left(\frac{1}{R_x} + \frac{1}{R_y} \right) = {} & \frac{\mu_1}{k} X_1(\zeta) - \frac{\mu_2}{k} X_2(\zeta) - P_1(t) + P_2(t) \\
& + \left[\left(\frac{\mu_2}{k} - \frac{\mu_1}{k} \right) \bar{U} - (\rho_1 - \rho_2)\hat{g} \right] \zeta.
\end{aligned}
\tag{10.11}
$$

We note that for $q\zeta \ll 1$, we may, to order ϵ, write $X_1(\zeta) \simeq -\frac{\gamma}{q}\epsilon \exp(iqx + \gamma t)$ and a similar expression for X_2. Note also that for $q\zeta \ll 1$, we find that the radius of curvature in the x-direction is given by $1/R_x \simeq -(\partial^2 \zeta/\partial x^2)$, and we have $R_y \sim b/2\cos\theta$, so that we may write

$$
(p_1 - p_2) = P_c - \sigma \frac{\partial^2 \zeta}{\partial^2 x},
\tag{10.12}
$$

where we have introduced the *capillary pressure* $P_c = 2\sigma \cos(\theta/b)$. It follows from Eq. (10.11) that in the limit $\epsilon \to 0$,

$$
P_2 - P_1 = P_c = 2\sigma \cos(\theta/b).
\tag{10.13}
$$

Thus the stationary pressure difference between the two fluids is given by the capillary pressure P_c. The remaining terms in Eq. (10.11) all contain a factor $\epsilon \exp(iqx + \gamma t)$, which may be eliminated, and we rewrite Eq. (10.11) as follows:

$$
\gamma \frac{\mu_2 + \mu_1}{k} - \left[\frac{\mu_2 - \mu_1}{k} \bar{U} - (\rho_1 - \rho_2)\hat{g} \right] q + \sigma q^3 = 0.
\tag{10.14}
$$

We shall assume that $q > 0$ throughout. It then follows from Eq. (10.14) that $\gamma > 0$, and the interface is unstable for $\mu_2 > \mu_1$ if $\bar{U} > U_c$, where the critical velocity is given by

$$
U_c = \frac{k\hat{g}(\rho_1 - \rho_2)}{\mu_2 - \mu_1}.
\tag{10.15}
$$

Generally, the condition for instability may be written

$$
\bar{U} - U_c - \frac{\sigma k}{\mu_2 - \mu_1} q^2 > 0,
\tag{10.16}
$$

or

$$
\lambda^2 > \frac{(2\pi)^2 \sigma k}{(\mu_2 - \mu_1)(\bar{U} - U_c)}.
\tag{10.17}
$$

Note that $U_c < 0$ for $\rho_2 > \rho_1$ when $\mu_2 > \mu_1$, and the system is unstable even at $\bar{U} = 0$. Also, in the absence of gravity effects, the interface is unstable at any velocity since $U_c = 0$.

We find from Eq. (10.16) that the interface is unstable for disturbances with a wavelength $\lambda = 2\pi/q$ satisfying the relation

$$\boxed{\lambda > \lambda_c = 2\pi \left(\frac{\sigma k}{(\mu_2 - \mu_1)(\bar{U} - U_c)} \right)^{1/2}.} \tag{10.18}$$

Ignoring gravity effects, we find that the interface is always stable if the driving fluid has a higher viscosity than the fluid being driven.

When the interface is unstable, we have a real *critical wavelength* λ_c, which may be used to write Eq. (10.14) for the *growth rate* γ in a dimension-less (scaled) form:

$$\gamma = \frac{1}{\tau} \left[\left(\frac{\lambda_c}{\lambda} \right) - \left(\frac{\lambda_c}{\lambda} \right)^3 \right]. \tag{10.19}$$

The unit of the growth rate γ is τ^{-1}, where the time scale τ is defined by

$$\tau = \frac{\lambda_c}{2\pi A(\bar{U} - U_c)}. \tag{10.20}$$

Here we use the viscosity ratio (Atwood ratio) introduced by Tryggvason and Aref (1983), given by

$$A = \frac{\mu_2 - \mu_1}{\mu_2 + \mu_1}. \tag{10.21}$$

For a fluid of negligible viscosity, as is the case when air displaces an oil or glycerol, we have $A \sim 1$, and when two fluids of identical viscosity displace each other, we have $A \to 0$. In the latter case, the characteristic time τ diverges, and the instability will take an infinite time to develop.

Eq. (10.19) suggests that we scale distance with λ_c and time with τ, giving the dimensionless time t' and the dimensionless distances x', z' given by[1]

$$x' = \frac{x}{\lambda_c}, \quad z' = \frac{z}{\lambda_c}, \quad \text{and} \quad t' = \frac{t}{\tau}. \tag{10.22}$$

It is easy to show, by setting $d\gamma/d\lambda = 0$ in Eq. (10.19), that the maximum growth rate is obtained by perturbations with a wavelength

$$\boxed{\lambda_m = \sqrt{3}\lambda_c.} \tag{10.23}$$

We therefore expect that in experiments in a Hele-Shaw channel of width W, an initially straight interface will develop *viscous fingers* with a characteristic period λ_m. Using the expression (10.2) for the permeability and assuming that the viscosity

---

[1] Tryggvason and Aref (1983) introduce similar dimensionless quantities related to ours by $t'_{TA} = t'/\sqrt{24}$ and $x'_{TA} = 2\pi x'/\sqrt{24}$.

of the driving fluid is negligible, $\mu_1 \ll \mu_2$, as is the case when glycerol is displaced by air, we find for a horizontal cell that we expect fingers with a period given by

$$\lambda_m = \pi b \sqrt{\frac{\sigma}{\bar{U}\mu}} = \frac{\pi b}{\sqrt{\text{Ca}}}. \tag{10.24}$$

Here we have introduced the dimensionless *capillary number* Ca defined by

$$\text{Ca} = \frac{\bar{U}\mu}{\sigma}, \tag{10.25}$$

which measures the ratio of viscous to capillary forces.

10.2.1 Saffmann–Taylor Instability

Linear stability analysis is linear in the sense that it is only valid up to linear order in the ratio of amplitude to wavelength. As it predicts exponential growth rates for individual wavelengths, it also predicts a wavelength λ_m that will ultimately dominate the growth process. So, provided the initial perturbation from which the instability evolves is sufficiently small and contains a range of wavelengths that include λ_m, this will be the wavelength that appears at the end of the linear phase. However, if another wavelength dominates sufficiently in the initial perturbation, the amplitude of λ_m may not have time to dominate before the system moves out of the linear regime.

Clearly, the condition for instability, Eq. (10.18), and the expression for the wavelength of growth, Eq. (10.23), remain valid in the special cases when either $\hat{g}$ or $\bar{U}$ vanishes.

When the cell is horizontal $\hat{g} = 0$, then $U_c = 0$, and the critical wavelength

$$\lambda_c = 2\pi \left(\frac{\sigma k}{(\mu_2 - \mu_1)\bar{U}} \right)^{1/2} \tag{10.26}$$

depends only on the viscosities, driving velocity and surface tension. The instability is named after Philip G. Saffmann and G. I. Taylor.

10.2.2 Rayleigh–Taylor Instability

When the driving velocity $\bar{U}$ vanishes, the instability is known as Rayleigh–Taylor instability, and the critical wavelength takes the form

$$\lambda_c = 2\pi \left(\frac{\sigma}{(\rho_2 - \rho_1)\hat{g}} \right)^{1/2}, \tag{10.27}$$

which does not depend on the viscosities anymore but only on the surface tension, the mass densities and the component of gravity along the cell. Note that for both special cases, a larger σ causes a larger wavelength.

10.3 Observations of Viscous Fingers*

Saffman and Taylor (1958) and Chuoke et al. (1959) not only developed the theory of viscous fingering in a Hele-Shaw channel; they also studied viscous fingering experimentally. We show fingering patterns for air displacing glycerol observed by Saffman and Taylor in Figure 10.3. The initial air–glycerol interface had irregularities at the start of the experiments. Note that the observed wavelength of approximately 2.2 cm is quite close to the wavelength of maximum instability $\lambda_m = \sqrt{3}\lambda_c = 2.1$ cm.

Similar fingers were observed by Chuoke et al. As shown in Figure 10.4, the finger structure was quite close to the most unstable wavelength λ_m.

Saffman and Taylor also made a detailed study of the shape of a single finger in a long horizontal channel, as shown in Figure 10.5.

The tendency for fingers to develop equidistant spacings irrespective of the channel width W suggests that one should study the hydrodynamics of a single finger with periodic lateral boundary conditions. Saffman and Taylor found that

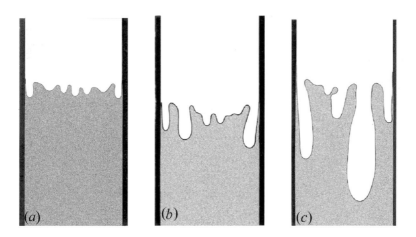

Figure 10.3 Viscous fingering in a vertical cell where air displaces glycerol from the top and downwards. $\bar{U} = 0.1$ cm/s and $\lambda_c = 1.2$ cm. (a) Early stage with observed average $\lambda \simeq 2.2$ cm. (b) Later stage: Fingers tend to space themselves. (c) Late stage: Longer fingers inhibit the growth of neighbors. (After Saffman and Taylor (1958).)

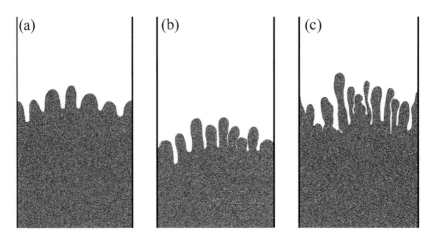

Figure 10.4 Water-glycerol solution (dark) $\mu_1 = 0.552$ poise, $\rho_1 = 1.21\,\mathrm{g/cm^3}$ moved upward and displaced oil $\mu_2 = 1.39$ poise, $\rho_2 = 0.877\,\mathrm{g/cm^3}$. The system was tilted an angle of $44°25'$. The bulk interfacial tension was $\sigma = 33\,\mathrm{dyne/cm}$, and the oil wetted the walls. The critical velocity was $U_c = 0.23\,\mathrm{cm/s}$. (a) $\bar{U} = 0.41\,\mathrm{cm/s}$, $\lambda_m = 3.5\,\mathrm{cm}$, observed $\lambda = 3.5\,\mathrm{cm}$. (b) $\bar{U} = 0.87\,\mathrm{cm/s}$, $\lambda_m = 2.6\,\mathrm{cm}$, observed $\lambda = 2.4\,\mathrm{cm}$. (c) $\bar{U} = 1.66\,\mathrm{cm/s}$, $\lambda_m = 1.6\,\mathrm{cm}$, observed $\lambda = 1.7\,\mathrm{cm}$. (After Chuoke et al. (1959).)

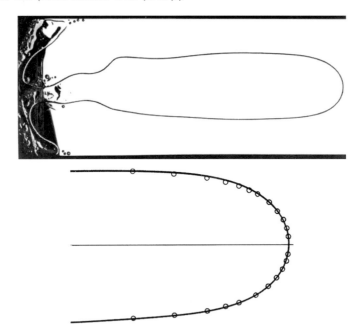

Figure 10.5 (a) An air finger advancing into glycerol in a Hele-Shaw channel $b = 0.09\,\mathrm{cm}$ and $W = 10.26\,\mathrm{cm}$. The finger started at a small bubble injected before the experiment. (b) Comparison of the shape of the finger tip in (a) and the theoretical expression indicated by $\circ$. (After Saffman and Taylor (1958).)

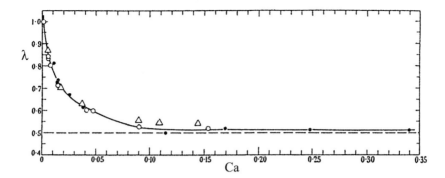

Figure 10.6 The relative width δ of a water finger advancing in oil as a function of the capillary number Ca $= U_{\text{tip}}\mu/\sigma$, where U_{tip} is the tip velocity. $\circ$ and $\triangle$ Shell Talpa $\mu = 4.5$ poise. $\bullet$ Shell Dalia $\mu = 0.30$ poise. (After Saffman and Taylor (1958).)

for negligible viscosity of the driving fluid, $\mu_1 \ll \mu_2$, and neglecting the influence of radius of curvature R_x on the pressure difference (that is, $R_x \gg b$), the bubble profile is given by (see Figure 10.5)

$$z = \frac{1-\delta}{\pi} \ln \left[\frac{1}{2} \left(1 + \cos \frac{\pi x}{\delta} \right) \right]. \tag{10.28}$$

Here the width of the finger is $W\delta$, so that δ is the fractional finger width. The velocity of the finger is $U^* = U\delta$. The relative width of the finger is, however, *not* given by their equations and is therefore arbitrary.

Saffman and Taylor concluded from their experiments on Hele-Shaw cells that

...as the speed of flow for any given fluid increases, δ rapidly decreases to $\delta = \frac{1}{2}$, and remains close to this value over a large range of speeds, till at high speeds of flow the tongue or finger of the advancing fluid itself breaks down and divides into smaller fingers.

They suggested that the shape of the meniscus depends only on $U_{\text{tip}}\mu/\sigma$, which is the capillary number Ca in Eq. (10.25) using the tip velocity U_{tip} as the characteristic velocity. As shown in Figure 10.6, the observed relative finger widths δ as function of the capillary number Ca fall on a single curve, giving a very satisfying data collapse for the experiments on single fingers propagating in a Hele-Shaw channel.

Saffman and Taylor had no explanation for the selection of $\delta = 0.5$ at high Ca. Also they note that an analysis of the stability of finger solutions given in Eq. (10.28) indicated that they are *unstable* for all values of δ. This instability of the finger solution was confirmed by McLean and Saffman (1981).

Pitts (1980) observed that the advancing finger left a thin layer of the viscous wetting fluid on the glass plates. Attempting to account for this effect, Pitts made an *ad hoc* assumption for the boundary conditions and found an equation for the finger shape given by Eq. (10.28) but with δ replacing $1 - \delta$ in front of the logarithm. He founds that this modified expression fitted the experimental results very well indeed. Pitts further made the assumption that the radius of curvature (for $\cos\theta = 1$) of the advancing meniscus is $R_y = b/2m$, with $m > 1$ instead of $m = 1$ as was assumed above. Also assuming that m is independent of the velocity $\bar{U}$, or the capillary number Ca, he found that his own experimental results and those by Saffman and Taylor (1958) were very well described by

$$\lambda \ln\left(\frac{\lambda}{1-\lambda}\right) = b\frac{\pi}{12}(m-1)\mathrm{Ca}^{-1}, \qquad (10.29)$$

when $m = 1.26$ is chosen. Unfortunately, no justification of the assumptions made by Pitts is available.

10.4 The Nonlinear Regime*

The development of viscous fingers in the Hele-Shaw cell beyond the initial (linear) stability analysis cannot be studied by analytical methods. Tryggvason and Aref (1983) solved the two-dimensional flow equations for the Hele-Shaw cell numerically. They represented the interface by a vortex sheet and computed the evolution of this vortex sheet using a variant of the vortex-in-cell method.

It is convenient in the numerical calculations to work with a Hele-Shaw channel of unit width and they used the scaled units $\tilde{x} = x/W$ and $\tilde{t} = tW/U_*$, where they use a velocity defined by $U_* = A(\bar{U} - U_c)$. They also introduced the non-dimensional surface-tension coefficient:

$$B = \frac{\sigma b^2}{12 U_* \bar{\mu} W^2} = \left(\frac{\lambda_c}{2\pi W}\right)^2 = \frac{1}{A\mathrm{Ca}^*}\left(\frac{b}{W}\right)^2. \qquad (10.30)$$

Here the average viscosity $\bar{\mu} = \frac{1}{2}(\mu_1 + \mu_2)$, and the capillary number Ca^* is a modification of Eq. (10.25), so that $\mathrm{Ca}^* = \bar{\mu}(\bar{U} - U_c)/\sigma$. However, as pointed out by Tryggvason and Aref (1983), the equations of motion for the interface are independent of the surface-tension coefficient B if one instead uses the scaling given in Eq. (10.22). Therefore their numerical results may be taken to be valid for any surface tension ratio, and the important parameter is the viscosity ratio A.

In their calculations, they chose to start with an interface consisting of an arbitrary collection of small-amplitude waves of various wavelengths. The surface tension was chosen such that the most unstable wavelength (according to linear stability theory) was about eight grid spacings. As seen from the second relation in Eq. (10.30),

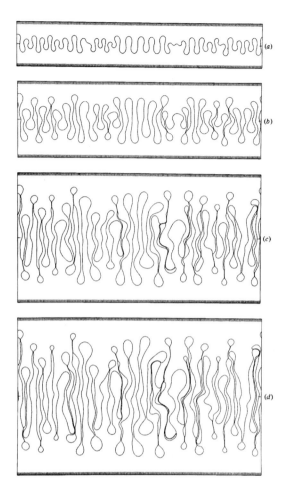

Figure 10.7 The evolution of the mixing layer with $A = 0$ and $B = 1.25 \, 10^{-5}$.
Note that this corresponds to gravity driven – not viscous – fingering. Note the
up-down symmetry. (After Tryggvason and Aref (1983).)

this is all that is needed in order to specify B. For the numerical results reproduced in
Figures 10.7–10.9, the important control parameter is therefore the viscosity ratio A.

When the viscosity ratio $A = 0$, we still have a fingering instability as shown
in Figure 10.7. This fingering instability is driven by the density difference of the
two fluids, and from Eqs (10.15) and (10.18) it follows that for this case, the critical
wavelength is $\lambda_c = 2\pi (\sigma / \hat{g}(\rho_2 - \rho_1))^{1/2}$. Therefore, this instability occurs only
if the upper fluid is denser than the lower fluid. The interface is very complex but
has an up-down symmetry that is in fact apparent in the equations when $A = 0$.

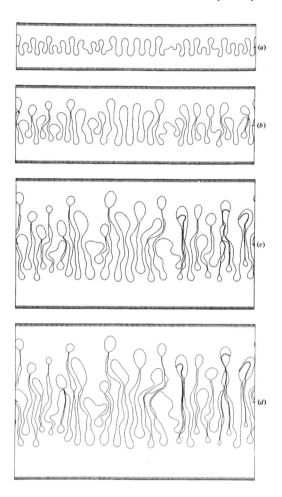

Figure 10.8 The evolution of viscous fingers with $A = 0.5$ and with the same initial conditions and value of B as in Figure 10.7. (After Tryggvason and Aref (1983).)

The fingers tend to become very thin and have rather small bubbles at the end. It is likely that some of these bubbles will eventually detach and propagate into the more viscous fluid. Such a detachment is not allowed by the program used by Tryggvason and Aref.

As the viscosity ratio is increased, one finds that the interface is less chaotic, and the driving low viscosity fluid penetrates more deeply into the high viscosity fluid. There is a slight tendency to splitting of the low viscosity fingers. On the other hand, fingers tend to merge and could possibly collapse to produce a reduced number of fingers with increasing time if such topology changes had been allowed

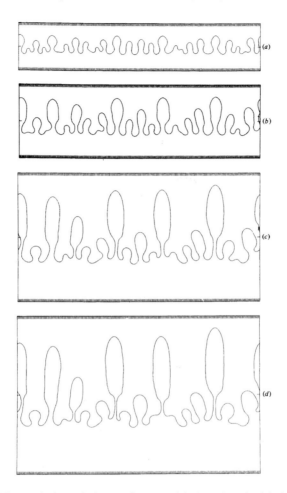

Figure 10.9 The evolution of viscous fingers with $A = 1$ and with the same initial conditions and value of B as in Figures 10.7 and 10.8. (After Tryggvason and Aref (1983).)

by the algorithm used. The numerical results in Figure 10.9 for $A = 1$ have a striking resemblance to the experimental results shown in Figures 10.3 and 10.5.

10.5 Experiments on Viscous Finger Dynamics*

In a set of elegant experiments, Maher (1985) studied the evolution of viscous fingers as a function of the viscosity ratio A. Maher used the unique features of the properties of binary mixtures near the critical unmixing temperature T_c. The binary mixture of isobutyric acid and water has at the critical composition a critical temperature $T_c = 26.12°C$ and separates into two immiscible phases, one rich in water and one rich in isobutyric acid. At the critical temperature, two phases can

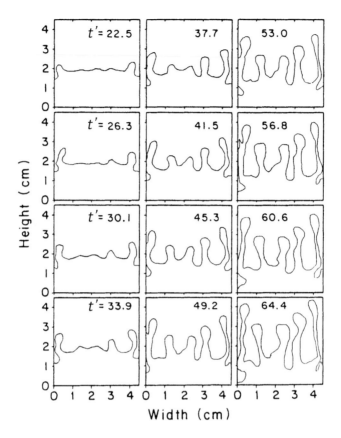

Figure 10.10 Time series of fingering patterns for isobutyric acid plus water. $A = 0.015$, $B = 6.6 \cdot 10^{-4}$. The dimensionless time, t', is indicated for each frame. (After Maher (1985).)

no longer be distinguished, and therefore we have $\rho_1 = \rho_2$, $\mu_1 = \mu_2$ and $\sigma = 0$. In terms of the reduced temperature $\epsilon = (T_c - T)/T_c$, the viscosity coefficient is $A = 0.053\,\epsilon^{\beta}$, and the interface tension coefficient is $B = 0.024\,\epsilon^{2\nu-\beta}$. The critical exponents are $\beta = 0.31$ and $\nu = 0.61$. Therefore, Maher obtained very small values of A by simply varying the temperature and thereby for the first time obtained viscous fingering results in this regime as shown in Figure 10.10.

The observed fingering pattern (shown in Figure 10.10) is similar to those obtained numerically by Tryggvason and Aref for $A = 0$. Also the expected variation of finger width λ_c with A and B was observed. Note again that the experimental results for $A \sim 0$ correspond to a gravity driven system. In fact the results in Figure 10.10 were obtained by inverting a small cell ($45 \times 45 \times 1$ mm^3) that had come to equilibrium at a temperature slightly below T_c.

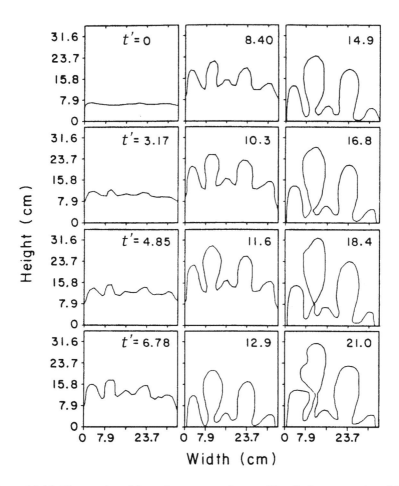

Figure 10.11 Time series of fingering patterns for paraffin oil plus water. $A = 0.93$, $B = 8.3 \cdot 10^{-4}$. (After Maher (1985).)

Maher also performed fingering experiments by displacing paraffin oil with water using $A = 0.93$ and $B = 8.3 \cdot 10^{-4}$. His results are shown in Figure 10.11. Again the experimental results and the numerical results obtained by Tryggvason and Aref for $A = 1$ are strikingly similar (see Figure 10.9).

Maher in addition observed that the dimensionless length of the mixing region $\theta' = \text{width}/\lambda_c$ depends on the dimensionless time $t' = t/\tau$ as follows:

$$\theta' = (t')^a. \tag{10.31}$$

The observed exponent $a = 1.6$ had some scatter but gave a good approximation for both the high and the low A values. The numerical results by Tryggvason and Aref for θ' vs. t' fall reasonably well on a power law with $a = 1.85$ for $A = 1$. For $A = 0$, their results give $a \sim 1.85$ for $t' < 25$ and then cross over to $a \sim 1.3$.

There is no theoretical prediction for the width of the mixing zone, and the result in Eq. (10.31) may not hold in detail. However, both theory and experiments on the Rayleigh–Benard and Taylor instability have shown that the amplitudes of the instabilities grow with time in a power-law fashion similar to Eq. (10.31), and with $a = 1.5$.

Exercises

10.1 **Linear stability of a ball on a hill top:**
We will consider a problem where the unstable variable is a single coordinate rather than the amplitude of a Fourier mode. The variable is the horizontal distance x of a ball sliding from the top of a parabolic hilltop of height $h(x) = -(x/R)^2$. We will also assume that the ball and the hilltop are immersed in a viscous fluid so that the resitance to motion is $-\alpha \dot{x}$. The mass of the ball is m so that the weight is mg. Explain that the equation of motion is

$$m\ddot{x} = -\alpha \dot{x} - h'(x)mg. \tag{10.32}$$

There is no rolling, only sliding of the ball, so we need not take account of the momemnt of inertia.

10.2 Assume that after a small initial pertubation x_0, $x(t)$ evolves as $x(t) = x_0 e^{\gamma t}$ and find an expression for γ.

10.3 Provide a physical explanation of this expression. What happens when the ball is in the stable potential well $h(x) = (x/R)^2$?

11

Displacement Patterns in Porous Media

When a non-wetting fluid displaces a wetting one, the process is called *drainage*. In the opposite case, when the invading fluid is wetting, the process is called *imbibition*. The wetting properties of the porous medium relative to the fluids determines this. The viscosity difference between the fluids is another important relation for the flow structures that emerge.

In this chapter we study displacement processes in disordered media. We will move from the simple case of viscous fingering to the more complex cases where capillary forces and gravity play important roles too. There is always a driving pressure, but in the simplest cases there may be only one other force affecting the hydrodynamics. The simplest case, perhaps, is when this other force is the capillary force. This happens when gravity and viscous forces are eliminated, typically by having a slow flow take place horizontally. The other case is the opposite one, where only viscous forces matter, in fast flows, that is.

In both these cases, the emergent structures lack a length scale; they are fractal. This makes continuum descriptions challenging, as for fractals, there is no critical length scale above which densities, like a fluid saturation, become scale-independent. The mass density of a fractal depends on length L as L^{D-d}, where $D - d$ is the difference between Euclidean and fractal dimensions.

Only when several forces combine, like capillary and gravitational forces, will a characteristic length scale emerge, and in this case, there will be a length scale above which the displacement patterns appear homogeneous. This is important for the upscaling problem where the variables on larger scales are always averaged densities.

11.1 Flow in Porous Media Dominated by Capillary Forces

In Section 4.4, it was pointed out that invasion percolation is a simple algorithmic model for slow drainage processes where the only relevant forces are the external

(a)   (b)

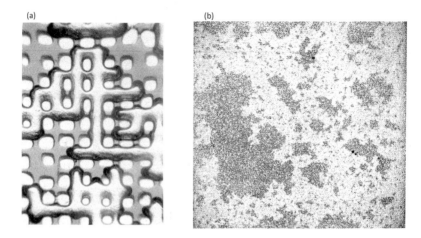

Figure 11.1 (a) A small section of the etched network model first used by
Lenormand and Zarcone (1985) (b) A larger section of the etched network model.
It shows displacement of the wetting fluid, paraffin oil (black), by the nonwetting
fluid, air, injected on the left-hand side of the network. On the right-hand side, a
semipermeable membrane prevents the non-wetting fluid from flowing outside.

driving pressure and the internal capillary pressures. Historically, this was first suggested by Wilkinson and Willemsen (1983) and de Gennes (1983). Later, Lenormand and Zarcone (1985) did experiments that confirmed this. Figure 11.1(a) and
(b) show the etched network model and the resulting fluid displacement patterns that
emerge. Figure 11.2 gives the result of box-counting, a fractal dimension between
1.80 and 1.83. This agrees well with the fractal dimension for invasion percolation
that was found in Section 4.4. In these experiments, Ca $\approx 10^{-6}$–10^{-8}, which is
very low.

A more recent experimental setup, which has proved exceptionally versatile for
a broad range of studies of flows in porous media, is shown in Figure 11.3 (Moura
et al., 2015). Here the porous model consists of a monolayer of glass beads of
diameter $a = 1$ mm that are spread randomly between two contact papers (Måløy
et al., 1985). The model is a transparent rectangular box of dimensions $L \times W$ and
thickness a, where L and W are typically 40 cm.

The porous medium is initially saturated with a viscous water–glycerol mixture.
To prevent bending of the model, a 2 cm thick glass plate and a 2 cm thick Plexiglas plate are placed on top of the model. To squeeze the beads and the contact
paper together with the upper plate, a Mylar membrane mounted on a 2.5 cm thick
Plexiglas plate, below the model, is kept under a constant pressure. Figure 11.4
shows the displacement pattern in colors that code for the invasion time.

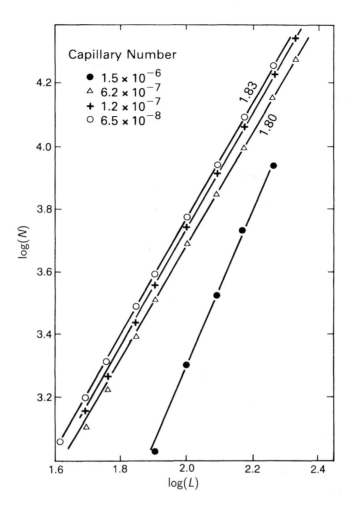

Figure 11.2 Results of box-counting where the box size is $L \times L$: N is the number of filled pore channels for different capillary numbers. The slopes 1.83 and 1.80 are least-squares fit to the data. (After Lenormand and Zarcone (1985).)

11.2 Flow in Porous Media Dominated by Viscous Forces

The opposite of the slow drainage regime where capillary forces dominate is the regime of fast displacement where viscous forces dominate. Then the capillary number is large, and we can neglect the surface tension effects. Also, the viscosity ratio between the displacing and displaced fluids becomes important.

If the displacing fluid is the less viscous one, the Saffman–Taylor instability described in Section 10.1 becomes relevant.

Figure 11.5 shows an experiment where this is the case. Here the less viscous fluid, air, displaces the more viscous epoxy. If some part of the air advances more

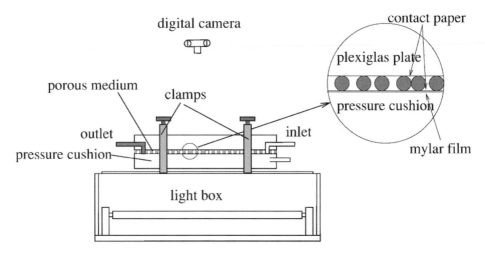

Figure 11.3 Sketch of the experimental setup with the light box for illumination, the porous model, and the digital camera. The porous medium is sandwiched between two contact papers and kept together with a "pressure cushion". (After Moura et al. (2015).)

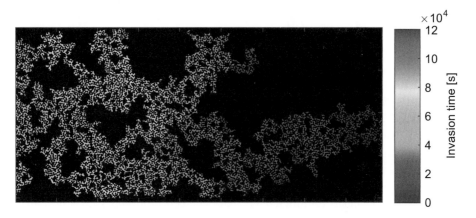

Figure 11.4 Spatio-temporal map of the invasion up to breakthrough (the average flow direction is from left to right). The color map shows the elapsed time for the invasion of a given pore (in seconds). The experiment lasts $\sim$33 h. (Figure courtesy of Marcel Moura.) (A black and white version of this figure will appear in some formats. For the colour version, refer to the plate section.)

closely towards the outer edge, it will create a locally larger pressure gradient in the epoxy, thus increasing the flow rate there. In other words, a local increase in the flow rate of the air will cause a further increase in the flow rate. This mechanism gives rise to rapidly protruding fingers that subsequently tend to split, thus forming the branching structure of Figure 11.5. For moderate Ca, the finger width would be given by λ_m of Eq. (10.24). However, as Ca becomes very large, $\lambda_m \to 0$, and there is nothing to define a characteristic scale for the patterns of Figure 11.5. For this reason the structures are fractal. In Figure 11.6, the fractal dimension is measured

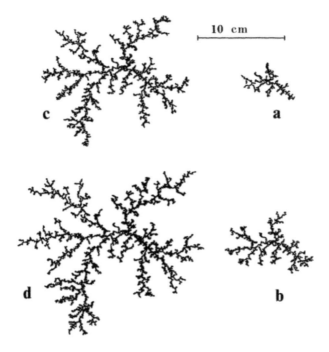

Figure 11.5 Air, shown as black, is injected in the center, displacing liquid epoxy that is contained in a porous medium of glass spheres arranged as in Figure 11.3. (After Måløy et al. (1985).)

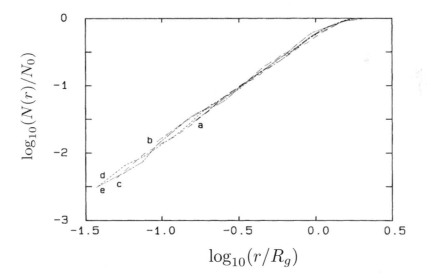

Figure 11.6 The normalized volume of the invasion structure of Figure 11.5 that is contained inside a circle of radius r. The Latin letters a–e label different experiments where the radius of gyration R_g ranges from 1.7 to 6.7 cm. (After Måløy et al. (1985).)

and found to lie around $D = 1.62$. This number may be compared to results of modeling, where growth happens at the location of the largest pressure graident.

11.3 Crossover from Capillary to Viscous Behavior

What happens when Ca is moderate and there are both capillary and viscous forces at play? There is an interesting answer to this question, and that is that it depends on the length-scale that the observation takes place on.

Using the experimental setup shown in Figure 11.3, it is possible to explore this. Figure 11.7 shows displacment patterns where the transition from capillary structures to viscous structures may be observed. At small flow velocities, the wide capillary finger structures with holes of different length scales may be seen. At larger flow rates, the fingers become narrower and exhibit the typical branching that is seen in Figure 11.5 too. However, even the pattern of the small flow rates shows branching; it just has wider branches.

This branch width corresponds to a crossover from capillary to visous behavior. The length (or width) scale at which this change takes place is called w_f and may be found by comparing the viscous pressure drop, Δp_μ, to the typical capillary pressure, σ/a. The condition is that at the pressure drop, $a \nabla P$ over a single pore is significantly smaller than σ/a. This means that a droplet of size a will not be mobilized by the viscous forces created by the flow. But at a larger length scale w_f,

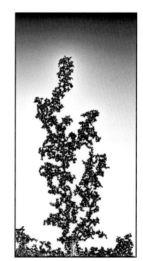

Figure 11.7 Viscous fingering experiment by Løvoll et al. (2004) where glycerol is displaced by air in a model of a porous medium. The gray tones in glycerol indicate the pressure, which is calculated separately. The capillary number decreases from the left figure to the right one.

the viscous forces will eventually equal the capillary pressure, and fluid clusters above this size will be mobilized by the surrounding flow. For this reason, fluid structures smaller than w_f will have much in common with the capillary dominated structures in Figure 11.4, while those larger than w_f should share features with those of Figure 11.5. Equating the two pressures,

$$\frac{\sigma}{a} = \frac{\mu U w_f}{\kappa} \tag{11.1}$$

gives the crossover-length

$$w_f = \frac{\kappa}{a\mathrm{Ca}}. \tag{11.2}$$

Figure 11.8 shows how the crossover takes place between a viscous regime with a fractal dimension that is roughly equal to that of Figure 11.6 and a capillary regime that is characterized by a fractal dimension that agrees with that of invasion percolation. The fact that the crossover takes place at $w_f/a\mathrm{Ca} = 1$ is easily understood. From Eq. (7.17), it is seen that the permeability may be estimated

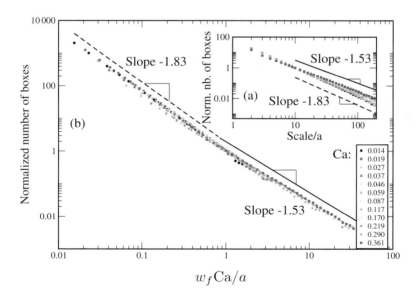

Figure 11.8 Small- and large-scale fractal dimension of the invasion cluster at various capillary numbers, determined by box-counting. (a) Insert: Number of boxes as function of the scale. (b) Data collapse for all capillary numbers after scaling the box size by Ca: Evidence of crossover scale $w_f \sim \mathrm{Ca}^{-1}$ separating a small-scale capillary fractal dimension 1.83 and a large viscous scale one with fractal dimension 1.53. (After Toussaint et al. (2005).)

roughly as $\kappa \sim a^2$, so that the crossover length $w_f \sim a/\text{Ca}$, which means that the crossover between capillary and viscous flow behavior happens when $w_f \text{Ca}/a \sim 1$, as the figure shows.

11.4 Displacement under the Effect of Gravity

In reality there is always gravity. This may be surpressed in a horizontal quasi-two-dimensional geometry. But the moment this geometry is tilted, the effect of gravity is re-introduced in a way that may be controlled by the tilt angle α. The experimental setup is shown in Figure 11.9, where the tangential component of gravity $g_t = g \sin \alpha$. This of course makes the lab-model more realistic as real three-dimensional systems are always affected by gravity. When α is large, g_t will also be large and cause a rather flat front. But as α becomes smaller, the front becomes wider. This is seen in Figure 11.10(a) and (b), which show how the displacement front evolves as air displaces a glycerol–water mixture under the effect of gravity.

In the following, we derive a prediction on how the front width depends on the gravity component g_t by comparing the tilted system with a horizontal ($\alpha = 0$) system where percolation has just happened. We note that in the horizontal system, the probability that a pore is invaded by air is just the critical percolation probability f_c, while it will vary with position in the tilted case. This is seen in Figure 11.10(a). As the percolating air front passes, it leaves behind a region of trapped glycerol clusters. These clusters do not come in all sizes up to the system size, as in the horizontal case, but rather they have a maximum size, corresponding to a larger volume fraction of air. So, away from the front on the side of the invaded air, $f > f_c$, while on the other side, $f = 0$. Somewhere in between, where the front is developing, $f = f_c$.

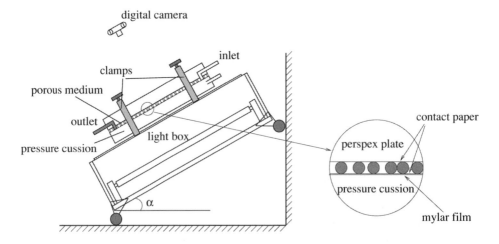

Figure 11.9 Sketch of the tilted experiment. (After Méheust et al. (2002).)

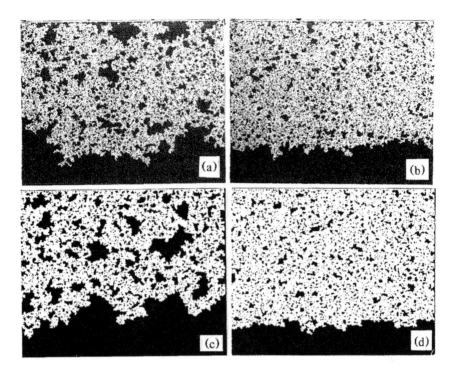

Figure 11.10 Slow displacement of a glycerol–water mixture by air in a two-dimensional porous medium at an angle to the horizontal of (a) $\alpha = 3°$ and (b) $\alpha = 11°$. (c) Numerical simulations of invasion percolation and a 400×400 lattice with Bo = 0.001 and (d) Bo = 0.01. (After Birovljev et al. (1991).)

We shall assume that (i) the size of the clusters is also the width of the front and (ii) that the cluster size is given by the correlation length $\xi(p)$ of *ordinary* percolation given by Eq. (4.15). These two assumptions may be summarized as

$$W \sim \xi \sim (f - f_c)^{-\nu}, \tag{11.3}$$

where f is the occupation probability at the top of the front, and we have used Eq. (4.39), that is, the value $\nu = 1.35$. Since the clusters that are trapped after the front has passed become isolated from further liquid exchange, they will remain fixed in size, as will f.

Now, the local value of f is related to the local capillary pressure via the normalized distribution of capillary threshold pressures $N(p_t)$, illustrated in Figure 11.11(b). The relation is

$$f(p) = \int_{-\infty}^{p} dp_t \, N(p_t). \tag{11.4}$$

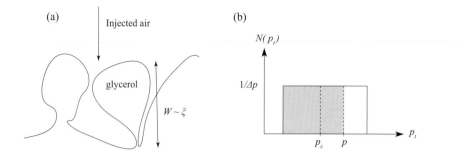

Figure 11.11 (a) A sketch of the front where the movement of the fluids traps a cluster of glycerol once the width W reaches the correlation length ξ. (b) The assumed distribution of capillary threshold pressures. Here p is the actual local pressure difference over the interface, and p_c is the global critical percolation pressure, that is, the pressure difference needed for percolation in a horizontal system.

Note that at the bottom of the front where gravity causes a larger glycerol pressure, the capillary pressure is smaller, and so is f. Correspondingly, at the top f is larger. The departure from f_c is then

$$f(p) - f_c = \int_{p_c}^{p} dp_t\, N(p_t) = \frac{p - p_c}{\Delta p}, \tag{11.5}$$

where Δp is the width of the $N(p_t)$ distribution, which, for simplicity, is taken to be flat. The hydrostatic pressure increase from the top of the front to the p_c location is only some fraction of W. But since a prefactor multiplying W will not alter the final scaling behavior, we simply write

$$p - p_c = g_t W \Delta\rho, \tag{11.6}$$

where $\Delta\rho > 0$ is just the density difference between air and glycerol. Inserting this in Eq. (11.5) gives

$$f(p) - f_c = \text{Bo}\frac{W}{a}, \tag{11.7}$$

where we have introduced the Bond number

$$\boxed{\text{Bo} = \frac{g_t a \Delta\rho}{\Delta p}.} \tag{11.8}$$

The Bond number[1] is quite simply the ratio of the hydrostatic pressure build-up over one pore to the typical pressure needed to move a fluid cluster. The reason for this is

---

[1] A slightly different definition using σ/a instead of Δp, and hence $\text{Bo} = \Delta\rho g_t a^2/\sigma$, is often employed in the literature. This number is generally of the same order of magnitude but misses the fact that the very narrow $N(p_t)$ distributions may produce narrower fronts.

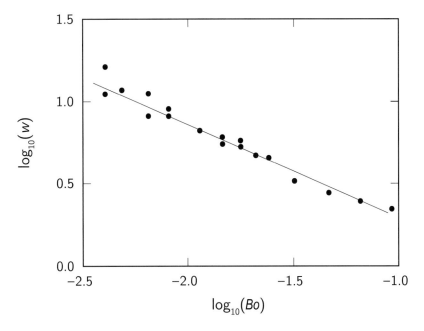

Figure 11.12 Front width as a function of Bond number Bo. The straight line has the predicted slope of -0.57. (After Birovljev et al. (1991).)

that the movement of a fluid cluster normally involves a pair of pores and requires a pressure difference corresponding to the fact that one pore must be invaded while the other is drained. The pressure needed for this is typically the difference between two pores, one of which is easily drained and one of which is easily invaded. This difference may be estimated as Δp.

Using Eq. (11.3) to replace $f(p) - f_c$ by $W^{-1/\nu}$ in Eq. (11.7) and solving for W gives

$$W = a \, \text{Bo}^{-\nu/(1+\nu)}, \qquad (11.9)$$

where $\nu/(1 + \nu) = 0.57$ if we use the above ν-value. Figure 11.12 shows that this prediction agrees well with experimental measurements.

Note that Eq. (11.9) implies that the front width diverges as $\alpha \to 0$ and Bo $\to 0$. This is what we would expect as the correlation length ξ diverges in invasion percolation for a horizontal system. Also, in the $\alpha \to 0$ limit, Bo $\to 0$, so from Eq. (11.7) we see that $f \to f_c$, as we know for the case of invasion percolation. Another significant observation from Eq. (11.9) is that $W \to 0$ as $\Delta p \to 0$ and Bo $\to \infty$. This is linked to the pairwise pore movement associated with the cluster displacments that define the front. Finally, the $N(p_t)$-distribution does not have to be flat as we have assumed in Figure 11.11(b). In fact, it makes little difference if the distribution is different; it will only change the interpretation of Δp slightly.

11.5 Steady State Multiphase Flow*

So far we have studied invasion processes that are transient: They start and they finish. Historically, these transient systems were also studied first and attracted a lot of interest, in part by virtue of their fractal nature. Systems that are in *steady states*, on the other hand, are in many respects far simpler. They have much in common with systems in statistical mechanical equilibrium, while the transient systems resemble non-equilibrium systems. It is therefore a historical oddity that the more challenging systems were studied so carefully without paying much attention to the simpler systems first.

Avraam and Payatakes (1995) were the first to treat steady states as a matter for serious investigation at the pore scale. They identified many key features of such systems. First, they do not have stationary interfaces, as clusters break up and move. This motion was coined *ganglia dynamics*, the moving clusters being the *ganglia*. They do, however, have stationary distributions of cluster sizes.

In the following we describe a more recent experiment that explores these concepts using the type of setup shown in Figure 11.3. In this case, however, it is modified to deal with a steadily injected mixture of two phases, as is illustrated in Figure 11.13. When this happens, clusters of wetting and non-wetting fluid move in a way that makes it impossible to distinguish drainage and imbibition processes. Figure 11.14 shows how the system evolves toward a steady state. In Figure 11.14, the water–glycerol mixture is of dark color, the air is bright white and glass beads surrounded by air may be indiscernible.

As the final steady state is approached, both capillary and viscous forces are at play. In fact, they compete so as to produce a critical size s^*, below which air clusters will become immobile and above which they will move but also break up. As the air-clusters break up, they also produce a range of smaller clusters, as is seen in Figure 11.14. In order to determine s^* and the relationship between flow rate and pressure drop, we note that just as in Section 11.3, there is a critical length scale where capillary and viscous forces balance each other. Now, this length scale is just $w_f = \sqrt{s^*}$. When the viscous pressure drop in the glycerol surrounding an air cluster, $w_f \nabla P$, exceeds the characteristic capillary pressure σ/a, the glycerol will invade the air clusters causing their break-up. Expressing the pressure gradient by the total pressure drop over the system Δp_L and the system length L, we may equate the pressures at hand by writing

$$\frac{\Delta p_L}{L} w_f = \frac{\sigma}{a}. \tag{11.10}$$

From this it is evident that

$$\frac{1}{w_f} \sim \Delta p_L. \tag{11.11}$$

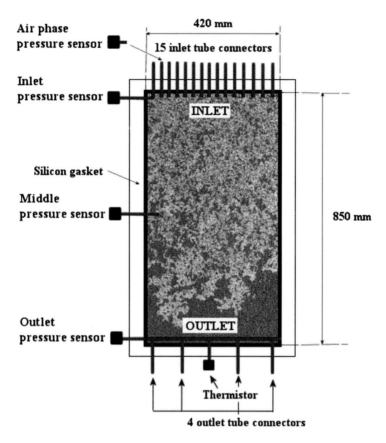

Figure 11.13 The experimental model. There are 15 independent inlet holes with attached pipes where we inject alternately the wetting and the nonwetting phase with syringe pumps. This leads to a mixing of the two phases inside the model two-dimensional porous medium, a random monolayer of glass beads, and a mix of the two phases flows out of the outlet channel at the opposite end of the system. (After Tallakstad et al. (2009).)

Now, we make the assumption that the permeability of the system is proportional to the number of channels available to the glycerol between the air clusters, or quite simply to the number L/w_f of clusters across the system (which may be assumed to be quadratic so that the length and width are equal). With this assumption, the flow rate becomes

$$Q \sim \frac{L}{w_f} \frac{\Delta p_L}{L} \sim \Delta p_L^2,$$ (11.12)

or, since Ca $\sim Q$,

$$\Delta p_L \sim \sqrt{\text{Ca}}.$$ (11.13)

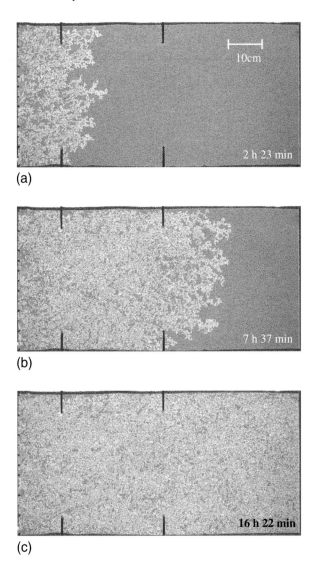

(a)

(b)

(c)

Figure 11.14 For a Ca $= 0.0079$ experiment, using the system of Figure 11.13, the system is shown at three different times. Both fluids are injected at the left-hand side; the outlet is at the right. The upper panel shows a sample in the early transient regime. The middle panel shows a later stage in the transient. The lower panel shows the fully developed steady state. (After Tallakstad et al. (2009).)

Equation (11.11) then tells us that $w_f \sim \mathrm{Ca}^{-1/2}$ and the critical cluster size

$$s^* \sim \frac{1}{\mathrm{Ca}}. \tag{11.14}$$

The situation is quite different from the transient case discussed in Section 11.3 because the permeability κ now depends on Δp_L. In fact, this breakdown of the

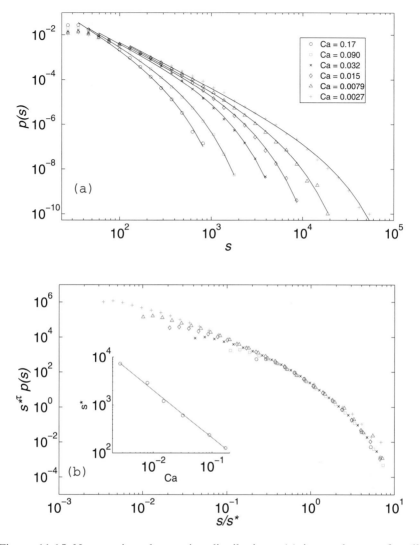

Figure 11.15 Non-wetting cluster size distribution $p(s)$ in steady state for all experiments. The cluster size s is measured in pixels; 1 pixel = 0.037 mm^2. The top figure shows normalized probability distributions. The dominating cutoff behavior is evident. In the lower figure, the horizontal and vertical axes are rescaled with $1/s^*$ and $(s^*)^\tau$, respectively, to obtain the data collapse. The Ca dependence of the cutoff cluster size s is shown in the inset. (After Tallakstad et al. (2009).)

simple linear (multiphase) Darcy law is quite remarkable. We will return to this problem in great detail in the next section.

In Figure 11.16 it is seen that Eq. (11.13) indeed gives a good description of the experimental behavior. It turns out that the data is well fitted by the function

$$p(s) \sim s^{-\tau} \exp(-s/s^*), \qquad (11.15)$$

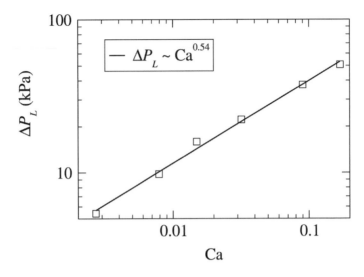

Figure 11.16 Average steady state pressure drop over the system of size L as a function of Ca, with the prediction of Eq. (11.13) represented by the full line. (After Tallakstad et al. (2009).)

where s^* is the critical cluster area. To get the τ-exponent, we note that the volume flux of air Q_a is an imposed quantity and must be equal to the accumulated volume flux of all clusters that move. The cross-sectional area of the whole system is $A = aL$, where a is the system thickness, and the cross-sectional area of a single cluster is $w_f a$. This means that the contribution of a single cluster of width w_f to the total volume flux of air is $a w_f U(s)$, where $U(s)$ is the average center of mass velocity of the cluster.

Further, the average number of clusters of extension $(w_f, w_f + dw_f)$ that intersects any given cross section A is given as $ds\, p(s) N w_f / L$, where N is the total number of clusters. In this expression, $ds\, p(s) N$ is the number of clusters in the given range, while w_f / L is the fraction that intersects A. Hence we can write

$$Q_a = \frac{aN}{L} \int_{a^2}^{\infty} ds\, s p(s) U(s), \qquad (11.16)$$

where we have used $s = w_f^2$ and the minimal cluster area a^2, which is the area of a single pore. Now, N is expected to be proportional to the system area L^2, so it does not derive from $p(s)$. When Q is increased, we expect more small clusters and fewer large ones. It turns out that these dependencies nearly cancel to produce a roughly constant N value.

To obtain $U(s)$, we make the plausible assumption that it is linear in Q_a. However, since clusters smaller than s^* are generally immobile, we should write

$$U(s) = f(s/s^*)\frac{Q_a}{A},$$ (11.17)

where $f(x)$ is some function that drops quickly to 0 below $x = 1$. Combining Eqs (11.15) and (11.17) in Eq. (11.16) then gives the total air flow

$$Q_a = Q_a \frac{aN}{V} \int_0^\infty ds \, s^{1-\tau} \exp(-s/s^*) f(s/s^*),$$ (11.18)

where $V = LA$, and we have replaced the lower integration limit a^2 by zero, since $f(x)$ anyway vanishes for $x < a^2/s^*$. Setting $x = s/s^*$ and cancelling Q_a on both sides gives

$$1 = s^{*(2-\tau)}\frac{aN}{V} \int_0^\infty dx \, x^{1-\tau} \exp(-x) f(x).$$ (11.19)

Since the left-hand side is s^*-independent, so must the right-hand side be, and it follows that

$$\boxed{\tau = 2.}$$ (11.20)

The solid lines in Figure 11.15 represent fits of Eq. (11.15) that have been optimized by tuning τ as a free parameter. The result, $\tau = 2.07 \pm 0.18$, agrees well with Eq. (11.20). The scaling form expressed in Eq. (11.15) also means that the product $y = p(s)s^\tau$ is a function only of $x = s/s^*$, and if we plot y as a function of x for experiments with various Ca-values, the result should fall on the same curve. This is indeed what is observed in Figure 11.15(b).

The distributions shown in Figure 11.15 do not change in time. The clusters continuously break up and merge on the scale up to s^*, but macroscopic properties like saturations do not change, save for fluctuations that decay as the averaging volume is increased. This is very much like the situation in equilibrium statistical mechanics where thermodynamic variables take on constant values up to some fluctuations that reflect the fact that the underlying microdynamics keeps moving the system around in phase space. This motion implies that the macroscopic behavior of the system is independent of microscopic details as these details are under continuous change.

11.6 Steady-State Flow in the Capillary Fiber Bundle Model*

The *capillary fiber bundle model*, first introduced by Scheidegger (1953, 1958), provides an excellent instrument for studying immiscible steady-state flow in porous media. We will in this section introduce this model and use it to study the non-linear Darcy law (11.12). We will indeed find flow rate being proportional to the pressure drop to some power, but – not surprisingly – not necessarily equal to two. We will also find that there is a crossover to linear behavior where the flow rate is proportional to the pressure drop at high pressure drops. A last important point is that we will be able to scale up the size of the model and answer the question, "What happens to the non-linear regime when it becomes large?"

11.6.1 Capillary Pipe Flow*

We considered in Section 9.5 a single bubble of length Δx moving in a capillary pipe of length L. The bubble is surrounded by fluid that is more wetting than the one constituting the bubble, but there is not complete wetting, that is, there are no wetting fluid films. Both fluids are incompressible. The pipe has a radius varying with the position x along the pipe, as given in Eq. (9.56). The capillary pressure on this bubble when it is at position x_b is given by Eq. (9.59).

We follow here Sinha et al. (2013). Suppose now we have N bubbles numbered $i = 1$ to N. Their midpoints are at positions $x_1 < x_2 < \cdots < x_N$, and they have lengths $\Delta x_1, \Delta x_2, \ldots, \Delta x_N$, with the constraint that $x_{i+1} - x_i > (\Delta x_i + \Delta x_{i+1})/2$, so that they do not overlap. The total capillary force exerted on this bubble train is the sum over all bubbles,

$$\Delta p_t = -\frac{4\sigma a}{r_0} \sum_{i=1}^{N} \sin\left(\frac{2\pi \Delta x_i}{l}\right) \sin\left(\frac{2\pi x_i}{l}\right); \tag{11.21}$$

see Eq. (9.59) for the force on a single a single bubble at x_b.

Let us now choose a point x_0 that follows the flow. That is, we will have $\pi r_0^2\, dx_0/dt = q$. The positions of all the bubbles will then be given by $x_i = x_0 + \delta x_i$, where $d\delta x_i/dt = 0$ since there are no relative motions of the bubbles. We may then rewrite Eq. (11.21) using the identity $\sin(\theta_1 + \theta_2) = \cos(\theta_1)\sin(\theta_2) + \sin(\theta_1)\cos(\theta_2)$:

$$\Delta p_t = -\frac{4\sigma a}{r_0}\left[\Gamma_c \sin\left(\frac{2\pi x_0}{l}\right) + \Gamma_s \cos\left(\frac{2\pi x_0}{l}\right)\right], \tag{11.22}$$

where

$$\Gamma_c = \frac{4\sigma\sqrt{1-a^2}a}{r_0} \sum_{i=1}^{N} \sin\left(\frac{2\pi \Delta x_i}{l}\right) \cos\left(\frac{2\pi x_i}{l}\right) \tag{11.23}$$

and

$$\Gamma_s = \frac{4\sigma\sqrt{1-a^2}a}{r_0} \sum_{i=1}^{N} \sin\left(\frac{2\pi \Delta x_i}{l}\right) \sin\left(\frac{2\pi x_i}{l}\right). \tag{11.24}$$

The Washburn equation for this bubble train is

$$q = -\frac{\pi r_0^4}{8\mu_{\text{eff}} L}[\Delta p - \Delta p_t], \tag{11.25}$$

where the effective viscosity is given by

$$\mu_{\text{eff}} = S_w \mu_w + S_n \mu_n. \tag{11.26}$$

The wetting and non-wetting saturations are given by

$$S_w = 1 - \sum_{i=1}^{N} \frac{\Delta x_i}{L} \tag{11.27}$$

and

$$S_n = \sum_{i=1}^{N} \frac{\Delta x_i}{L}, \tag{11.28}$$

respectively.

We turn the Washburn equation (11.25) into an equation of motion for x_0, finding

$$\frac{dx_0}{dt} = \frac{r_0^2}{8L\mu_{\text{eff}}} \left[|\Delta p| - \Gamma_s \sin\left(\frac{2\pi x_0}{l}\right) - \Gamma_c \cos\left(\frac{2\pi x_0}{l}\right) \right]. \tag{11.29}$$

We define

$$\gamma = \sqrt{\Gamma_s^2 + \Gamma_c^2}. \tag{11.30}$$

By non-dimensionalizing Eq. (11.29) and defining as in Section 9.5 a dimensionless angle $\Theta = 2\pi x_0/L$ and a dimensionless time $\tau = \gamma t \pi r_0^2/(4L^2 \mu_{\text{eff}})$, we find

$$\frac{d\Theta}{d\tau} = \left(\frac{|\Delta p|}{\gamma}\right) - \frac{\Gamma_s}{\gamma} \sin(\Theta) - \frac{\Gamma_s}{\gamma} (\cos \Theta). \tag{11.31}$$

We define

$$\tan(\Theta_\gamma) = \frac{\Gamma_s}{\Gamma_c}, \tag{11.32}$$

so that we may write

$$\frac{\Gamma_s}{\gamma} \sin(\Theta) + \frac{\Gamma_s}{\gamma} \cos(\Theta) = Cos(\Theta - \Theta_\gamma). \tag{11.33}$$

We then redefine the angle Θ,

$$\Theta - \Theta_\gamma - \frac{\pi}{2} \rightarrow \Theta, \tag{11.34}$$

giving

$$\frac{d\Theta}{d\tau} = \left(\frac{|\Delta p|}{\gamma}\right) - \sin(\Theta), \tag{11.35}$$

which is identical to the equation of motion of a single bubble, see Eq. (9.63).

We calculate the period, T_τ in the same way as we did for the single bubble, (9.64),

$$T_\tau = \int_0^{T_\tau} d\tau = \int_0^{2\pi} \frac{d\Theta}{(d\Theta/d\tau)} = \int_0^{2\pi} \frac{d\Theta}{|\Delta p|/\gamma - \sin(\Theta)}. \tag{11.36}$$

This integral diverges as $|\Delta p|/\gamma \to 1$ from above – this divergence being the saddle-node bifurcation, and we find

$$T_\tau = \frac{2\pi\gamma}{\sqrt{\Delta p^2 - \gamma^2}}. \tag{11.37}$$

The average angular speed, $\langle d\Theta/d\tau \rangle$, we find as in Eq. (9.65):

$$\left\langle \frac{d\Theta}{d\tau} \right\rangle = \frac{2\pi}{T_\tau} = \frac{1}{\gamma}\sqrt{\Delta p^2 - \gamma^2}. \tag{11.38}$$

Reverting to the original variables q and t, we thus have

$$\langle q \rangle = -\frac{\pi r_0^4}{8\mu_{\text{eff}}L}\, \text{sign}(\Delta p) \begin{cases} \sqrt{\Delta p^2 - \Delta p_t^2} & \text{if } |\Delta p| > \Delta p_t, \\ 0 & \text{if } |\Delta p| \leq \Delta p_t, \end{cases} \tag{11.39}$$

where we have renamed γ the threshold pressure

$$\Delta p_t = \gamma. \tag{11.40}$$

11.6.2 From a Single Capillary Fiber to the Capillary Fiber Bundle*

Suppose we put together K capillary fibers to form a capillary fiber bundle. In the following discussion, we follow Roy et al. (2019a). Each fiber is of the type we have just discussed in the preceding section. We assume that each fiber has its own threshold pressure drawn from a uniform distribution on the interval $[0, \Delta p_m]$. The corresponding cumulative probability is

$$P(p_t) = \begin{cases} 0, & \Delta p_t \leq 0, \\ \frac{\Delta p_t}{\Delta p_M}, & 0 < \Delta p_t \leq \Delta p_M, \\ 1, & \Delta p_t > \Delta p_M. \end{cases} \tag{11.41}$$

We now imagine an ensemble of such capillary fiber bundles, each containing K fibers. We may then use order statistics, see Eq. (3.30), to order the thresholds after averaging over the ensemble. We find

$$P(\Delta p_{t,(k)}) = \frac{k}{K + 1}, \tag{11.42}$$

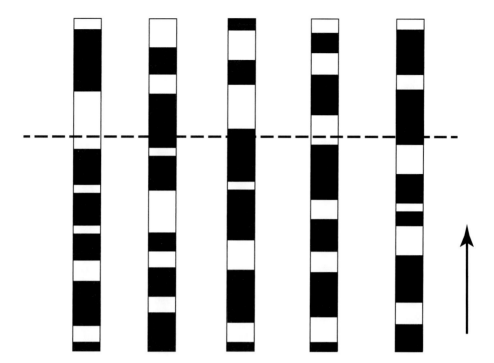

Figure 11.17 The capillary pipe model: K parallel capillary pipe filled with a more wetting fluid (white) wherein there are bubbles of less wetting fluid (black). There is a pressure difference Δp driving the fluids. When calculating the average flow rate vs. pressure difference relations, two averaging processes must be done. The first one is with respect to time. We envision this done by averaging over what passes through the broken line, representing a virtual cut through the model. The second averaging is over the pipe. This is illustrated by each pipe in the figure having its distribution of bubble sizes. Each pipe, however, has the same saturation, and hence, effective viscosity.

where $1 \le k \le K$. Thus, we have

$$\Delta p_{t,(k)} = \Delta p_M \frac{k}{K+1}. \tag{11.43}$$

The average volumetric flow rate through the capillary fiber bundle is then

$$\langle Q \rangle = \sum_{k=1}^{(K+1)a/\Delta p_M} \bar{q}\left(\Delta p_{t,(k)}\right), \tag{11.44}$$

where $a = \min(|\Delta p|, \Delta p_M)$, and $|\Delta p| \ge \frac{\Delta p_M}{K+1}$. We now assume $K \to \infty$, turning the sum into an integral,

$$\langle q \rangle = \frac{\langle Q \rangle}{K} = -\frac{r_0^4 \Delta p_M \operatorname{sign}(\Delta p)}{8\pi \mu_{\mathrm{av}} L} \int_0^{a/\Delta p_M} dx \sqrt{\left(\frac{|\Delta p|}{\Delta p_M}\right)^2 - x^2}, \tag{11.45}$$

for $|\Delta p| \geq \frac{\Delta p_M}{K+1} \to 0$. We do the integral and find

$$\langle q \rangle = -\frac{r_0^4}{32\mu_{av}L}\left|\frac{\Delta p}{\Delta p_M}\right|\Delta p, \tag{11.46}$$

when $|\Delta p| \leq \Delta p_M$ and

$$\langle q \rangle = -\frac{r_0^4\Delta p_M\,\mathrm{sign}(\Delta p)}{16\pi\,\mu_{eff}L}\left[\sqrt{\left(\frac{|\Delta p|}{\Delta p_M}\right)^2 - 1} + \left(\frac{|\Delta p|}{\Delta p_M}\right)^2\arcsin\left(\frac{\Delta p_M}{|\Delta p|}\right)\right], \tag{11.47}$$

when $|\Delta p| > \Delta p_M$. In the limit $|\Delta p| \gg \Delta p_M$, Eq. (11.47) gives

$$\langle q \rangle = -\frac{r_0^4}{8\pi\,\mu_{eff}L}\Delta p, \tag{11.48}$$

which is the Darcy law for the fiber bundle – which is close to the Hagen–Poiseuille equation for a pipe, see Eq. (5.18).

Now, Eq. (11.46) is very interesting. We have in fact seen a similar equation before. Eq. (11.12) describes the effective Darcy law for a steady-state immiscible two-phase flow. Let us therefore check whether this behavior is generic or dependent on the specific threshold distribution we chose, Eq. (11.41).

We now assume a cumulative threshold probability

$$P(\Delta p_t) = \begin{cases} 0, & \Delta p_t \leq \Delta p_m, \\ \frac{\Delta p_t - \Delta p_m}{\Delta p_M - \Delta p_m}, & \Delta p_m < \Delta p_t \leq \Delta p_M, \\ 1, & \Delta p_t > \Delta p_M, \end{cases} \tag{11.49}$$

by moving the lower cutoff of the threshold distrbution from zero to a finite value Δp_m.

In this case, Eq. (11.42) yields the ordered threshold sequence

$$\Delta p_{t,(k)} = \Delta p_m + (\Delta p_M - \Delta p_m)\frac{k}{K+1}. \tag{11.50}$$

In the $K \to \infty$ limit, Eq. (11.44) now gives

$$\langle q \rangle = -\frac{r_0^4(\Delta p_M - \Delta p_m)\,\mathrm{sign}(\Delta p)}{8\pi\,\mu_{eff}L}$$

$$\int_{\Delta p_m/(\Delta p_M - \Delta p_m)}^{b/(\Delta p_M - \Delta p_m)} dx\sqrt{\left(\frac{|\Delta p|}{\Delta p_M - \Delta p_m}\right)^2 - x^2}, \tag{11.51}$$

where $b = \max(|\Delta p|, \Delta p_M)$.

When we assume $\Delta p_m \leq |\Delta p| \leq \Delta p_M$, we find

$$\langle q \rangle = -\frac{r_0^4(\Delta p_M - \Delta p_m)\,\mathrm{sign}(\Delta p)}{32\pi\,\mu_{\mathrm{eff}}L}\left(\frac{|\Delta p|}{\Delta p_M - \Delta p_m}\right)^2$$

$$\times\left[\pi - 4\left(\frac{\Delta p_m}{|\Delta p|}\right)^2\sqrt{\left(\frac{|\Delta p|}{\Delta p_m}\right)^2 - 1} - 2\,\mathrm{arccot}\left(\frac{2\sqrt{(|\Delta p|/\Delta p_m)^2 - 1}}{2(|\Delta p|/\Delta p_m)^2}\right)\right].$$

(11.52)

To lowest order in $(|\Delta p| - \Delta p_m)$, this expression behaves as

$$\langle q \rangle = -\frac{r_0^4\,\mathrm{sign}(\Delta p)}{3\sqrt{2}\pi\,\mu_{\mathrm{eff}}L}\frac{\sqrt{\Delta p_m}}{(\Delta p_M - \Delta p_m)}(|\Delta p| - \Delta p_m)^{3/2},$$

(11.53)

as $|\Delta p| \to \Delta p_m$. Hence, the similarity with Eq. (11.12) is gone. However, the *mechanism* responsible for the non-linear behavior of the Darcy law is evident. The number of "open" – that is, flowing – capillaries increases with the pressure difference $|\Delta p|$, the mechanism already detailed in Section 11.5.

When we assume $|\Delta p| > \Delta p_M$, we find

$$\langle q \rangle = -\frac{r_0^4\Delta p}{16\pi\,\mu_{\mathrm{eff}}(\Delta p_M - \Delta p_m)L}$$

$$\times\left[\Delta p_M\sqrt{1 - \left(\frac{\Delta p_M}{|\Delta p|}\right)^2} + |\Delta p|\arcsin\left(\frac{\Delta p_M}{|\Delta p|}\right)\right.$$

$$\left. -\Delta p_m\sqrt{1 - \left(\frac{\Delta p_m}{|\Delta p|}\right)^2} - |\Delta p|\arcsin\left(\frac{\Delta p_m}{|\Delta p|}\right)\right].$$

(11.54)

As $|\Delta p| \gg \Delta p_M$ and $|\Delta p| \gg \Delta p_m$, this expression reverts to linear dependency on Δp, that is, Darcy-type behavior.

11.6.3 Scaling the Threshold Pressure*

What happens to the threshold pressure Δp_t as $L \to \infty$? In order to answer this question, we assume that the number of bubbles, N, diverges in this limit, but in such a way that the bubble density $n = N/L$ remains constant.

The threshold pressure Δp_t for a single capillary is defined in Eqs (11.23), (11.24), (11.30) and (11.40).

Focus on the sum occurring in Γ_s, Eq. (11.23):

$$\sum_{i=1}^{N}\sin\left(\frac{2\pi\,\Delta x_i}{l}\right)\sin\left(\frac{2\pi\,\delta x_i}{l}\right).$$

(11.55)

If the relative positions δx_i are randomly placed, for example, by a Poisson process, the term $\sin(2\pi\,\delta x_i/l)$ will change sign randomly. There will also be random sign changes coming from the term $\sin(2\pi\,\Delta x_i/l)$ depending on the length of the bubbles. We are then dealing with a random walk, and we have

$$\left|\sum_{i=1}^{nL}\sin\left(\frac{2\pi\,\Delta x_i}{l}\right)\sin\left(\frac{2\pi\,\delta x_i}{l}\right)\right| = C\sqrt{nL}, \qquad (11.56)$$

where C is some positive constant, see Section 8.1. Thus, we find

$$\Gamma_s = C\frac{4\sigma\sqrt{1-a^2}a}{r_0}\sqrt{nL}. \qquad (11.57)$$

Likewise, we have

$$\Gamma_c = C'\frac{4\sigma\sqrt{1-a^2}a}{r_0}\sqrt{nL}. \qquad (11.58)$$

From Eq. (11.22), we then find

$$\frac{\Delta p_t}{L} \propto \sqrt{\frac{n}{L}} \rightarrow 0, \qquad \text{as } L \rightarrow \infty. \qquad (11.59)$$

This has important consequences for the behavior of the capillary fiber bundles. We considered two threshold distributions, given by Eqs (11.41) and (11.49). In both cases the lower and upper limiting thresholds Δp_m and Δp_M scale as $\sqrt{L}$, whereas $|\Delta p|$ scales as L in order to keep the pressure gradient $\Delta p/L$ constant. Thus, in the limit $L \rightarrow \infty$, the expressions *all revert to the linear Darcy behavior.*

This limit is in fact the continuum limit. Our conclusion is that the capillary fiber bundle models reverts to the Darcy behavior,

$$\langle q \rangle = -\frac{r_0^4}{8\pi\,\mu_{\text{eff}}}\frac{\partial P}{\partial x}, \qquad (11.60)$$

in the continuum limit.

In Section 11.5 where we studied a two-dimensional Hele-Shaw cell filled with static glass beads, there was no threshold pressure, see Eq. (11.12). The reason for this is that both fluids percolate simultaneously through the porous medium. This phenomenon, both phases percolating, may be modeled using the capillary fiber bundle model by introducing a distribution in the number N of bubbles in the fibers, rather than having N fixed for each fiber. If there is at least one fiber that does not contain any bubbles, the threshold Δp_t will be zero since there will be flow as soon as there is a pressure difference Δp.

We note that percolation of both fluid phases is even more pronounced in three dimensions than it two. Hence, the typical situation in three dimensions will be one with zero threshold pressure, $\Delta p_t = 0$.

However, if there is no simultaneous percolation of the two immiscible fluids, there will be a non-zero threshold pressure Δp_t since any path from inlet to outlet will pass through fluid interfaces. How does Δp_t scale with the system size? As we will now argue, exactly the same mechanism that made the capillary fiber bundle model linear as $L \to \infty$ is also at work in the full two- or three-dimensional case. However, there is *another* mechanism present that leads to non-linearities in the latter case, which is *not* present in capillary fiber bundle: When the pressure difference across the porous medium is increased, fluid clusters that were stuck begin to move. This mechanism lacks in the capillary fiber bundle, where each capillary fiber either flows or does not flow. This mechanism is responsible for the non-linearity, as described in Section 11.5.

Let us now generalize the arguments we used for the capillary fiber bundle model to the two- or three-dimensional case. Choose some path through the porous medium starting at the inlet and ending at the outlet. Let us assume we are starting the path at a point where the surrounding fluid is wetting. This path will pass through a number of interfaces. Let us start a counter at zero. Every time we pass from the more wetting fluid to the less wetting fluid, we add a +1, and every time we pass through an interface from the less wetting fluid to the more wetting fluid, we add a -1. What will the counter end at when we reach the outlet. It will be +1 if we end at a point surrounded by the less wetting fluid and 0 if we end surrounded by the more wetting fluid. Hence, the interfaces will always come in pairs as with the less wetting bubbles in the capillary pipe system. We can sum up the capillary forces for each less wetting "bubble" (which are now clusters) as we did for the bubbles in the capillary pipe. They will almost cancel, but fluctuations remain, and these will be responsible for the non-linearities as they were in the capillary bubble. These fluctuations will be randomly positive or negative. Hence, the capillary forces along any path through the system will therefore behave as a random walk, so that the sum will scale as the square root of the length of the path. This means that the threshold pressure – which is the sum along the path giving the smallest value – will increase more slowly than L, resulting in

$$\frac{\Delta p_t}{L} \to 0, \qquad \text{as } L \to \infty, \qquad (11.61)$$

as for the capillary fiber bundle model, see Eq. (11.59). Figure 11.18 shows Δp_t as a function of L based on a dynamic network simulator that models the glass

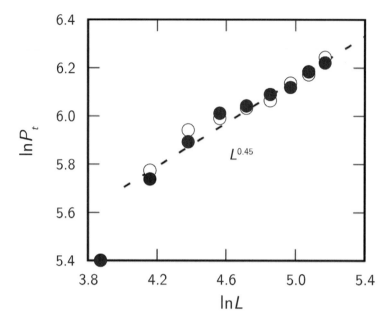

Figure 11.18 The glass bead-filled Hele-Shaw cell of Tallakstad et al. (2009) may be modeled using a dynamic network simulator (Aker et al., 1998; Sinha et al., 2021) that tracks the motion of the interfaces between the immiscible fluids. Using a square lattice of size $L \times L$, Roy et al. (2019b) have measured the threshold pressure Δp_t as a function of the size of the network L, both when the pressure difference $\Delta p/L$ (green circles) or the volumetric flow rates Q_w/L and Q_n/L are kept constant (purple circles). In both cases they find $\Delta p_t \sim L^{0.45}$.

bead-filled Hele-Shaw cell. It behaves as predicted. Hence, the threshold pressure Δp_t may be ignored for large systems, whether there is simultaneous percolation of both the immiscible fluids or not.

11.7 Mean Field Theory for Steady-State Immiscible Two-Phase Flow*

We now return to the non-linear Darcy equation (11.12). Here we will derive this equation using a very powerful technique to analyze the flow: *mean field theory*. "Mean field theory" is a term from theoretical physics. The engineering community will know this approach as homogenization.

Imagine the porous medium as a network of links. Here is how this approach works: 1. Pick out a link in the porous structure. 2. Replace all the other links by averaged links, all identical. 3. Average over this one remaining link. 4. Demand consistency; that is, the averaged link should behave in the same way as all the other already averaged links.

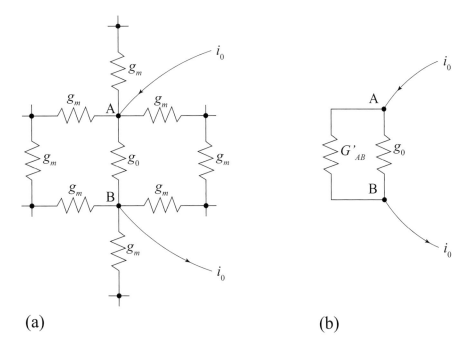

(a)   (b)

Figure 11.19 (a) A square network where all conductors have been replaced by average ones with conductance g_m, except for one that retains its original conductance g_0. We inject and extract a current i_0 at nodes A and B by placing voltages V_A and V_B, respectively. (b) The equivalent network, where G'_{AB} is the conductance of the network between nodes A and B without the conductance g_0. (From Kirkpatrick (1973).)

We now put flesh on this skeleton. We do so by basing ourselves on the mean-field theory developed by Kirkpatrick (1973) to deal with conductivity near the percolation threshold. He considered a network of electrical resistors with conductances (i.e., inverse resistance) g drawn from some probability distribution $p(g)$. We assume that the network is an infinite lattice with coordination number z, that is, each node has z conductors connected to it. Our goal is to replace the original lattice with links having their own conductances with another lattice where all the links have the same conductance, g_m, such that the conductance of the network as a whole remains the same after the replacement. Hence, we wish to answer the following question: What is g_m when we know $p(g)$?

For illustrative purposes, we use a square lattice oriented as shown in Figure 11.19(a). A uniform electrical field is set up in the vertical direction so that there is a voltage difference between each row equal to V_m when all links have been replaced by links with the mean field conductance g_m. However, we have left *one* link between nodes A and B with its original conductance g_0. Since $g_0 \neq g_m$, the voltage across this link is not V_m. In order to compensate for this, an extra current i_0 passed between nodes A and B so that the total current passing through this link is the same as in all the other links. Hence, we have

$$g_m V_m = g_0 V_m + i_0. \tag{11.62}$$

In order to accomplish this, a voltage difference V_0 between nodes A and B needs to be set up. If G'_{AB} is the conductance of the lattice between nodes A and B with the direct link between them removed, see Figure 11.19(b), then V_0 is given by

$$V_0(g_0 + G'_{AB}) = i_0. \tag{11.63}$$

We now calculate G'_{AB}. In order to do this, we replace the conductor with conductance g_0 by a conductor with conductance g_m between nodes A and B. We set up a voltage V between these two nodes, resulting in a current i being injected into and extracted from the network here. The current passing through the direct link between A and B is $2i/z$. This gives us what V must be, $V g_m = (2i/z)$, and thus G_{AB} since $V G_{AB} = i$. Hence, $G_{AB} = (z/2)g_m$. This in turn, gives us G'_{AB}, $G'_{AB} + g_m = G_{AB}$, or

$$G'_{AB} = \left(\frac{z}{2} - 1\right) g_m. \tag{11.64}$$

We now use this expression to eliminate i_0 between Eqs (11.62) and (11.63), and we find

$$V_0 = V_m \frac{g_0 - g_m}{g + \left(\frac{z}{2} - 1\right) g_m}. \tag{11.65}$$

Now, finally, comes the consistency step. We need that $\langle V_0 \rangle = 0$ when averaging over the disorder. Using Eq. (11.65), we then get the final expression

$$\left\langle \frac{g_0 - g_m}{g + \left(\frac{z}{2} - 1\right) g_m} \right\rangle = 0. \tag{11.66}$$

Hence, we need to solve this equation with respect to g_m – and we have the answer we are looking for.

We will now apply this method to the immiscible two-phase flow problem (Sinha and Hansen, 2012). We assume a lattice with coordination number z in which two immiscible fluids are flowing. Each link in the network obeys the constitutive equation, (11.39), relating time-averaged flow rate to the pressure difference Δp_L and a threshold Δp_t (and thus a cumulative probability $P(\Delta p_t)$ such that $dP/d\Delta p_t = p(\Delta p_t)$) for a capillary pipe containing a bubble train. We will assume that the disorder in the network comes from a statistical distribution of the thresholds, $p(\Delta p_t)$.

We need the mobilities of the links, which correspond to the conductances in the electrical case. They are given by

$$
\frac{\mathrm{d}\langle q \rangle}{\mathrm{d}(\Delta p_L)} = \frac{\pi r_0^4}{8 \mu_{\mathrm{eff}} l} m(\Delta p_L, \Delta p_t)
$$

$$
= \frac{\pi r_0^4}{8 \mu_{\mathrm{eff}} l}
\begin{cases}
\dfrac{|\Delta p_L|}{\sqrt{\Delta p_L^2 - \Delta p_t^2}} & \text{if } |\Delta p_L| > \Delta p_t, \\
0 & \text{if } |\Delta p_L| \leq \Delta p_t,
\end{cases}
\tag{11.67}
$$

where l is the length of a link. Eq. (11.66) then becomes

$$
\int_0^\infty \mathrm{d}P \, p(P) \frac{m(\Delta p_L, P) - M}{m(\Delta p_L, P) + (z/2 - 1)M} = 0,
\tag{11.68}
$$

where M is the mean-field mobility. We solve this equation with respect to M. We combine the equation with Eq. (11.67),

$$
\int_0^\infty \mathrm{d}p' \, p(p') \frac{|\Delta p_L| - M\sqrt{|\Delta p_L|^2 - p'^2}}{|\Delta p_L| + \left(\frac{z}{2} - 1\right) M\sqrt{|\Delta p_L|^2 - p'^2}} = 0.
\tag{11.69}
$$

We now assume a uniform distribution on the interval $[0, \Delta p_M]$, Eq. (11.41), and for simplicity, we assume a square lattice so that the coordination number $z = 4$. We then have the equation

$$
\int_0^a \frac{\mathrm{d}p'}{\Delta p_M} \frac{|\Delta p_L| - M\sqrt{|\Delta p_L|^2 - p'^2}}{|\Delta p_L| + M\sqrt{|\Delta p_L|^2 - p'^2}} - \int_a^{\Delta p_M} \frac{\mathrm{d}p'}{\Delta p_M} = 0,
\tag{11.70}
$$

where again $a = \min(|\Delta p_L|, \Delta p_M)$. The first thing we do is to determine the flow threshold. This is the value $\bar{p}_t$ of $|\Delta p_L|$ for which M disappears. So, setting $M = 0$ in Eq. (11.70), we find

$$
\frac{\bar{p}_t}{\Delta p_M} - 1 + \frac{\bar{p}_t}{\Delta p_M} = 0,
\tag{11.71}
$$

since we must have that the threshold $\bar{p}_t$ must be in the range where $|\Delta p_L| \leq \Delta p_M$. Hence, from Eq. (11.71) we have

$$
\bar{p}_t = \frac{\Delta p_M}{2}.
\tag{11.72}
$$

We now calculate the integral (11.70) assuming $|\Delta p_L| < \bar{p}_t = \Delta p_M/2$ so that $M > 0$. We find

$$
-\frac{|\Delta p_L|}{\Delta p_M}\left[1 - \frac{\pi}{M} + \frac{\pi - 2\arctan(M/\sqrt{1 - M^2})}{M\sqrt{1 - M^2}}\right] - 1 + \frac{|\Delta p_L|}{\Delta p_M} = 0.
\tag{11.73}
$$

We expand this expression to linear order in M to find

$$\frac{|\Delta p_L|}{\Delta p_M}\left[1 - \frac{\pi}{2}M + \mathcal{O}(M^2)\right] = 1 - \frac{|\Delta p_L|}{\Delta p_M}. \tag{11.74}$$

Hence, we have

$$M = \frac{4}{\pi} - \frac{2\Delta p_M}{\pi|\Delta p_L|}. \tag{11.75}$$

We assume $|\Delta p_L|$ is close to but larger than the threshold value $\bar{p}_t$, so that we expand this expression in terms of $\bar{p}_t - |\Delta p_L|$. We find to lowest order

$$M = \frac{8}{\pi\,\Delta p_M}\,(\bar{p}_t - |\Delta p_L|). \tag{11.76}$$

We now consider the case when $|\Delta p_L| > \Delta p_M$. Eq. (11.70) then becomes

$$\int_0^{\Delta p_M/|\Delta p_L|} dx\,\frac{1 - M\sqrt{1-x^2}}{1 + M\sqrt{1-^{2}}} = 0. \tag{11.77}$$

We find after doing the integral that this equation becomes

$$-\frac{\Delta p_M}{|\Delta p_L|} + \frac{2}{M}\arctan\left(\frac{\Delta p_M}{|\Delta p_L|}\right)$$

$$+\frac{2}{M\sqrt{1-M^2}}\left[\arctan\left(\frac{M}{\sqrt{1-M^2}}\frac{\Delta p_M}{|\Delta p_L|}\right)\right.$$

$$\left.- \arctan\left(\frac{1}{\sqrt{1-M^2}\left[1 - \left(\frac{\Delta p_M}{|\Delta p_L|}\right)^2\right]}\frac{\Delta p_M}{|\Delta p_L|}\right)\right] = 0. \tag{11.78}$$

We are interested in the asymptotic limit when $|\Delta p_L| \gg \Delta p_M$. Hence, eq. (11.78) becomes to lowest order

$$\frac{1-M}{1+M}\frac{\Delta p_M}{|\Delta p_L|} = 0, \tag{11.79}$$

which has as a solution that

$$M = 1. \tag{11.80}$$

After integrating Eq. (11.76), we have the behavior of $\langle q\rangle$, the flow in any link in the three ranges $|\Delta p_L| < \bar{p}_t$; $|\Delta p_L| \geq \bar{p}_t$, but close to $\bar{p}_t = \Delta p_M/2$; and lastly $|\Delta p_L| > \Delta p_M$:

$$\langle q \rangle = -\frac{r_0^4}{8\mu_{\mathrm{eff}}l} \begin{cases} 0 & \text{if } |\Delta p_L| < \bar{p}_t, \\ \frac{4}{\pi \Delta p_M} \, \mathrm{sign}(\Delta p_L) \left(|\Delta p_L| - \bar{p}_t^2\right)^2 & \text{if } \bar{p}_t \le |\Delta p_L| \ll \Delta p_M, \\ \Delta p_L & \text{if } \Delta p_M \ll |\Delta p_L|. \end{cases}$$

$$(11.81)$$

Hence, we see that the flow in the links has the behavior seen in the experiments, see Section 11.5, except for the appearance of a non-zero threshold $\bar{p}_t$.

Mean field theory is an *approximate* technique. It focuses on one link and replaces all the others by identical links. This destroys the cluster structure in the network. The reader should remember that the heuristic derivation of the square Darcy law (11.12) given in Section 11.5 was based on there being clusters of all sizes. The mean-field derivation sees only a single link.

Exercises

11.1 Invasion percolation with and without trapping:
In this problem we will look at the effect of "trapping" of the defending fluid. This effect is caused by incompressibility of the defending fluid, which keeps clusters that have been cut off from the rest of the fluid from decreasing further.

Apply the invasion percolation algorithm on a square lattice, and measure the fractal dimension D_0 of the percolation cluster by the box-counting metod.

11.2 Modify this algorithm to avoid trapping: This is done by identifying perimeters that have been closed by the growth of the occupied sites. Once a perimeter is closed, no further growth inside it is allowed.

11.3 Measure the new fractal dimension D_{tr}, and explain why $D_{tr} < D_0$.

11.4 Conductivity at the percolation threshold:
In Section 4.6, we described the conductivity of the random resistor network near the percolation threshold as following

$$G \sim \Theta(p - p_c)(p - p_c)^t,$$

where Θ is the Heaviside step function and p_c the percolation threshold. Use the Kirkpatric mean field theory described in Section 11.7 to calculate the conductivity exponent t for a square lattice where the bond percolation threshold is $p_c = 1/2$.

11.5 Generalize the result to a lattice with coordination number z; that is, every node has z bonds attached to it.

12

Continuum Descriptions of Multiphase Flow

The displacement of one fluid by another in porous media is of central interest in oil production, ground water management and many industrial processes. Often for practical purposes, the scales over which this displacement occurs are in the kilometer range. Still, the relevant physics controlling this process is happening at the scale of the pores, which typically is in the μm range. There are some nine orders of magnitude between this scale and the kilometer scale.

In 1972, the condensed matter physicist Philip W. Anderson wrote a very influential article named "More is Different" (Anderson, 1972).[1] The essence of Anderson's comment is that the description of nature is hierarchical with respect to (length) scale. Each scale has its appropriate description, which differs from those appropriate for the other scales. The description at one level is anchored in the one below in scale and acts as input to the one above in scale. No scale – or description – is more fundamental than any other. This philosophy is very different from the reductionist approach that has prevailed in physics throughout the twentieth century: the smaller, the more fundamental. Within this framework, elementary particle physics would be more fundamental than atomic physics, which in turn would be more fundamental than chemistry. The point Anderson is making is that elementary particle physics simply is not appropriate for describing the world at the level at which chemistry operates. In the same way, molecular dynamics is not the appropriate tool to describe the behavior of the water in a cup. Thermodynamics and hydrodynamics are.

This discussion is highly relevant for porous media. A reservoir that covers a volume of a kilometer cubed will contain some 10^{24} pores. To describe the motion of fluids in such a structure based on flow at the pore scale is no more appropriate

---

[1] We also recommend the article by Schweber in *Physics Today* (1993) for an even wider perspective.

than using molecular dynamics in the case of the water in a cup (which incidentally will contain some 10^{24} water molecules). We need to find another description that works at this scale.

The obvious choice to describe multiphase flow in porous media at large scales is to treat the porous medium as a continuum and construct appropriate differential equations. But, we have to ask ourselves, is this at all possible? Before discussing an answer, we need set the scene.

Continuum variables

A continuum description will require the introduction of new variables. These variables will have little or no value at the pore scale. However, we will have used variables that resemble those that we will introduce here. This sometimes induces confusion as the variables have different contents on different scales.

In order to keep the complexity down, we will consider two immiscible fluids, one more wetting, the other less wetting with respect to the porous matrix. We will refer to them as "w" and "n" in the following.

All the variables will be fields. That is, if $\mathbf{x}$ is a point in the porous medium at the continuum scale, then $\phi(\mathbf{x})$ is the porosity at $\mathbf{x}$. $S_w(\mathbf{x})$ will be the wetting saturation and $S_n(\mathbf{x})$ the non-wetting saturation, so that

$$S_w(\mathbf{x}) + S_n(\mathbf{x}) = 1. \tag{12.1}$$

We will have a mass density

$$\rho(\mathbf{x}) = \phi(\mathbf{x})(S_w(\mathbf{x})\rho_w + S_n(\mathbf{x})\rho_n), \tag{12.2}$$

where ρ_w and ρ_n are the densities of the two immiscible fluids. There will be a pressure field $p(\mathbf{x})$. The fluids will move with Darcy velocities $\mathbf{U}_w(\mathbf{x})$ and $\mathbf{U}_n(\mathbf{x})$, and their combined velocity is

$$\mathbf{U}(\mathbf{x}) = \mathbf{U}_w(\mathbf{x}) + \mathbf{U}_n(\mathbf{x}). \tag{12.3}$$

We will also work with the seepage velocities

$$\mathbf{v}(\mathbf{x}) = \frac{1}{\phi(\mathbf{x})}\mathbf{U}(\mathbf{x}),$$

$$\mathbf{v}_w(\mathbf{x}) = \frac{1}{\phi(\mathbf{x})S_w(\mathbf{x})}\mathbf{U}_w(\mathbf{x}), \tag{12.4}$$

$$\mathbf{v}_n(\mathbf{x}) = \frac{1}{\phi(\mathbf{x})S_n(\mathbf{x})}\mathbf{U}_n(\mathbf{x}),$$

and we have that

$$\mathbf{v}(\mathbf{x}) = S_w(\mathbf{x})\mathbf{v}_w(\mathbf{x}) + S_n(\mathbf{x})\mathbf{v}_n(\mathbf{x}). \tag{12.5}$$

It will become clear later on why we have defined two classes of velocities.

We will also define the fractional flow rates $f_w(\mathbf{x})$ and $f_n(\mathbf{x})$ so that

$$f_w(\mathbf{x}) + f_n(\mathbf{x}) = 1. \tag{12.6}$$

Both the Darcy velocities and the fractional flow rates need precise definitions. We do this by introducing the Representative Elementary Volume.

Representative Elementary Volume

The variables we have just described only make sense at the continuum level. In order to fill them with physical contents, we need to connect them to the description at the level below (in the language of "More is different" (Anderson, 1972)): the pore level. The way to do this is through the *Representative Elementary Volume*, the REV for short, which we first met at the end of Chapter 11. However, there is a difference between the REV we met there and the REV we will construct here. In that chapter, we constructed a REV for the pore-level description. We will now construct a REV for the continuum level.

We have our porous medium. We pick a point $\mathbf{x}$ somewhere in it. Around this point we single out a volume just large enough so that the porous medium may be seen as a continuum, but not larger. This is the REV. It has a volume $V(\mathbf{x})$, a pore volume $V_p(\mathbf{x})$ and a porosity

$$\phi(\mathbf{x}) = \frac{V_p(\mathbf{x})}{V(\mathbf{x})}. \tag{12.7}$$

But, what do we mean by the "continuum limit"? Suppose v_p is the volume of a typical pore. Then, in the continuum limit, the ratios between the volume and pore volume of the REV and v_p diverge,

$$\frac{V}{v_p} \to \infty, \quad \text{and} \quad \frac{V_p}{v_p} \to \infty. \tag{12.8}$$

The pore volume may be divided into a wetting volume $V_w(\mathbf{x})$ and a non-wetting volume $V_n(\mathbf{x})$ so that

$$V_w(\mathbf{x}) + V_n(\mathbf{x}) = V_p(\mathbf{x}), \tag{12.9}$$

and we define the saturations as

$$S_w(\mathbf{x}) = \frac{V_w(\mathbf{x})}{V_p(\mathbf{x})} \quad \text{and} \quad S_n(\mathbf{x}) = \frac{V_n(\mathbf{x})}{V_p(\mathbf{x})}. \tag{12.10}$$

The mass of wetting fluid in the REV is $M_w(\mathbf{x})$ and of the non-wetting fluid $M_n(\mathbf{x})$. We define their densities with respect to pore space:

$$\rho_w(\mathbf{x}) = \frac{M_w(\mathbf{x})}{V_p(\mathbf{x})} \quad \text{and} \quad \rho_n(\mathbf{x}) = \frac{M_n(\mathbf{x})}{V_p(\mathbf{x})}. \tag{12.11}$$

The REV will have a thermodynamic internal energy associated with the fluids. We use the grand potential for this, $\Upsilon(\mathbf{x})$, and define the pressure as (Kjelstrup et al., 2019)

$$p = -\frac{\Upsilon(\mathbf{x})}{V_p(\mathbf{x})}. \tag{12.12}$$

The reader should at this point recognize that the variables occurring on the right-hand sides of Eqs (12.7)–(12.12) are all *extensive variables*. The usual definition of this concept is that they are all proportional to the volume $V(\mathbf{x})$. A better definition states that they are *additive*: If we combine REVs, the variables from each add up to form the new combined value.

The variables on the left-hand sides of the equations are *intensive*. They are insensitive to the size of the REV in the sense that if we change the size of the REV (within reason), their values do not change. In the continuum limit, we work only with intensive variables acting as fields – that is, they depend on the position $\mathbf{x}$.

Note that REVs may overlap. Indeed, this is an important aspect. We may choose a neighboring point $\mathbf{x} + \delta\mathbf{x}$ where $\delta\mathbf{x}$ is much smaller than the size of the REVs. This allows us, for example, to define derivatives of the intensive, that is, continuum variables.

The observant reader may now ask, where are the velocities in this discussion? In order to define those, we need to bring in another aspect in the description of the REV. We need to orient the REV. Suppose our REV is shaped like a cylinder with axis length L and areas A at both ends. (We stop writing explicitly the position of the REV $\mathbf{x}$ from now on, assuming it is implicitly understood.) Hence, the volume of the REV is given by $V = AL$.

The transversal area A will consist of matrix and pores. The pores will cover a trasversal pore area A_p given by

$$A_p = \frac{V_p}{L} = \frac{\phi V}{L} = \phi A. \tag{12.13}$$

The transversal pore area may be divided into that which is covered by the wetting fluid, A_w, and that which is covered by the non-wetting fluid, A_n, so that

$$A_w + A_n = A_p. \tag{12.14}$$

By combining these definitions with the definition of the saturations, Eq. (12.10), we have

$$S_w = \frac{A_w}{A_p} \quad \text{and} \quad S_n = \frac{A_n}{A_p}. \tag{12.15}$$

We now *orient* the REV (making sure it is small enough) so that its axis is parallel to the local streamline and there is no flow through the side walls. The fluids enter at the bottom area and leave through the top. The flow is parallel to the axis of the cylinder. There is a volumetric flow rate of wetting fluid Q_w entering the REV through the "bottom" of the REV. If we assume the fluid to be imcompressible, the same volumetric flow rate leaves the REV through the "top" area. Likewise there is a volumetric flow rate Q_n of incompressible non-wetting fluid passing through the ends of the cylinder. The total volumetric flow rate is given by

$$Q_w + Q_n = Q. \tag{12.16}$$

We may now define the seepage velocities

$$v_w = \frac{Q_w}{A_w}, \quad v_n = \frac{Q_n}{A_n}, \tag{12.17}$$

and we have

$$Q = A_w v_w + A_n v_n. \tag{12.18}$$

Hence, the seepage velocities are the average velocities of each fluid species in the pores. The average seepage velocity v is defined as

$$v = \frac{Q}{A_p}. \tag{12.19}$$

Combining this equation with the previous one gives Eq. (12.5).

We choose the direction of the velocity vectors according to the orientation of the REV, so that $v \to \mathbf{v}$, $v_w \to \mathbf{v}_w$ and $v_n \to \mathbf{v}_n$.

We define the Darcy velocities as

$$U = \frac{Q}{A}, \quad U_w = \frac{Q_w}{A} \quad \text{and} \quad U_n = \frac{Q_n}{A}. \tag{12.20}$$

So, why do we distinguish between the seepage and the Darcy velocities? The Darcy velocities have the advantage that they do not change even if the porosity ϕ or the saturations S_w and S_n change. They are the intensive equivalents to the volumetric flow rates, being the volumetric flow rate per area. The seepage velocities have the advantage that they are intuitive, and the formulas we will use explicitly bring out their dependencies.

We also distinguish between intensive and extensive variables for those associated with the flow. The intensive variables do not depend on the sizes of the end areas of the REV, whereas the extensive variables do. The velocities are intensive, and the volumetric flow rates are extensive.

Lastly, we define the fractional flow rates

$$f_w = \frac{Q_w}{Q} \quad \text{and} \quad f_n = \frac{Q_n}{Q}.$$ (12.21)

There is a subtlety here that in fact is important. In dealing with the velocities, we have based ourselves on a different kind of extensivity than when dealing with the other quantities. Here we allow for the change of the areas A, A_p, A_w and A_n while keeping the length of the REV fixed. These areas, in addition to Q, Q_w and Q_n are extensive with respect to such deformations of the REV, but not with respect to volume changes. We will return to this important point in Section 12.4.

*But Can We Do This?**

We posed earlier on in this chapter the question, "Is it at all possible to define a continuum limit?" We now address this.

Let us first make the following observation. Suppose we start with a porous medium that is saturated with a wetting fluid. We then very slowly inject a non-wetting fluid. The drainage process that ensues is one of invasion percolation. When the injection process is finished, the non-wetting fluid will have formed a singly connected but fractal volume. Hence, the volume of the injected wetting fluid scales as

$$V_w \sim \left(\frac{V}{v_p}\right)^{D/d},$$ (12.22)

where $D < d$ is the fractal dimension, and d is the dimension of the porous medium, that is, three in most cases. The pore volume, which is not fractal, behaves as

$$V_p \sim \left(\frac{V}{v_p}\right)^1,$$ (12.23)

when the volume of a typical pore is used as scale. The wetting saturation, defined in eq. (12.10), thus behaves as

$$S_w = \frac{V_w}{V_p} = \frac{\left(\frac{V}{v_p}\right)^{D/d}}{\left(\frac{V_p}{v_p}\right)} = \left(\frac{V}{v_p}\right)^{D/d-1}.$$ (12.24)

Since $D/d - 1 < 0$, we have that

$$S_w \to 0$$ (12.25)

in the continuum limit $V/v_p \to \infty$. Hence, the saturation in the continuum limit is zero, that is, exactly the same value as before the injection process.

The lesson from this example is that fractals belong to the pore-level description, but not to the continuum description, where they are irrelevant. Fractals that are generated by instabilities at the pore level are irrelevant at the continuum level.

Let us now consider a capillary pipe of length L filled with bubbles. We studied such a situation in great detail in Section 11.6.1. However, now consider a distribution of bubbles that is generated by the following algorithm: Generate a random walk $\tilde{y} = \tilde{y}(x)$ where $0 \le x \le L$ and $\tilde{y}(0) = 0$. Orient the random walk by transforming $\tilde{y}(x) \to y(x) = \tilde{y}(x) - [\tilde{y}(L) - \tilde{y}(0)]x/L$ so that $y(L) = y(0) = 0$. Let the line segments for which $y(x) > 0$ represent the wetting fluid and the line segments for which $y(x) < 0$ represent the non-wetting fluid. The points where $y(x) = 0$ represent the interfaces between the two fluids.

By construction, the saturations are given by $S_w = S_n = 1/2$ in this system when $L \to \infty$, that is, the continuum limit. The interfaces, on the other hand, form a fractal set with fractal dimension $D = 1/2$, so that their number scales as $L^{1/2}$ (Hansen et al., 1994). Hence, as L diverges in the continuum limit, the number of interfaces per length L falls off as $1/\sqrt{L}$. The length distribution $p(\Delta)$ of both the wetting and the non-wetting bubbles (which are one-dimensional clusters) will behave as $p(\Delta) \sim \Delta^{-3/2}$, making the average bubble size $\langle \Delta \rangle$ diverge as $\sqrt{L}$. Still, the average bubble size relative to L, $\langle \Delta \rangle / L \sim 1/\sqrt{L}$, vanishes in the $L \to \infty$ limit.

This simple example tells us that bubbles and clusters are concepts that belong to the pore level description and not to the continuum level description. All that is left of the bubble structure are the saturations S_w and S_n in the continuum limit.

It should be noted that we constructed this example in such a way that the average bubble size would diverge. The price for this was that the number of interfaces per REV volume L would disappear. In general, this does not have to be the case. In fact, in the typical situation, it is quite the opposite. The influence of the interfaces will therefore enter the continuum description. This, as we shall see, will be through the capillary pressure function, p_c.

So, to answer the question posed in the title of this section, "But Can We Do This?" That is, can we describe multiphase flow in porous media fully in the continuum limit? The answer is yes. However, we have to remember that many concepts that make sense at the pore level have no meaning at the continuum level. For some, this is because their effects become vanishly small in this limit so that they may be ignored, for example, fractals. Or the concepts have to be substituted by other concepts, such as the concept of interfaces at the pore level being substituted by the capillary pressure function at the continuum level.

12.1 Generalizing the Darcy Law to Two-Phase Flow: The Continuum Limit*

We have already encountered the Darcy equation for single phase flow, Eq. (7.8):

$$\mathbf{U} = -\frac{k}{\mu} \nabla p, \tag{12.26}$$

where we ignore gravity. Can we generalize this expression to two-phase flow?

We saw that this is far from trivial in Section 11.5. There we found that flow rate for immiscible two-phase flow under steady-state conditions becomes essentially quadratic in the pressure drop, see Eq. (11.12). In this section, we will discuss what consequences this has for constructing a Darcy law for immiscible two-phase flow in the continuum limit.

*12.1.1 Multiphase Darcy Equation in Continuum Limit**

In Chapter 11, we discussed how the Darcy equation, which is an equation describing single fluid flow in porous media, may be generalized to the immiscible two-phase flow problem. We first met this problem in Section 11.5, where experiments using a two-dimensional Hele-Shaw cell gave that the flow rate turned out to be proportional to the square of the pressure drop, see Eq. (11.12). This behavior, over a finite range of pressure drops, was confirmed through a mean field calculation, see Eq. 11.81 in Section 11.7.

We may write the Dacy equation for immiscible two-phase flow in porous media under *steady-state conditions* as

$$\mathbf{U} = -M(S_w, \mu_w, \mu_n, \mathrm{Ca}) \nabla p, \tag{12.27}$$

where $M(S_w, \mu_w, \mu_n, \mathrm{Ca})$ is the mobility. This is a *constitutive equation* relating velocity $\mathbf{U}$ to the driving force Δp.

Note that Eq. (12.27) has an underlying assumption that there are no macroscopic gradients in the saturation, that is, $|\nabla S_w| = 0$. We will generalize this expression to the case that $|\nabla S_w| \neq 0$ in Section 12.4.3.

*The Darcy Equation in the High-Capillary Number Limit**

What happens when the capillary number is so high that the capillary forces do not influence the flow at the pore scale but are still strong enough to prevent the fluids from becoming fully miscible? This question was recently studied numerically by Sinha et al. (2019). They found the "standard" Darcy equation (7.8), that is, a mobility

$$M(S_w, \mu_w, \mu_n, \text{Ca}) = \frac{k}{\mu_{\text{eff}}(S_w)}, \tag{12.28}$$

but with an effective viscosity obeying the equation

$$\mu_{\text{eff}}(S_w)^\alpha = \mu_w^\alpha S_w + \mu_n^\alpha (1 - S_w). \tag{12.29}$$

We recognize this expression for the effective viscosity when $\alpha = 1$. It is appropriate for bubbles moving in a capillary pipe, see Section 9.5. However, in a two- or three-dimensional porous medium, Sinha et al. found α to vary between -0.5 and 0.6 depending on the structure of the bubbles compared to the pore geometry.

12.2 Relative Permeabilites

Think steady-state flow. The saturation is constant throughout the porous medium. What would be the simplest possible model for two-phase flow under such conditions? Here is how Wyckoff and Botset (1936) thought more than 80 years ago. Suppose, I am the wetting fluid. What do I see? I see the matrix and the other fluid. I am therefore constrained to flow in the pore space that is left available to me after the non-wetting fluid has taken its part. The Darcy equation that I obey is the single fluid one since I am a single fluid, but with the permeability reduced due to the presence of the other fluid. Hence, according to Eq. (7.8), I obey

$$\mathbf{U}_w = -\frac{k \, k_{rw}(S_w)}{\mu_w} \Delta p, \tag{12.30}$$

where we have ignored buoyancy. The factor $k_{rw}(S_w)$ is the *wetting relative permeability*, and we have that $0 \leq k_{rw}(S_w) \leq 1$. Now, take the mantle of the non-wetting fluid. By precisely the same argument, we have

$$\mathbf{U}_n = -\frac{k \, k_{rn}(S_w)}{\mu_n} \Delta p. \tag{12.31}$$

Here $0 \leq k_{rn}(S_w) \leq 1$ is the *non-wetting relative permeability*. We form the sum $\mathbf{U} = \mathbf{U}_w + \mathbf{U}_n$, and the generalized Darcy equation becomes

$$\mathbf{U} = -\left[\frac{k \, k_{rw}(S_w)}{\mu_w} + \frac{k \, k_{rn}(S_w)}{\mu_n}\right] \nabla p. \tag{12.32}$$

Equations (12.30), (12.31) and (12.32) are examples of *constitutive equations*, relating motion to driving forces.

12.2.1 Saturation Gradients

There is a key assumption here for these three equations, (12.30), (12.31) and (12.32) to apply: There are no saturation gradients. If there were, currents would be set up in the porous medium to level them out. The driving force behind these currents: Capillary suction. This led Leverett (1941) to propose a generalization of the Wyckoff and Botset equations by introducing a *capillary pressure field* $p_c = p_c(S_w)$ in addition to splitting the single pressure p into a pressure in the wetting fluid p_w and a pressure in the non-wetting fluid, so that

$$p_c = p_w - p_n, \tag{12.33}$$

modeled on the theory of capillary rise, see Chapter 9. The original equations, (12.30) and (12.31), then become

$$\mathbf{U}_w = -\frac{k\, k_{rw}(S_w)}{\mu_w}\Delta p_w, \tag{12.34}$$

and

$$\mathbf{U}_n = -\frac{k\, k_{rn}(S_w)}{\mu_n}\Delta p_n. \tag{12.35}$$

By eliminating p_w using (12.33) in Eq. (12.34) and adding this equation and Eq. (12.35), we generalize Eq. (12.32) to

$$\mathbf{U} = -\left[\frac{k\, k_{rw}(S_w)}{\mu_w} + \frac{k\, k_{rn}(S_w)}{\mu_n}\right]\nabla p_n + \frac{k\, k_{rw}(S_w)}{\mu_w}\frac{dp_c}{dS_w}\nabla S_w. \tag{12.36}$$

Thus, two things have been accomplished: 1. We have added a saturation gradient term to the equation for $\mathbf{U}$, and 2. we have provided a physical reasoning of why it should be the way it is.

The first accomplishment is fine. However, the reasoning behind the second one is weak. It is fine to define a pressure p_w and a different pressure p_n when we are at the pore level looking at an interface and we can go in and measure the pressures on both sides of it with a pressure gauge. However, this is a theory on the continuum level. The concept of interfaces is *not defined at this level*. We may measure the pressure in a point; $p = p(\mathbf{x})$ is well defined. However, we cannot specify whether we are measuring the pressure in the wetting or the non-wetting fluid. Remember, the saturations S_w and S_n are *continuous* functions. When $S_w(\mathbf{x})$ and $S_n(\mathbf{x})$ are neither 0 nor 1, which pressure are we then measuring at $\mathbf{x}$, $p_w(\mathbf{x})$ or $p_n(\mathbf{x})$?[2]

---

[2] This is vaguely reminiscent of the concept of partial pressure in thermodynamics. It can be *defined* as the pressure times the mole fraction of the gas, but it cannot be measured directly.

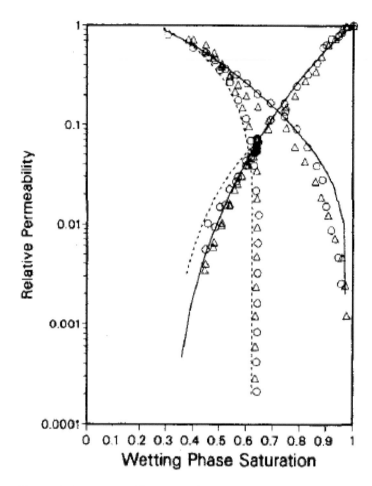

Figure 12.1 Relative permeability curves for Berea sandstone based on water–oil (continuous curves), oil–gas (∘) and water–gas (△). (From Oak et al. (1990).)

We imagine hearing protests from some readers at this point: But, you *can* measure p_w and p_n directly by inserting a filter in the probe with pores that wet either of the two fluids. The problem with this is that this constitutes a pore-scale measurement and not a continuum scale measurement.

We have now presented *relative permeability theory*, which is the leading – in fact the only – engineering tool for dealing with immiscible two-phase flow in porous media. It requires the measurement of three functions, $k_{rw}(S_w)$, $k_{rn}(S_w)$ and $p_c(S_w)$. These measurements are done routinely in laboratories all over the world under the label *special core analysis*, often abbreviated to SCAL or SPCAN.

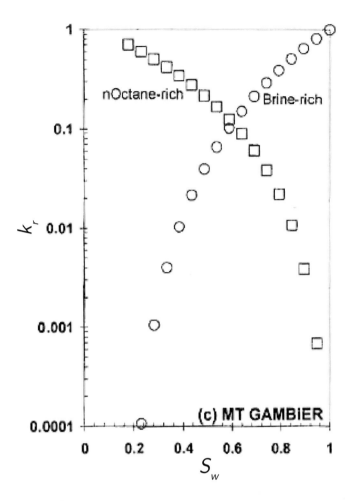

Figure 12.2 Relative permeability curves for Mt. Gambier limestone based on the injection of brine and n-octane. (From Olafuyi et al. (2008).)

We show in Figures 12.1 and 12.2 the relative permeability curves for Berea and Mt. Gambier limestone (Oak et al., 1990; Olafuyi et al., 2008). In Figure 12.1, different immiscible fluids have been used (Oak et al., 1990). If the relative permeability curves were a property of the porous medium alone and not the immiscible fluids, these curves should have fallen on top of each other.

A common correlation for the relative permeabilities is the Corey model (Honarpour et al., 1986),

$$k_{rw} = (S_w^*)^4, \tag{12.37}$$

$$k_{rn} = k_{rn}^0 (1 - S_w^*)^2 (1 - (S_w^*)^2), \tag{12.38}$$

where

$$S_w^* = \frac{S_w - S_{w,irr} - S_{n,r}}{1 - S_{n,r} - S_{w,irr}}. \tag{12.39}$$

Here $S_{w,irr}$ is the irreducible wetting fluid saturation – the smallest possible value for S_w, and $S_{n,r}$ is the residual non-wetting fluid saturation – the smallest possible value for S_n. Hence, $0 \leq S_w^* \leq 1 - S_{n,r}$. The parameter k_{rn}^0 is the end-point non-wetting relative permeability, and we have $0 < k_{rn}^0 \leq 1$. The Corey model is often generalized to Corey-type models,

$$k_{rw} = (S_w^*)^{N_w}, \tag{12.40}$$
$$k_{rn} = k_{rn}^0(1 - S_w^*)^{N_n}, \tag{12.41}$$

where N_w and N_n are exponents in the range from 1 to 6.

Figure 12.3 shows the capillary pressure curves measured for Berea, Bentheimer sandstone and Mt. Gambier limestone (Olafuyi et al., 2008). The resulting curves, measured with different standard techniques, are compared with two much-used models for the capillary pressure, the Brooks–Corey model and the van Genuchten model (Brooks and Corey, 1966; van Genuchten, 1980). The Brooks–Corey model for capillary pressure is

$$p_c = p_c^0(S_w^*)^{-1/\lambda}, \tag{12.42}$$

where the exponent λ is around 1–2. The van Genuchten model is

$$p_c = p_c^0[(S_w^*)^{-1/m} - 1]^{1-m}. \tag{12.43}$$

A typical value for m is 0.5.

We will not go into the derivations of these models. Rather, we refer the interested reader to the huge literature that exists on them.

The reader should note that relative permeability theory is popular because it is not bad. At low flow rates, it does a pretty good job.

12.3 Continuity Equations

In order to construct a closed set of differential equations, the generalized Darcy equations are not enough. Let us count variables. We have the saturations S_w and S_n, but their sum is one – so they count as one. Then we have the pressures p_w and p_n. That makes three. Lastly, we have the two velocities $\mathbf{U}_w$ and $\mathbf{U}_n$. So, there are five variables in total. The generalized Darcy equations, (12.34) and (12.35), and the capillary pressure relation (12.33) make three equations. We need two more for the five variables $\mathbf{U}_w$, $\mathbf{U}_n$, S_w, p_w and p_n.

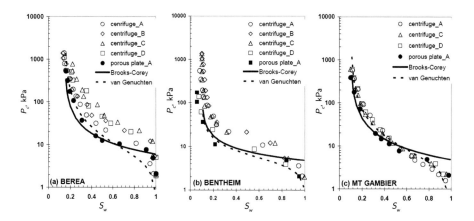

Figure 12.3 Capillary curves for Berea sandstone, Bentheimer sandstone and Mt. Gambier limestone based on different methods (see source paper for which methods) and compared to the Brooks–Corey model and the van Genuchten model. (From Olafuyi et al. (2008).)

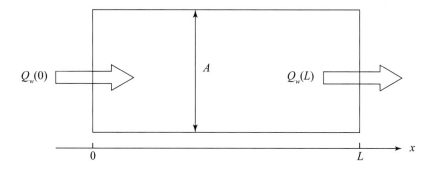

Figure 12.4 The REV is oriented along the x-axis. The volumetric flow rate is $Q(0)$ at $x = 0$ and $Q(L)$ at $x = L$. We must have that $Q(0) = Q(L)$ if the fluids are incompressible, so that whatever enters the REV also leaves it.

We assume that the fluids are incompressible. In terms of our oriented REV, this means that the amount of fluid that enters it per unit time, $Q(0)$, (see Figure 12.4) has to match the amount of fluid that leaves it per unit time, $Q(L)$. We must have that $Q(0) - Q(L) = A(U(0) - U(L)) = -AL(\partial U/\partial x) = -V(\partial U/\partial x) = 0$. We divide by the volume and generalize to three dimensions,

$$\nabla \cdot \mathbf{U} = 0. \tag{12.44}$$

Hence, we have the fourth equation. We note that in terms of the average seepage velocity, (12.19), this equation becomes

$$\nabla \cdot (\phi \mathbf{v}) = 0. \tag{12.45}$$

We follow the same prescription for the wetting fluid, taking into account that the saturation may change. Hence, with reference to the REV in Figure 12.4, a volume of the wetting fluid $Q_w(0)$ enters the REV per unit time, and a volume $Q_w(L)$ leaves it per unit time. Hence, we must have $dV_w/dt = V\phi(\partial S_w/\partial t) = Q_w(0) - Q_w(L) = -AL(\partial U/\partial x)$, or simplified, $\phi(\partial S_w/\partial t) + \partial U/\partial x = 0$. In three dimensions, this becomes

$$\phi\frac{\partial S_w}{\partial t} + \nabla \cdot \mathbf{U}_w = 0. \tag{12.46}$$

– and the fifth equation is ours. In terms of the seepage velocities, it becomes

$$\phi\frac{\partial S_w}{\partial t} + \nabla \cdot (\phi S_w \mathbf{v}_w) = 0. \tag{12.47}$$

If we now use these five equations to eliminate the velocities – the two generalized Darcy equations, (12.34) and (12.35), the capillary pressure Eq. (12.33) and the two continuity equations, (12.44) and (12.46) – we will end up with two convection–diffusion equations for p_w and S_w, which we proceed to solve according to the boundary conditions we have.

The characteristics of the porous medium and the fluids enter through the three functions $k_{rw} = k_{rw}(S_w)$, $k_{rn} = k_{rn}(S_w)$ and $p_c = p_c(S_w)$.

12.3.1 Buckley–Leverett displacement

Let us investigate a very interesting consequence of the continuity equations, (12.44) and (12.46): The creation of shock fronts when the fluids move. This is known as Buckley–Leverett theory.

Buckley and Leverett (1942) start their famous paper, "Crude oil has no inherent ability to expel itself from the pores of the reservoir rocks in which it is found; rather, it must be forcibly ejected or displaced by the accumulation of other fluids." It is the ejection process that creates the shock.

The starting point is the one-dimensional version of Eq. (12.46),

$$\phi\frac{\partial S_w}{\partial t} + \frac{\partial U_w}{\partial x} = 0. \tag{12.48}$$

We introduced the fractional flow rate f_w in Eq. (12.21) in Chapter 12. By noting that $Q_w = AU_w$ and $Q = AU$, we may write the fractional flow as

$$f_w = \frac{U_w}{U}. \tag{12.49}$$

Hence, Eq. (12.48) becomes

$$\phi\frac{\partial S_w}{\partial t} + \frac{\partial(U f_w)}{\partial x} = 0. \tag{12.50}$$

Since Eq. (12.44) is

$$\frac{\partial U}{\partial x} = 0, \tag{12.51}$$

we may transform (12.50) into

$$\frac{\partial S_w}{\partial t} + U_0 \frac{\partial f_w}{\partial x} = 0, \tag{12.52}$$

where

$$U_0 = \frac{Q}{\phi A}. \tag{12.53}$$

We now make the key assumption: We assume that the fractional flow f depends on all the other parameters in the problem *through the saturation S only*, that is,

$$f(S_w, p, \ldots) = f(S_w(p, \ldots)). \tag{12.54}$$

This leads to

$$\frac{\partial f(S_w)}{\partial x} = \left(\frac{\mathrm{d}f}{\mathrm{d}S_w}\right) \frac{\partial S_w}{\partial x}. \tag{12.55}$$

The piston flow velocity U_0 is the velocity expected if there is no smearing of the front, that is, the invading fluid displaces the defending fluid without any mixing in a purely one-dimensional displacement process. With this notation, we arrive at the *Buckley–Leverett equation*

$$\boxed{\frac{\partial S_w}{\partial t} + U(S_w)\frac{\partial S}{\partial x} = 0,} \tag{12.56}$$

where the front velocity at saturation S is

$$\boxed{U(S_w) = U_0 \left(\frac{\mathrm{d}f_w}{\mathrm{d}S_w}\right).} \tag{12.57}$$

Figure 12.5 shows the fractional flow and $\mathrm{d}f_w/\mathrm{d}S_w$ as a function of the wetting saturation for a case discussed by Buckley and Leverett (1942) where the viscosity ratio between the oil and water is $\mu_o/\mu_w = 2$.

From Eq. (12.57) it is possible to propagate an initial saturation profile $S_w(x, t_0)$ to a new profile $S(x, t + \Delta t)$ using the expression

$$\Delta x = \frac{Q \Delta t}{\phi A} \frac{\mathrm{d}f(S)}{\mathrm{d}S}. \tag{12.58}$$

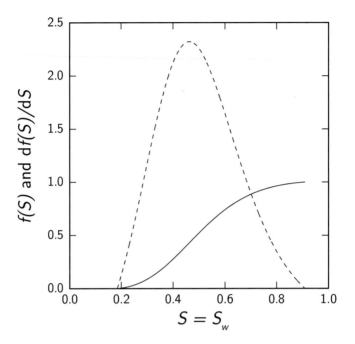

Figure 12.5 Effect of water saturation $S = S_w$, on the fractional flow $f = f_w$ (full line) and the Buckley–Leverett front velocity $U(S_w)/U_0 = df_w/dS_w$ (dashed line). The curves were calculated on the basis of the relative permeability curves given in Buckley and Leverett (1942) with a viscosity ratio between the oil and water $\mu_o/\mu_w = 2$.

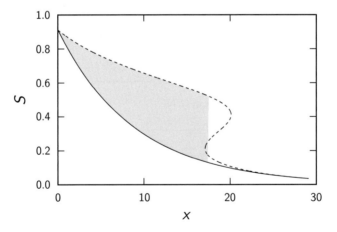

Figure 12.6 The initial water saturation $S(x_w, t_0)$ as a function of position (full line). After a time step Δt such that $Q\Delta t/\phi A = 5$, the saturation profile (dashed curve) is propagated by Eq. (12.57) to the unstable profile (dashed curve). Mass balance limits the profile (shaded region), and a shock front at x_f results.

As shown in Figure 12.6, the initial saturation is propagated to form an unstable distribution in which $S_w(x)$ becomes multiple valued. This ambiguity is resolved by insisting on continuity:

$$\frac{Q\Delta t}{\phi A} = \int_0^{x_f} dx \, [S_w(x, t_0 + \Delta t) - S(x, t_0)], \qquad (12.59)$$

and a shock front arises at x_f. Of course, one may solve the differential Eq. 12.56 numerically, also including capillary effects and gravity, but the Buckley–Leverett shock survives in general.

12.4 Euler Scaling Theory*

In relative permeability theory, two pressures, p_w and p_n are defined, and the difference between them is the capillary pressure function, see Eq. (12.33). This makes sense at scales below the continuum scale. However, the role played by the capillary pressure is in reality to introduce forces driving the fluids to flow caused by gradients in the saturation: Capillary forces push the non-wetting fluid away in favor of the wetting fluid. For this to happen at the continuum scale, a gradient in S_w must be present. Hence, one obtains exactly the same physics by doing away with ∇p_n and ∇p_w, replacing them by ∇p and ∇S_w, and stating that the pressure p is the single number you measure at the point in the porous medium you have chosen. We have already anticipated this in Eq. (12.36), where we have used the gradient in the saturation explicitly.

We will now describe a recently advanced theory relating the seepage velocities $\mathbf{v}_w$, $\mathbf{v}_n$ to another velocity pair of which $\mathbf{v}$ is one (Hansen et al., 2018). We could of course have expressed eveything that follows in terms of the Darcy velocities $\mathbf{U}_w$, and $\mathbf{U}_n$, but refrain from doing so for the simple reason that we wish the text to be as close in notation to the existing literature on this subject.

But first, why go beyond relative permeability? Is what we have not good enough? No, it is not. Relative permeability theory works pretty well in a window of small velocities, but as soon as the non-linearities set in, when the capillary number that appears in the mobility in Eq. (12.27) becomes a relevant parameter, the theory looses all predictive power. When a non-linear Darcy-type law, as in Eq. (12.27), predicts a Darcy velocity $\mathbf{U}$, or equivalently, a seepage velocity $\mathbf{v} = \mathbf{U}/\phi$, how can one split it into the velocities of each fluid species, $\mathbf{v}_w$ and $\mathbf{v}_n$, in any unique way? There is no unique way to determine what the relative permeabilities k_{rw} and k_{rn} should be. The *Euler scaling theory* attempts to remedy this.

In order to start our review of this theory, we need to return to the REV, which we described in Section 12. The REV, which we oriented in the flow direction, had a volume V, a length L and a transverse cross-sectional area A. All flow entered or left the REV through the transversal areas. We orient our coordinate system so that the x-axis points along the axis of the REV.

Let us now choose some transverse cut through the REV. It has an area A. A part $A_p = \phi A$ of this area is covered by the pores; the rest, $(1 - \phi)A$, is covered by the matrix. A volumetric flow rate $Q = Q_w + Q_n$ passes through the transverse pore area A_p. A part $A_w = S_w A_p$ is filled with the wetting fluid, and an area $A_n = S_n A_p = (1-S_w)A_p$ is filled with the non-wetting fluid, so that $A_p = A_w + A_n$. There is a pressure p associated with the cut. This pressure will be the same everywhere on the cut.[3] However, it is not the pressure helping to drive the flow; it is the pressure gradient, $\partial p / \partial x$, which is orthogonal to the cut. A second driving force comes from a gradient in the saturation, $\partial S_w / \partial x$, which also must be orthogonal to the cut. As a consequence, A_w and A_n change along the x-axis in the REV. However, their sum $A_p = A_w + A_n$ does not. We have that[4]

$$Q = Q\left(A_w, A_n, p, \frac{\partial p}{\partial x}, \frac{\partial S_w}{\partial x}\right). \tag{12.60}$$

Since we are dealing with incompressible fluids, Q cannot depend on p, so that p may be eliminated.

Let us now change the area A by changing the size of the REV, *but without changing L*. Hence, we change

$$A \to \lambda A \quad \text{and} \quad L \to L. \tag{12.61}$$

This results in

$$A_w \to \lambda A_w \quad \text{and} \quad A_n \to \lambda A_n. \tag{12.62}$$

The volumetric flow rate Q is an extensive (=additive) variable in A_w and A_n so that we have

$$Q\left(\lambda A_w, \lambda A_n, \frac{\partial p}{\partial x}, \frac{\partial S_w}{\partial x}\right) = \lambda Q\left(A_w, A_n, \frac{\partial p}{\partial x}, \frac{\partial S_w}{\partial x}\right). \tag{12.63}$$

Now, take the derivative with respect to λ of this expression and set $\lambda = 1$. We obtain

$$A_w \left(\frac{\partial Q}{\partial A_w}\right)_{A_n} + A_n \left(\frac{\partial Q}{\partial A_n}\right)_{A_w} = Q. \tag{12.64}$$

We have left out the dependence of the two gradients, $\partial p / \partial x$ and $\partial S_w / \partial x$, as they are kept constant throughout these manipulations. We see that the two partial derivatives have the units of velocity, and we name them the *thermodynamic velocities*,

---

[3] Remember we are in the continuum limit. Fluctuations on the pore level are invisible.
[4] It should be noted that p and $\partial p / \partial x$ are independent variables, as is the case for S_w and $\partial S_w / \partial x$.

$$\tilde{v}_w = \left(\frac{\partial Q}{\partial A_w}\right)_{A_n} \tag{12.65}$$

and

$$\tilde{v}_n = \left(\frac{\partial Q}{\partial A_n}\right)_{A_w}. \tag{12.66}$$

Equation (12.64) may then be written

$$Q = A_w \tilde{v}_w + A_n \tilde{v}_n. \tag{12.67}$$

The scaling property (12.63) makes Q a *homogeneous function of degree one*, and the step from this equation to the next is the contents of the Euler theorem for homogeneous functions.

Now, compare Eq. (12.67) to Eq. (12.18), $Q = A_w v_w + A_n v_n$. Does this mean that $\tilde{v}_w = v_w$ and $\tilde{v}_n = v_n$? No. The most general relation between them we may write down is

$$\tilde{v}_w = v_w + S_n v_m, \tag{12.68}$$
$$\tilde{v}_n = v_n - S_n v_m, \tag{12.69}$$

where v_m was named the *co-moving velocity* by Hansen et al. (2018).

It is easy to convince oneself that the velocities v, v_m, v_w, v_n, $\tilde{v}_w$ and $\tilde{v}_n$ are homogeneous functions of order zero. That is,

$$v(\lambda A_w, \lambda A_n) = v(A_w, A_n), \tag{12.70}$$

and likewise for the other velocities. This means that the velocities can only depend on A_w and A_n through their ratio, $A_w/A_n = S_w/(1 - S_w)$. Hence, we must have

$$v(A_w, A_n) = v(S_w), \tag{12.71}$$

and likewise for the other velocities.

Let us now make a *change of variables* from (A_w, A_n) to (S_w, A_p), that is,

$$S_w = \frac{A_w}{A_w + A_n}, \tag{12.72}$$
$$A_p = A_w + A_n, \tag{12.73}$$

or inverted

$$A_w = S_w A_p, \tag{12.74}$$
$$A_n = (1 - S_w) A_p. \tag{12.75}$$

We now calculate

$$
\begin{aligned}
\left(\frac{\partial Q}{\partial S_w}\right)_{A_p} &= A_p \frac{dv}{dS_w} \\
&= \left(\frac{\partial A_w}{\partial S_w}\right)_{A_p} \left(\frac{\partial Q}{\partial A_w}\right)_{A_n} + A_n \left(\frac{\partial Q}{\partial A_n}\right)_{A_w} \left(\frac{\partial Q}{\partial A_n}\right)_{A_w} \\
&= A_p \tilde{v}_w - A_p \tilde{v}_n.
\end{aligned}
\tag{12.76}
$$

Hence, dividing by A_p gives

$$
\frac{dv}{dS_w} = \tilde{v}_w - \tilde{v}_n = v_w - v_n + v_m,
\tag{12.77}
$$

where on the right-hand side we have used Eqs (12.68) and (12.69).

We now take the derivative of Eq. (12.5) with respect to S_w to find

$$
\begin{aligned}
\frac{dv}{dS_w} &= \frac{d}{dS_w} [S_w v_w + (1 - S_w) v_n] \\
&= v_w - v_n + S_w \frac{dv_w}{dS_w} + (1 - S_w) \frac{dv_n}{dS_w}.
\end{aligned}
\tag{12.78}
$$

Subtracting Eq. (12.77) from this equation leaves us with

$$
v_m = S_w \frac{dv_w}{dS_w} + (1 - S_w) \frac{dv_n}{dS_w}.
\tag{12.79}
$$

We may see Eqs (12.5) and (12.79) as transformation from a (v_w, v_n) velocity basis to a (v, v_m) velocity basis,

$$
\mathbf{v} = S_w \mathbf{v}_w + S_n \mathbf{v}_n,
\tag{12.80}
$$

$$
\mathbf{v}_m = S_w \mathbf{v}'_w + S_n \mathbf{v}'_n,
\tag{12.81}
$$

where $v' = dv/dS_w$. We have here written the velocities as vectors. These velocities are intensive variables, and we may therefore get rid of any reference to the REV and its properties.

Equations (12.80) and (12.81) may be inverted. We do this by using Eq. (12.77),

$$
S_n(v' - v_m) = S_n v_w - S_n v_n = S_n v_w - (v - S_w v_w) = v_w - v,
\tag{12.82}
$$

or

$$
v_w = v + S_n(v' - v_m),
\tag{12.83}
$$

and similarly for v_n. Our end result is therefore

$$
\mathbf{v}_w = \mathbf{v} + S_n(\mathbf{v}' - \mathbf{v}_m),
\tag{12.84}
$$

$$
\mathbf{v}_n = \mathbf{v} - S_w(\mathbf{v}' - \mathbf{v}_m).
\tag{12.85}
$$

Thus, knowing $\mathbf{v} = \mathbf{v}(S_w, \nabla p, \nabla S_w)$ and $\mathbf{v}_m = \mathbf{v}_m(S_w, \nabla p, \nabla S_w)$ will allow us to calculate $\mathbf{v}_w = \mathbf{v}_w(S_w, \nabla p, \nabla S_w)$ and $\mathbf{v}_n = \mathbf{v}_n(S_w, \nabla p, \nabla S_w)$.

What has been achieved here is quite remarkable. The two pairs of equations, (12.80) and (12.81), and (12.84) and (12.85), make the mapping between $\mathbf{v}$ and $\mathbf{v}_w$ and $\mathbf{v}_n$ unique; $(\mathbf{v}, \mathbf{v}_m) \Leftrightarrow (\mathbf{v}_w, \mathbf{v}_n)$. The definition of the co-moving velocity $\mathbf{v}_m$ makes this possible.

*12.4.1 The Co-Moving Velocity for the Capillary Fiber Bundle Model**

Hansen et al. (2018) worked out the co-moving velocity for the capillary fiber bundle model. In the bubble train model, see Section 11.6.1, it turns out that it is simply

$$v_m = v'. \tag{12.86}$$

This is obvious from Eqs (12.84) and (12.85) since we must have that $v_w = v_n = v$. More interesting is the case when some of the fibers are so thin that only the wetting fluid is able to enter them. The irreducible wetting saturation is in this case given by the area covered by these small capillaries, $S_{w,irr}$. We assume the flow velocity in these capillaries is v_s, whereas the flow velocity in the large capillaries is given by v_l. We then find

$$v_m = \frac{S_{w,irr}}{S_w}(v_l - v_s) + \frac{\mathrm{d}v_l}{\mathrm{d}S_w}. \tag{12.87}$$

*12.4.2 Relation to Relative Permeability Theory**

Relative permeability theory is not bad as long as velocities are small. Let us work backwards from the relative permeability curves shown in Figures 12.1 and 12.2 and determine the co-moving velocity v_m from these curves.

We combine Eqs (12.34) and (12.35) with Eq. (12.81) and find

$$v_m = \mu_w v_0 \left[\frac{S_w}{\mu_w} \frac{\mathrm{d}}{\mathrm{d}S_w} \left(\frac{k_{rw}}{S_w} \right) + \frac{S_w}{\mu_n} \frac{\mathrm{d}}{\mathrm{d}S_w} \left(\frac{k_{rn}}{S_n} \right) \right], \tag{12.88}$$

where we have defined the velocity scale

$$v_0 = -\frac{k}{\phi \mu_w} \frac{\partial p}{\partial x}. \tag{12.89}$$

We also have

$$v' = \mu_w v_0 \frac{\mathrm{d}}{\mathrm{d}S_w} \left[\frac{k_{rw}}{\mu_w} + \frac{k_{rn}}{\mu_n} \right]. \tag{12.90}$$

We show the results for the data from Oak et al. (1990) in Figure 12.7 and the data from Olafuyi et al. (2008) in Figure 12.8.

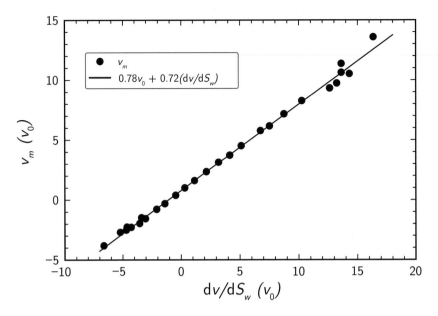

Figure 12.7 The co-moving velocity v_m from the relative permeability data of Oak et al. (1990), see Figure 12.1.

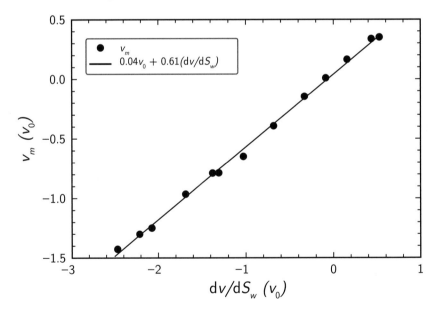

Figure 12.8 The co-moving velocity v_m from the Mt. Gambier limestone relative permeability data of Olafuyi et al. (2008), see Figure 12.2.

We find that in both cases,

$$v_m = av_0 + bv'$$ (12.91)

fits the data very well. Roy et al. (2021) have analysed a number of additional experimental relative permeability data sets with similar results. Hence, we see that $v_m \to 0$ and $v \to 0$. Furthermore, we expect $v_w = v_n = v$ when $v \to \infty$, that is, in the high-capillary number limit. From Eqs (12.84) and (12.85) we then expect $v_m \to v'$. Hence, we must have that $a \to 0$ and $b \to 1$ in this limit.

Let us now make the assumption that

$$v_m = bv', \tag{12.92}$$

and assume the generalized Darcy equations, (12.34) and (12.35), are valid. This means that we have

$$\frac{S_w}{\mu_w} \frac{d}{dS_w} \left(\frac{k_{rw}}{S_w} \right) + \frac{S_n}{\mu_n} \frac{d}{dS_w} \left(\frac{k_{rn}}{S_n} \right) = b \frac{d}{dS_w} \left[\frac{k_{rw}}{\mu_w} + \frac{k_{rn}}{\mu_n} \right]. \tag{12.93}$$

Since this expression is to be valid for all combinations of μ_w and μ_n, the equation splits in two,

$$S_w \frac{d}{dS_w} \left(\frac{k_{rw}}{S_w} \right) = b \frac{d}{dS_w} k_{rw}, \tag{12.94}$$

and

$$S_n \frac{d}{dS_n} \left(\frac{k_{rw}}{S_n} \right) = b \frac{d}{dS_n} k_{rn}, \tag{12.95}$$

where in the last equation we have used that $S_n = 1 - S_w$. We solve these two equations, finding

$$k_{rw} = k_{rw}^0 S_w^{1/(1-b)}, \tag{12.96}$$

and

$$k_{rn} = k_{rw}^0 (1 - S_w)^{1/(1-b)}. \tag{12.97}$$

Comparing with the Corey-type relative permeabilities, Eqs (12.40) and (12.41), we find that the exponents N_w and N_n obey

$$N_w = N_n = \frac{1}{1-b}. \tag{12.98}$$

With the exponents N_w and N_n in the range 1 to 6, we find that b is in the range 0 to $5/6 \approx 0.83$ (Roy et al., 2021).

*12.4.3 Euler Scaling Theory under Non-steady Conditions**

Equations (12.84) and (12.85) give the velocity for each fluid species $\mathbf{v}_w$ and $\mathbf{v}_n$ once the constitutive equations for $\mathbf{v}$ and $\mathbf{v}_m$ are given. The constitutive equation for $\mathbf{v}$ we argued in Section 12.1.1 would have the form given by Eq. (12.27) in steady state. We will now generalize this equation to non-steady state, that is, when there are saturation gradients present:

$$\mathbf{v} = -\tilde{M}(S_w, \mu_w, \mu_n, \text{Ca}) \nabla p - \tilde{N}(S_w, \mu_w, \mu_n, \text{Ca}) \nabla S_w, \tag{12.99}$$

where $\tilde{N}(S_w, \mu_w, \mu_n, \text{Ca})$ is a new mobility related to the saturation gradient. The sign in front of the new term signals that the wetting fluid is drawn toward lower saturations due to capillary suction. We note that it may change, for example, under mixed wetting conditions.

The mobility $\tilde{M}$ is related to the one defined in Eq. (12.27) by $M = \tilde{M}/\phi$ since we are here dealing with the seepage velocities rather than the Darcy velocities.

We now write $\tilde{N}(S_w, \mu_w, \mu_n, \text{Ca})$ as

$$\tilde{N}(S_w, \mu_w, \mu_n, \text{Ca}) = \tilde{M}(S_w, \mu_w, \mu_n, \text{Ca}) \frac{dp_c}{dS_w}, \tag{12.100}$$

so that we may transform Eq. (12.99) into

$$\mathbf{v} = -\tilde{M}(S_w, \mu_w, \mu_n, \text{Ca}) \nabla (p - p_c). \tag{12.101}$$

Hence, we have introduced – or rather *defined* – the capillary pressure $p_c = p_c(S_w, \mu_w, \mu_n, \text{Ca})$ in a natural way.

We now derive the most general form that the constitutive equation for the co-moving velocity $\mathbf{v}_m$ may take. At equilibrium, we have that $\mathbf{v}_w = 0$ and $\mathbf{v}_n = 0$ in addition to $\mathbf{v} = 0$.

From Eqs (12.84) and (12.85), we see that we must have

$$\frac{d\mathbf{v}}{dS_w} - \mathbf{v}_m \to 0 \tag{12.102}$$

as the system approaches equilibrium.

From Eq. (12.101) we have that

$$\frac{d\mathbf{v}}{dS_w} = -\frac{d\tilde{M}}{dS_w} \nabla (p - p_c) + \tilde{M} \frac{d^2 p_c}{dS_w^2} \nabla S_w. \tag{12.103}$$

The seepage velocity **v** vis zero when

$$\nabla p = \nabla p_c = \frac{dp_c}{dS_w} \nabla S_w. \tag{12.104}$$

Hence, we have

$$\frac{d\mathbf{v}}{dS_w} = \tilde{M} \frac{d^2 p_c}{dS_w^2} \nabla S_w \tag{12.105}$$

from Eq. (12.103). The most general form for $\mathbf{v}_m$ that is compatible with Eq. (12.102) and taking Eq. (12.105) into account is

$$\mathbf{v}_m = -\tilde{B}\nabla(p - p_c) + \tilde{M}\frac{\mathrm{d}^2 p_c}{\mathrm{d}S_w^2}\nabla S_w, \tag{12.106}$$

where $\tilde{B} = \tilde{B}(S_w, \mu_w, \mu_n, \mathrm{Ca})$ is the co-moving mobility.

It should be noted that $\mathbf{v}_m$ is *non-zero* at equilibrium when $\mathbf{v}_w = \mathbf{v}_n = \mathbf{v} = 0$; it is

$$\mathbf{v}_m = \tilde{M}\frac{\mathrm{d}^2 p_c}{\mathrm{d}S_w^2}\nabla S_w, \tag{12.107}$$

thus compensating $\mathrm{d}\mathbf{v}/\mathrm{d}S_w$ in this limit.

In the range where it makes sense to use relative permeabilities, we compare Eq. (12.91) with Eq. (12.106) remebering that $\nabla S_w = 0$ since we have assumed steady-state, finding

$$\tilde{B} = a\frac{k}{\mu_w} + b\frac{\mathrm{d}\tilde{M}}{\mathrm{d}S_w}. \tag{12.108}$$

As we see, the relative permeability framework fits nicely into the Euler scaling theory. However, the latter is much more general.

So, here is the calculational scheme using the Euler theory. We need the constitutive equations for $\mathbf{v}$ and $\mathbf{v}_m$, (12.101) and (12.106) (this latter being surprisingly simple in practice, see Figures 12.8 and 12.8). In addition we need the two continuity Eqs (12.44) and (12.47). They form a closed set of equations.

Exercises

12.1 **Relative permeability in the limit of zero surface tension:**
The relative permeability constitive Eqs (12.34) and (12.35), and the pressure eq. (12.33) when combined with the continuity Eqs (12.44) and (12.45), form a closed set that will predict how the saturation and pressure fields develop. There are three functions that describe the properties of the porous medium and the fluids, $k_{rw} = k_{rw}(S_w)$, $k_{rn} = k_{rn}(S_w)$ and $p_c = p_c(S_w)$. Suppose now we turn off the capillary forces, $p_c(S_w) \to 0$. The Darcy velocity $\mathbf{U}$ is then given by Eq. (12.32).

12.2 If we let the viscosities of the two fluids approach each other, that is, $\mu_w - \mu_n \to 0$, what would the relative permeabilities $k_{rw} = k_{rw}(S_w)$ and $k_{rn} = k_{rn}(S_w)$ need to be?

12.3 If the two viscosities are different, $\mu_w \neq \mu_w$, what can you say about the relative permeabilities then?

13

Particle Simulations of Multiphase Flows

Flow in porous media is one of those physical phenomena that take place over a broad range of length-scales. Wetting phenomena and boundary effects are direct results of solid–fluid interactions at the molecular level, while ground water flow is governed by geological and meteorological systems that may span several 100 kilometers. For this reason, the problem of modeling, connecting the scales in a physical way, yet keeping an adequate resolution, is very challenging. In this chapter, we look at how this challenge may be met by means of particle based numerical techniques.

Such techniques are built on a small scale physical description and are often both closer to reality and more flexible in terms of including the right effects. We shall proceed from the simplest possible model, the random walker, toward the more detailed realism of molecular dynamics before we reintroduce simplification of the particle dynamics via the lattice gas and lattice Boltzmann models. The latter models are governed by the Navier–Stokes equations by virtue of their underlying conservation laws.

We show in Figure 13.1 an example of immiscible two-phase flow in a sandstone, simulated using the lattice Boltzmann model.

13.1 Random Walks and the Simulation of the Advection–Diffusion Equation

As we noted in Chapter 8, random walkers are described by the diffusion equation, and, when they are made to move along with a given flow field, the advection–diffusion Eq. (8.48). Such particle dynamics is often referred to as *Brownian dynamics*.

The random walk model is also known as Brownian dynamics, which consists of a number of particles, labeled k, that are propagated by simple random shifts to their positions. Random walks may be on a lattice or consist of steps in continuous

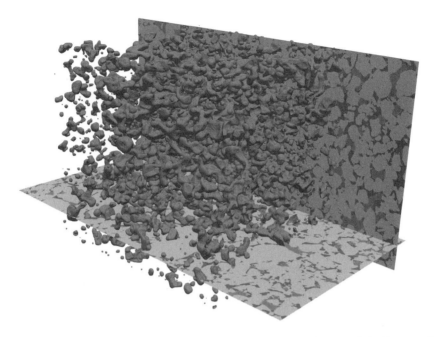

Figure 13.1 Pore scale simulation of two-phase flow, directly on a digital image of a sandstone sample, using a lattice Boltzmann method. The picture shows residual non-wetting fluid phase distribution in the pore space after an imbibition process. The non-wetting fluid clusters occupy the larger pore bodies of the model. (Figure courtesy Thomas Ramstad.) (A black and white version of this figure will appear in some formats. For the colour version, refer to the plate section.)

space. In Brownian dynamics, the steps are of random size and off-lattice. The only difference from the random walk is that now the random step lengths are directly related to a time step. The algorithm for updating the position of particle k at timestep n may be written

$$\mathbf{r}_k \rightarrow \mathbf{r}_k + \mathrm{d}\mathbf{r}_k(n), \quad \text{where} \quad \mathrm{d}\mathbf{r}_k(n) = \mathbf{u}(\mathbf{r}_k)\mathrm{d}t + \mathrm{d}\mathbf{W}(n), \quad (13.1)$$

and $\mathbf{u}(\mathbf{x})$ is the background velocity field. Here $\mathrm{d}\mathbf{W}$ is sometimes called a *Wiener process*, defined by the condition $\langle \mathrm{d}W_i(n)\mathrm{d}W_j(m) \rangle = \sigma_c^2 \delta_{nm}\delta_{ij}\mathrm{d}t$. The steps $\mathrm{d}W_i$ are thus uncorrelated and sampled from a Gaussian distribution. In a simulation, $\mathrm{d}t$ is always finite. The walk may be realized by choosing

$$\mathrm{d}W_i(n) = x(n)\frac{\sigma_c\sqrt{\mathrm{d}t}}{d}, \quad (13.2)$$

where x is a Gaussian random variable with $\langle x^2 \rangle = 1$ and d is the spatial dimension. Note that if $\mathbf{u} = 0$, the process is purely diffusive, and we may write

$$\mathbf{r}_k = \sum_{n=1}^{N} \mathrm{d}\mathbf{W}(n) \quad (13.3)$$

Figure 13.2 The evolution of 60 000 random walkers. Time increases from left to right.

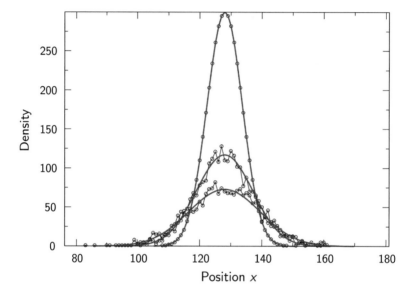

Figure 13.3 Density profiles corresponding to the same random walker algorithm as in Figure 13.2. However, the random walkers are initialized as a strip with translational symmetry in the y-direction, and a Gaussian density profile in the x-direction. The density profiles are averaged in the y-direction which extends 200 sites. The full lines show the solution of the one-dimensional diffusion equation. Note the noise in the measured graphs. (A black and white version of this figure will appear in some formats. For the colour version, refer to the plate section.)

for the N first timestep up to $t = N\mathrm{d}t$. Taking the square and the average yields

$$\langle \mathbf{r}_k^2 \rangle = \sum_{nm} \langle \mathbf{W}(n) \cdot \mathrm{d}\mathbf{W}(m) \rangle = N\mathrm{d}t\, \sigma_c^2 = \sigma_c^2 t, \qquad (13.4)$$

where we have used that all components of the steps $\mathrm{d}\mathbf{W}_n$ are independent. This shows that the molecular diffusivity $D_m = \mathrm{d}\sigma_c^2/2$, which should be compared to Eq. (8.14).

Figure 13.2 shows the evolution of a finite set of random walkers that spread diffusively. Note that, even though each random walker has no preference to move either way, their density spreads out irreversibly. Figure 13.3, which displays the result of a density variation only in the x-direction, shows how their collective behavior conforms to Eq. (8.36). Note how fluctuations on this single simulation are introduced on top of the smooth analytic predictions for the ensemble average.

13.2 Molecular Dynamics Simulations

Molecular dynamics (MD) (Berendsen, 2007) is a real attempt to model actual molecular motion. It uses potentials that seek to mimic real molecular interactions, at least two-body interactions. The particles are moved forward in time by integrating Newton's second law. As Newton's third law is respected as well, momentum is conserved, and therefore MD is a way to simulate small scale hydrodynamics. The space and time scales that may be simulated are, of course, strongly limited by the molecular level of detail. However, as computer resources grow, the simulations reach ever further toward the macro-scale. In this context, MD simulations are used to study the flow in pores at the nanometer scale, the motion of large polymers and surface wetting effects.

13.2.1 Lennard-Jones Potential and the Velocity Verlet Algorithm

At least two ingredients are needed for an MD simulation, the intermolecular potential energy (which gives the forces) and an accurate numerical scheme to integrate Newton's second law forward in time. For a given particle or molecule i at position $\mathbf{r}_i$ and with mass m, Newtons second law takes the form

$$m \frac{d^2 \mathbf{r}_i}{dt^2} = -\sum_j \frac{\partial V(r_{ij})}{\partial \mathbf{r}_i}, \tag{13.5}$$

where i and j label the particles, r_{ij} is their separation, and $V(r)$ is the potential energy of two particles a distance r apart. A popular choice for $V(r)$ is the *Lennard-Jones potential*, which is an approximation to the interatomic potential in noble gases. Between such single atom molecules there are only *van der Waals forces* and the core repulsion due to the Pauli exclusion principle. This repulsion may be represented by a positive term in the potential energy that behaves as $1/r^{12}$. The van der Waals forces are caused by the interaction between the electric dipole moments of the atoms: The fluctuation in one dipole moment sets up a field that induces a dipole moment in the other atom. The potential energy goes as the dipole moment times the field. Since one dipole moment is proportional to the electric field caused by the other dipole, which goes as $1/r^3$, the potential goes as $1/r^6$. The van der Waals force is always attractive, so this contribution is negative. Grouping these terms, we may write

$$\boxed{V_{LJ} = 4\epsilon \left(\left(\frac{\sigma}{r}\right)^{12} - \left(\frac{\sigma}{r}\right)^6 \right),} \tag{13.6}$$

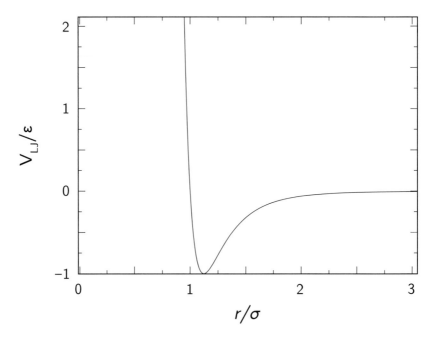

Figure 13.4 The Lennard-Jones potential as given in Eq. (13.6).

where ϵ sets the energy scale and σ the length scale. In Figure 13.4, both the core repulsion and region of attraction (where $V_{LJ}(r) < 0$) may be seen.

Now, in order to integrate Eq. (13.5), we have to introduce a finite timestep dt, and it makes an enormous difference how the discretization error behaves with dt. A convenient choice is the *velocity Verlet algorithm*, which updates the particle position $\mathbf{r}$ and velocity $\mathbf{v}$ by the scheme

$$\mathbf{r}(t + dt) = \mathbf{r}(t) + \mathbf{v}(t)dt + \frac{1}{2}\mathbf{a}(t)dt^2, \tag{13.7}$$

$$\mathbf{v}(t + dt) = \mathbf{v}(t) + \frac{1}{2}(\mathbf{a}(t) + \mathbf{a}(t + dt))dt. \tag{13.8}$$

The acceleration $\mathbf{a} = -\nabla V(\mathbf{r})$ is calculated before and after the update of $\mathbf{r}(t)$ in order to get both $\mathbf{a}(t)$ and $\mathbf{a}(t + dt)$. The discretization error in the velocity Verlet algorithm scheme is of order dt^3. The $\mathbf{r}$ update contains the correct terms up to order dt^2, and so the correction term is of third order. The same is the case for the $\mathbf{v}$-update since

$$\frac{1}{2}(\mathbf{a}(t) + \mathbf{a}(t + dt))dt = \mathbf{v}'(t)dt + \frac{1}{2}\mathbf{v}''(t)dt^2 + \mathcal{O}(dt^3). \tag{13.9}$$

For this algorithm to work, it is important that the potential only depends on $\mathbf{r}$ and not on $\mathbf{v}$.

The $\mathcal{O}(dt^3)$ error may accumulate at each timestep, so that the final error after a time t or $N = t/dt$ steps is of order dt^2. As dt is reduced, the number of time steps is increased, and round-off errors, which also accumulate at every step, may eventually dominate. Consequently there is an optimal dt-value that minimizes the final error.

A good verification that the code works is total energy conservation. However, this test may only be used for the case when one is actually simulating a system where the energy is intended to be conserved. This normally requires fixed particle number and volume. In the next section, we discuss fixed temperature systems, where the energy is not conserved. However, when coding such systems, it may be useful to go via intermediate stages where the energy *is* conserved.

Molecular dynamics code is often used for other purposes than simulation of molecules. In fact, by replacing $V_{LJ}(r)$, the molecular dynamics code may be used to simulate larger particles such as sand grains. In this case frictional forces are normally introduced, in which case energy conservation cannot be expected to hold.

13.2.2 Thermostats in Molecular Dynamics

Controlling the temperature or temperature gradients in molecular dynamics is often needed and requires some manipulation of the mean kinetic energy of the particles. A brute force method is to use the equipartition principle and measure the mean energy $\langle (m/2)v^2 \rangle = k_B T d/2$, where d is the dimension. Then the temperature may be reset by rescaling all velocities as $\mathbf{v} \to \mathbf{v}\sqrt{T_0/T}$, where T_0 is the target temperature.

Another approach, which is more continuous, is the application of a weak Langevin force, F_L. This force describes the fluctuating fluid force on suspended Brownian particles and enters into the *Langevin equation*

$$m\frac{d\mathbf{v}}{dt} = \mathbf{F}_L = -\alpha\mathbf{v} + \mathbf{F}'(t), \qquad (13.10)$$

where M is the mass of the particle, α is a friction coefficient and the fluctuating force F' satisfies the following correlations:

$$\langle F_i'(t)F_j'(0) \rangle = A\delta_{ij}\delta(t) \quad \text{and} \quad \langle F_i'v_j \rangle = 0. \qquad (13.11)$$

With discrete time, the Dirac delta function is most conveniently taken to be

$$\delta(t) = \begin{cases} \frac{1}{dt}, & \text{when} \quad |t| < dt/2, \\ 0 & \text{otherwise.} \end{cases} \qquad (13.12)$$

Multiplying Eq. (13.10) by the factor $e^{-(\alpha/m)t}$, and integrating, we get the formal solution

$$\mathbf{v}(t) = \int_{-\infty}^{t} dt' \, e^{-(\alpha/m)(t-t')} \frac{\mathbf{F}'(t')}{m}. \tag{13.13}$$

This equation may be used to calculate the mean kinetic energy

$$\frac{m}{2} \langle \mathbf{v}^2 \rangle = \frac{1}{2m} \int_{-\infty}^{t} \int_{-\infty}^{t} dt' \, dt'' \, e^{-(\alpha/m)(2t-t'-t'')} \langle \mathbf{F}'(t') \cdot \mathbf{F}'(t'') \rangle. \tag{13.14}$$

Now, using Eq. (13.11), we are left with

$$\frac{k_B T}{2} d = \frac{m}{2} \langle \mathbf{v}^2 \rangle = \frac{A}{4\alpha} d, \tag{13.15}$$

which means that

$$A = 2\alpha k_B T. \tag{13.16}$$

In a simulation, the Langevin force $\mathbf{F}_L$ may be realized by setting the force

$$\boxed{F_i' = \sqrt{\frac{2\alpha k_B T}{dt}} x \qquad \text{with } \langle x^2 \rangle = 1,} \tag{13.17}$$

where x is a random variable with $\langle x \rangle = 0$. If α is taken to be small, the perturbation represented by $\mathbf{F}_L$ will be correspondingly weak. The fact that other potential forces act on the particles will not alter the resulting temperature.

Another famous and useful thermostat is known as the *Nosé–Hoover thermostat* (Nosé, 1984).

13.2.3 Using Molecular Dynamics to Simulate Flow in Porous Media

Using molecular dynamics to simulate small scale flow in porous media, it is necessary to impose boundary conditions that allow for a flow through a given geometry. A convenient choice is periodic boundary conditions and an added force in the direction of the desired flow velocity. Two phases may be accounted for by introducing two particle species labeled $I = 1, 2$ and modifying the interaction strength ϵ in Eq. (13.6) to vary with the species, so that the interaction energy becomes

$$\boxed{V_{LJ} = 4\epsilon_{IJ} \left(\left(\frac{\sigma}{r}\right)^{12} - \left(\frac{\sigma}{r}\right)^6 \right),} \tag{13.18}$$

where $\epsilon_{12} = \epsilon_{21}$ by Newton's third law. If $\epsilon_{12} < \epsilon_{11}$ and ϵ_{22}, it may give rise to phase separation and surface tension between the phases.

Also, by introducing stationary wall particles $I = 3$, the wetting properties of the walls may be tuned by varying ϵ_{I3}.

As an example of how molecular dynamics is applied, we include a short description of a simulation study by Wu et al. (2013). In order to understand how droplets may be driven by thermal gradients along nanochannels, they carried out molecular

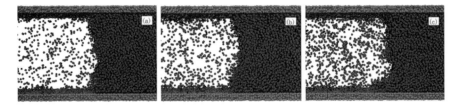

Figure 13.5 The liquid–vapor interface of a droplet confined in a nanochannel. The wall particles are red, the fluid particles blue. The liquid-to-vapor density ratio is about 37 in (a), 18 in (b) and 11 in (c), corresponding to the liquid-vapor coexistence temperatures 0.80 ϵ/k_B, 0.85 ϵ/k_B and 0.90 ϵ/k_B, respectively. Simulations by Wu et al. (2013). (A black and white version of this figure will appear in some formats. For the colour version, refer to the plate section.)

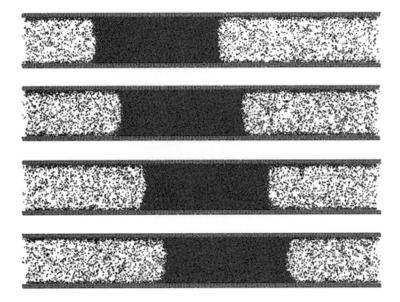

Figure 13.6 Snapshots of the liquid–vapor interface of a moving droplet confined in a nanochannel with a thermal gradient along it. (Wu et al., 2013). (A black and white version of this figure will appear in some formats. For the colour version, refer to the plate section.)

dynamics simulations where the temperature was set to allow for a liquid–vapor phase separation. A similar and earlier study was carried out by Wold and Hafskjold (1999). In the these simulations, there was an interaction between the fluid molecules and the molecules that made up the walls of the nanochannels that caused the liquid to wet the walls. This wetting angle would depend on temperature, and, hence, thermal gradients along the channel would create a driving force for the flow. This interface at different temperatures is illustrated in Figure 13.5, while the actual displacement of the droplet is shown in Figure 13.6.

13.3 Lattice Gas Model for Hydrodynamics

The *lattice gas models* were introduced in the late 1980s and represented somewhat of a new paradigm for the simulation of hydrodynamic flows. This paradigm was coined *bottom-up simulation*, as opposed to *top-down*. The bottom-up approache is characterized by the fact that it starts with a simulation algorithm on the molecular or particle level, and then one derives the Navier–Stokes equation from the physical description at this level. The conventional top-down approach, on the other hand, starts from the conservation laws that are described in the macroscopic Navier–Stokes equations and works downward to smaller scales by discretizing these.

The *lattice gas cellular automata* form a basis for the subsequent lattice Boltzmann models that are discussed in this chapter. The book by Rivet and Boon (2001) is a good introduction to the theory of lattice gases, which is described more briefly in this section, where we also shows how lattice gases are connected with the lattice Boltzmann and lattice BGK (Bhat-nagar, Gross and Krook) models.

Two fluids with quite different microscopic interactions may still have the same macroscopic behavior. The reason for this is that the form of the macroscopic equations of motion, which describe this behavior, only depend on the conservation laws obeyed by these interactions and not on their detailed form. This was one of the main motivations for the introduction of the *Lattice Gas Automaton* (LGA) as a method to simulate fluid flow. The LGA models the fluid as a large number of particles with interactions that conserve mass and momentum. While the aim of molecular dynamics is to simulate the real microworld, the microworld of the LGA is completely fictitious. Still, realistic macroscopic behavior is recovered when space and time averages are performed.

A *cellular automaton* is a system of discrete variables on a lattice, which is updated according to some simple and local rule. Beside the LGA, famous examples include *Conway's Game of Life* (Berlekamp et al., 1982) and the cellular automata for the critical dynamics of sandpiles introduced by Bak et al. (1987) a few decades ago. Another example is the model introduced by Hardy, de Pazzis and Pomeau (HPP) in 1976. This model, which is also a model for hydrodynamics, is known as an LGA; it succeeds in describing soundwaves, and, if two particle species are introduced, diffusion, but not hydrodynamics.

The simplest and first lattice gas model that produced realistic hydrodynamic behavior was introduced by Frisch, Hasslacher and Pomeau (FHP) in 1986 and works on a triangular lattice. This lattice is crucial for the isotropy (Frisch et al., 1986, 1987) and also for the isotropy of the flow dynamics. With the square lattice of the HPP model, the lattice directions are visible in the physical flow field. The FHP model describes particles with discrete velocities that move in discrete space and time. The particles all have the same speed, which means that conservation of

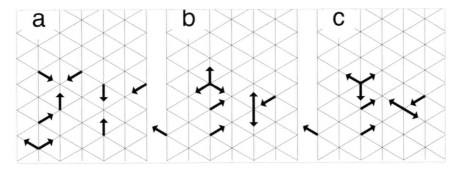

Figure 13.7 Figure (a) shows the initial state, figure (b) shows the state after propagation and figure (c) shows the state after the two- and three-body collisions. The directions of the velocities after the two-body collision is chosen by a random number.

mass and conservation of energy are the same thing. As a result, the concept of temperature has no independent non-trivial meaning within the FHP model.

Figure 13.7 shows the two basic steps of the LGA, *propagation* followed by *collisions*. The particles reside on the lattice sites and are represented by arrows according to their velocities. There can be at most one particle per direction, and there are no interactions between particles except at the lattice sites.

In addition to the internal evolution rules, boundary conditions must be implemented. The simplest choice is the periodic boundary condition, which gives global conservation of both mass and momentum – since these quantities are conserved locally. In order to introduce solid walls, one must introduce new collisions that prevent particles from moving out across the boundaries. These might be either of the bounce-back type, which send particles back into the direction from which they came, or of the mirror reflection type, where only one component of the particle's momentum is changed. When particle velocities are averaged, the effect of the *bounce-back collision* is a hydrodynamic no-slip boundary condition (Lavallee et al., 1991). The mixing of bounce-back and *mirror collisions* gives a boundary condition where the average flow velocity is non-zero at the wall. In practice, the most common boundary condition is the bounce-back one. In simulations this condition is easily implemented by the introduction of a "solid-matrix" that takes on the value 0 away from the boundary sites and 1 on these sites. Wherever the value 1 in encountered, the normal collision step is replaced by a step where all velocities are reversed. An obvious virtue of all lattice gas and lattice Boltzmann models is the ability to deal with highly complex boundaries.

As mention in Section 6.5, the boundary conditions on solid walls represent an independent part of the physical description relative to the equations of motion, which describes the interior part of the fluid away from the walls. For real fluids it is common to assume a vanishing velocity – a no-slip condition – at the walls. However, this is only a good approximation when the length scales involved are well above the mean free path. At smaller scales, which are relevant in a lattice gas, finite slip velocities become visible.

In order to introduce a body-force, like gravity, an additional collision step is needed that puts momentum into the system. This can be done in several ways, one of which is to flip particle velocities at a few randomly chosen sites into the direction of the forcing.

If there were no collisions, all the particles would move independently, and the lattice gas would have a trivial and non-physical behavior. With the two-body collisions shown in Figure 13.7, particles interact in a way that conserves mass and momentum. But with the *two-body collisions* alone there are additional conservation laws: The quantity defined as the number of particles moving in a chosen direction minus the number of particles moving in the opposite direction on a given line of the lattice, is also conserved – on every line. This is known as a *spurious conservation law,* which, in this case, leads to anisotropies. The three-body collisions, which are also shown in Figure 13.7, break this additional conservation law, and with the minimum set of two- and three-body collisions the lattice gas behaves isotropically.

13.3.1 Statistical Mechanics of the Lattice Gas

In the simplest lattice gas model, illustrated in Figure 13.7, the state at a single site (at position $\mathbf{x}$ at time t) is given by the six occupation numbers $n_i(\mathbf{x}, t) = 0$ or 1, which are simply the particle numbers in direction i. There can be at most six particles per site, and they must all move in different directions. The time development of the n_i's corresponding to Figure 13.7 is given by

$$n_i(\mathbf{x} + \mathbf{c}_i, t + 1) = n_i(\mathbf{x}, t) + \Omega_i\left(\{n_i(\mathbf{x}, t)\}\right), \qquad (13.19)$$

which is known as the Boltzmann equation for the LGA. Here, and in what follows, we will take take time to be measured in units of the timestep and space to be measured in units of the lattice constant. The term $\Omega_i(\{n_i(\mathbf{x}, t)\})$ is the change in n_i due to collisions, and it can be written as a simple algebraic expression of the n_i's.

The full detailed state of the lattice is given by the set of all n_i's on all sites. In simulations on a computer, all this information is stored and updated. However, the quantities of physical interest are the (ensemble or time) averaged mass and momentum densities. The question we address here is how the detailed description of Eq. (13.19) can be reduced to a set of equations that describe only the dynamics of

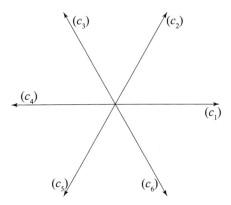

Figure 13.8 The particle velocities c_i are numbered from 1 to 6 and have unit length.

the averaged density and velocity fields. This is not a trivial task, and the derivations leading to the fluid dynamic equations of motion involve several assumptions that are only partly qualified.

We shall proceed to examine the conceptual content of this derivation. However, for the sake of simplicity, some of the technical points, which are elaborated by Frisch et al. (1987), are left out. The boolean variables n_i taking only the values 0 or 1 are strongly fluctuating quantities. Their ensemble average $N_i = \langle n_i \rangle$, however, is expected to vary much more slowly, as will the averaged mass and momentum densities per site, ρ and $\rho\mathbf{u}$, which are defined as

$$\rho = \sum_i N_i, \tag{13.20}$$

$$\rho\mathbf{u} = \sum_i \mathbf{c}_i N_i, \tag{13.21}$$

respectively. The velocity vectors $\mathbf{c}_i$ are defined in Figure 13.8. For the conservation laws to hold, the mass and momentum contained in the n_i's on the left- and right-hand sides of Eq. (13.19) must be the same. This is reflected in the following property of the collision term:

$$\sum_i \Omega_i = \sum_i \mathbf{c}_i \Omega_i = 0. \tag{13.22}$$

The starting point for the derivation of the equations of motion for ρ and $\rho\mathbf{u}$ will be the equilibrium values N_i^{eq} of the particle distributions, which we will obtain following classical arguments of statistical mechanics:

As seen from Figure 13.7, the lattice gas is not deterministic. In the two-body collision, there are two possible output states. In general an input state s on a given site, which is the set of the six n_i's, will become a state s' with probability $A(s \rightarrow s')$ in the collision step. The probability A is normalized according to

$$\sum_{s'} A(s \rightarrow s') = 1, \tag{13.23}$$

and we will only consider the case where *semi-detailed balance* is satisfied (Frisch et al., 1987). This condition is given by the equation

$$\sum_{s} A(s \rightarrow s') = 1, \tag{13.24}$$

which states that provided every input configuration s is equally likely, every output configuration s' will occur with equal probability as well.

On a lattice with periodic boundary conditions, both the overall momentum $\mathbf{Q}$ and the total particle number $\mathcal{N}$ are conserved. A state S of the lattice is specified by giving all the configurations $s(\mathbf{x}, t)$. From Eq. (13.24) it is possible to show that an ensemble of lattices, all with the specified values of $\mathcal{N}$ and $\mathbf{Q}$, will be stationary. In other words, if the states are distributed evenly in the phase space of all states S, they will remain so over time. All the states S must have the same values of the conserved mass and momentum $\mathcal{N}$ and $\mathbf{Q}$. It follows that the probability of finding a given state S is $1/W(\mathcal{N}, \mathbf{Q})$, where $W(\mathcal{N}, \mathbf{Q})$ is the number of states S with $\mathcal{N}$ particles and momentum $\mathbf{Q}$.

A cell is a single direction i at a lattice site. Consider now this cell to be in statistical equilibrium with the rest of the system. The cell is then like a system described in the grand canonical ensemble, except for the overall conserved energy being replaced by an overall conserved momentum $\mathbf{Q}$. The probability $p(n_i)$ of finding n_i particles in this cell is then proportional to the number of states available to rest of this system, that is,

$$p(n_i) \sim W(\mathcal{N} - n_i, \mathbf{Q} - n_i \mathbf{c}_i), \tag{13.25}$$

where W is a rapidly varying function of $(\mathcal{N}, \mathbf{Q})$. Expanding the more slowly varying logarithm of W, we obtain

$$\ln p(n_i) \sim \ln W(\mathcal{N} - n_i, \mathbf{Q} - n_i \mathbf{c}_i)$$
$$\sim -\frac{\partial \ln W}{\partial \mathcal{N}} n_i - \frac{\partial \ln W}{\partial Q_\alpha} n_i c_{i\alpha}, \tag{13.26}$$

where summation over the repeated cartesian index α is implied. Exponentiation of this expression gives the probability

$$p(n_i) \sim \exp[-(h + \mathbf{q} \cdot \mathbf{c}_i) n_i], \tag{13.27}$$

where we have defined the n_i-independent quantities

$$h = \frac{\partial \ln W}{\partial \mathcal{N}},$$ (13.28)

$$q_\alpha = \frac{\partial \ln W}{\partial Q_\alpha}.$$ (13.29)

The *equilibrium distribution* is now given as

$$N_i^{\text{eq}} = \frac{\sum_{n_i=0,1} n_i \, p(n_i)}{\sum_{n_i=0,1} p(n_i)} = [1 + \exp(h + \mathbf{q} \cdot \mathbf{c}_i)]^{-1}.$$ (13.30)

The quantities h and $\mathbf{q}$ may be identified as Lagrangian multipliers that may be determined in terms of ρ and $\rho\mathbf{u}$ by the constraints given by Eqs (13.20) and (13.21). It is not possible, however, to solve these equations exactly for h and $\mathbf{q}$, and one must resort to Taylor expansions in u_α. Due to the parity symmetry[1] of the triangular lattice, the expression for N_i^{eq} given in Eq. (13.30) must be invariant under the simultaneous replacements $\mathbf{c}_i \rightarrow -\mathbf{c}_i$ and $\mathbf{u} \rightarrow -\mathbf{u}$. It follows that h and $\mathbf{q}$ must be even and odd functions of u_α, respectively. By using this fact and carrying out a Taylor expansion of Eq. (13.30) to second order in u_α, N_i^{eq} can be determined as a function of ρ and $\mathbf{u}$ by the relations (13.20) and (13.21). The result is

$$N_i^{\text{eq}} = \frac{\rho}{6} \left(1 + 2\mathbf{c}_i \cdot \mathbf{u} + 4g(\rho)Q_{i\alpha\beta}u_\alpha u_\beta \right),$$ (13.31)

where

$$Q_{i\alpha\beta} = c_{i\alpha}c_{i\beta} - \frac{1}{2}\delta_{\alpha\beta}$$ (13.32)

and

$$g(\rho) = \frac{3 - \rho}{6 - \rho}.$$ (13.33)

It can be shown that the tensor $Q_{i\alpha\beta}$ satisfies the equation

$$\sum_i Q_{i\alpha\beta} = 0,$$ (13.34)

which can be viewed as a completeness relation for the $\mathbf{c}_i$'s. Since the set $\mathbf{c}_i$ contains three linearly independent subsets that all span space, it is possible to show that

$$\sum_i c_{i\alpha}c_{i\beta} = 3\delta_{\alpha\beta}.$$ (13.35)

---

[1] For every lattice vector $\mathbf{c}_i$ there is also the opposite vector $-\mathbf{c}_i$.

13.3.2 Hydrodynamic Equations

When the lattice gas is in a homogeneous state, the values of the N_i's will be given by the equilibrium distribution in Eq. (13.31). If there are spatial variations in ρ and $\mathbf{u}$, but over a typical length L that is much larger than the lattice constant, the system will evolve in time through long range transport of mass and momentum to a global homogeneous equilibrium. The quantity

$$\epsilon = \frac{1}{L} \qquad (13.36)$$

is normally referred to as the *Knudsen number*, which is the ratio of the mean free path (1 in our case) and the characteristic length over which hydrodynamic quantites vary. Since particles move at unit speed on the lattice, the typical time for temporal variations will thus in general be the same as that for spatial variations. The time needed to reach a local equilibrium is of the order of the mean free time $\tau \sim 1$, and hence, the time for relaxation to global equilibrium is much larger than τ.

The local values of the N_i's are given by the equilibrium values plus some correction terms that depend only on the gradients of ρ and $\mathbf{u}$. With the above assumption, these gradients will be of order ϵ, and we can write

$$N_i(\mathbf{x}, t) = N_i^{\text{eq}}(\rho(\mathbf{x}, t), \mathbf{u}(\mathbf{x}, t)) + N_i^{\text{neq}}, \qquad (13.37)$$

where N_i^{neq} to leading order is some linear function of the gradients in ρ and $\mathbf{u}$. Since a gradient in u_α may be estimated as $u_\alpha / L \sim \epsilon u_\alpha$, we have that the correction term $N_i^{\text{neq}} \sim \epsilon$.

All the macroscopic equations of motion follow from Eqs (13.19) and (13.22), which together express the conservation of mass and momentum. In order to get these conservation relations in the form of differential equations, a Taylor expansion in ϵ is needed. This type of expansion is known as a *Chapman–Enskog expansion*. We shall do this for the expansion of the mass conservation relation to first order in ϵ to get the continuity equation. However, we will only quote the equations resulting from the second order expansions of the relation for conservation of momentum.

Summing the average of Eq. (13.19) over the six lattice directions, we obtain the exact relation

$$\sum_i N_i(\mathbf{x} + \mathbf{c}_i, t + 1) - N_i(\mathbf{x}, t) = 0. \qquad (13.38)$$

Taylor-expanding the above difference around $(\mathbf{x}, t)$, it follows from the assumption on the scale of variations in ρ and $\mathbf{u}$ that the derivatives of N_i will be of order ϵ. Expanded to first order in ϵ, the left-hand side of Eq. (13.38) takes the form

$$\sum_i (\mathbf{c}_i \cdot \nabla + \partial_t) N_i(\mathbf{x}, t) = 0. \qquad (13.39)$$

Dropping the N_i^{neq} part, which gives rise to terms of higher order in ϵ only, we can write

$$\sum_i (\mathbf{c}_i \cdot \nabla + \partial_t) N_i^{\text{eq}}(\mathbf{x}, t) = 0. \tag{13.40}$$

Substituting the form of N_i^{eq} given in Eq. (13.31), and using the identity Eq. (13.34), we get the desired continuity equation

$$\partial_t \rho + \nabla \cdot (\rho \mathbf{u}) = 0. \tag{13.41}$$

To summarize, the assumptions needed to get Eq. (13.41) only involves the assumption of a scale separation between the lattice constant and L and the existence of a local equilibrium given in the form of Eq. (13.31).

Additional assumptions are needed in order to get the incompressible Navier–Stokes equations. The strongest of these is that the characteristic temporal length is not ϵ^{-1}, but ϵ^{-2}. In the following we will examine the conditions under which this will be the case.

In a real gas or fluid, variations in ρ will propagate rapidly as sound waves, whereas variations in $\mathbf{u}$ at constant ρ will spread more slowly through the diffusion of momentum. By dropping a bomb on a windy day, one certainly creates density variations in the air – but these variations will attenuate much faster than will the wind. On the macroscopic scale, the relaxation of ρ and $\mathbf{u}$ will thus be governed by different physical mechanisms acting on different timescales. The question is whether the same conclusion can be drawn from the knowledge of the LGA rules and of some carefully specified initial state of the system.

In Figure 13.9, a simple initial state of a lattice gas is shown. The scale of spatial variations in the conserved quantities is $L \gg 1$, and to leading order in ϵ, N_i will be given by the equilibrium expression of Eq. (13.31).

Consider first the case in which there is a density difference $\Delta \rho = \rho_1 - \rho_2 > 0$ between regions 1 and 2 of the lattice. The probability of having a particle arrive at the dashed line from above will then be larger than the probability of having a particle arrive from below. There will thus be a mass current ρu_y from region 1 to region 2 of the lattice. Since the mass is conserved, the increase of mass ΔM_2 in region 2 will be given solely by this current. ΔM_2 can be estimated as

$$\Delta M_2 \sim L \left(\mathbf{e}_y \cdot \sum_i \mathbf{c}_i N_i^{\text{eq}} \right) t \tag{13.42}$$

$$\sim L u_y t,$$

where $\mathbf{e}_y$ is the unit vector in the y-direction, and t is the time. The last equation follows from Eqs (13.31) and (13.34) and the fact that odd order moments of the $\mathbf{c}_i$'s vanish due to the parity symmetry of the lattice.

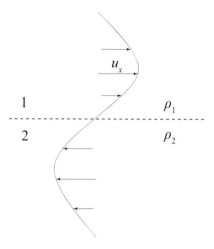

Figure 13.9 A simple initial state of a lattice gas.

After the characteristic relaxation time t_ρ has elapsed, $\Delta M_2 \sim L^2 \rho$. It follows from Eq. (13.42) that $t_\rho \sim L/u_y \sim \epsilon^{-1}$.

Now, consider the case in which the density field has been allowed to relax over a time t_ρ, but there is still a velocity difference $\Delta u = u_1$ between regions 1 and 2. The density will then be constant across the dashed line, and there will be no transport of mass in the y-direction. As with the increase of mass in region 2, the increase of momentum ΔQ_2 in region 2 will be given completely by the momentum current across the dashed line. At $\mathbf{x} = 0$, the current can be estimated by the average numbers of particles arriving from neighboring sites through the relation

$$\Delta Q_2 \sim L \left[(N_2^{eq}(\mathbf{c}_5) - N_3^{eq}(\mathbf{c}_6)) - (N_6^{eq}(\mathbf{c}_3) - N_5^{eq}(\mathbf{c}_2)) \right] t$$
$$\sim L \rho t \, \delta u,$$

(13.43)

where the last equation is again obtained from Eq. (13.31), and δu is the average change in the velocity across the dashed line. Since this difference is over a distance of order unity, $\delta u \sim \nabla u \sim u_1/L$, and Eq. (13.43) takes the form

$$\Delta Q_2 \sim \rho u_1 t.$$

(13.44)

This means that the time t_u corresponding to the relaxation of the momentum, $\Delta Q_2 \sim L^2 \rho u_1$, is given as $t_u \sim L^2/u_1 \sim \epsilon^{-2}$, which is the result we set out to show.

Although this result suggests the possibility that density variations will be attenuated on *one* timescale leaving the velocity field to evolve on another, longer scale, it is not obvious that it can be taken over to more complicated flow fields in general.

By requiring that the velocity u be much smaller than unity,

$$u \ll 1, \tag{13.45}$$

the incompressible Navier–Stokes equations may now be derived from Eqs (13.19) and (13.22) using the same procedure as that leading to the continuity equation (Frisch et al., 1987). The amount of algebra required is, however, considerably greater. The result, which is obtained by this procedure of keeping terms to second order in ϵ and second order in u, is

$$\nabla \cdot \mathbf{u} = 0, \tag{13.46}$$

$$\partial_t \mathbf{u} + g(\rho)\mathbf{u} \cdot \nabla \mathbf{u} = -\frac{1}{\rho}\nabla P + \nu\nabla^2\mathbf{u}, \tag{13.47}$$

where the pressure

$$P = \frac{\rho}{2}(1 - g(\rho)u^2), \tag{13.48}$$

and the kinematic viscosity ν depends on the choice of collision rules through Ω_i. This is the same situation as for a real fluid: The form of the Navier–Stokes equations results from the conservation laws alone. But the viscosity depends on the particular form of the microscopic interactions.

The viscosity of a lattice gas will in general decrease as the number of collisions is increased. The reason for this is that increasing the number of different collisions means reducing the mean free path. But the number of ways to rearrange the particles within the six-velocity model is quite limited. One way to deal with that limitation is to add a rest particle, that is, a particle with zero velocity, and as many mass- and momentum-conserving collision rules as possible. This gives rise to the so called *FHP-3 model*, which is illustrated in Figure 13.10.

The presence of the $g(\rho)$ factor, which is given in Eq. (13.33), causes Eqs (13.47) and (13.46) to differ slightly from the true Navier–Stokes equation (6.39). As a result of the fact that $g \neq 1$, Eqs. (13.47) and (13.46) will not be invariant under the *Galilean transformation*

$$\mathbf{x} \to \mathbf{x}' = \mathbf{x} - \mathbf{v}t,$$
$$t \to t' = t, \tag{13.49}$$

where $\mathbf{v}$ is the velocity of some new frame reference relative to the lattice. The lack of *Galilean invariance* can be understood from the fact that there is only one particle speed. For this reason, the existence of a finite flow velocity $\mathbf{u}$ means that there will be fewer particles moving in the direction perpendicular to the flow than along it. In a real gas, or in a model with a continuum of velocities, a macroscopic flow can (and will) be created by adding the velocity $\mathbf{u}$ to all the single particle velocities.

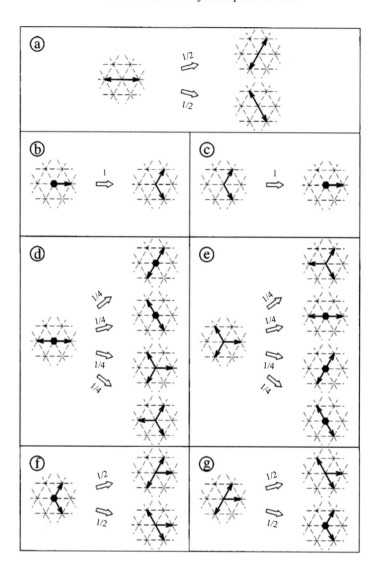

Figure 13.10 A selection of the FHP-3 collision rules, as illustrated by Rivet and Boon (2001). Rest particles are shown as dots, and the numbers are the probability of picking the given outcome of a collision. The full set of rules may be obtained by rotating those illustrated.

This means that the distribution of particle velocities will be the same in any frame of reference where the flow velocity is locally zero. This is not the case in a lattice gas.

Provided $g(\rho) > 0$, the problem of the lack of Galilean invariance can be solved by multiplying Eq. (13.47) by the factor $g(\rho)$ and absorbing it in the scaled velocity $\mathbf{u}' \equiv g(\rho)\mathbf{u}$ and pressure $P' \equiv g(\rho)P$. The correction terms corresponding to the derivatives of $g(\rho)$ are of higher order in ϵ than the remaining terms and can therefore be neglected. The equation for $\mathbf{u}'$ and P' will thus be exactly the incompressible Navier–Stokes equations.

However, the rate of mass transport is $\rho\mathbf{u}$, and not $\rho\mathbf{u}'$. This presents no problem when there is only one particle species present, which is the simplest case. But if more than one particle species is introduced, the physical velocity is set by the local velocity of the interfaces between regions of different particles, which will equal $\mathbf{u}$.

From a programmer's point of view, the lattice gas model is particularly attractive since it consists of very simple updatings of a *boolean field* given by the n_i's. The representation of a configuration s only requires 6 bits, and it is thus possible to store the configurations of 5 sites in a single 32-bit integer. For instance, the configuration of one particle with velocity $\mathbf{c}_2$ and one with velocity $\mathbf{c}_5$ will be stored as the number $2^2 + 2^5$, or this number left-shifted by an integer multiple of 6. The collision step will consist of feeding the integer corresponding to the state s into a table of length 64 to get the integer corresponding to the output state s'. The propagation step is carried out by shifting the elements of the array containing the s's. The high speed of these operations on a computer accounts for the high computational efficiency of the LGA. However, the intrinsic fluctuations in lattice gases require a significant amount of averaging in space or time to obtain smooth hydrodynamic fields.

13.3.3 The Isotropy of the Lattice Gas

Why is the triangular lattice necessary, and in fact sufficient, to guarantee that the flow fields are insensitive to the orientation of the underlying lattice? While the expansion leading up to the mass conservation Eq. (13.41) involved moments of $\mathbf{c}_i$ only up to second order, the expansion leading to Eq. (13.47) involved moments of $\mathbf{c}_i$ up to fourth order, that is, the tensor

$$T_{\alpha\beta\gamma\delta} = \sum_i c_{i\alpha}c_{i\beta}c_{i\gamma}c_{i\delta}. \tag{13.50}$$

The second order moments of $\mathbf{c}_i$ form an *isotropic tensor* both for the triangular and the square lattice. This is seen by considering the tensor as a map taking two vectors to a number. Eq. (13.35) shows that this map is simply the inner product of these vectors and therefore does not depend on the orientation of the underlying vectors $\mathbf{c}_i$. This is true for the square lattice as well, for which the lattice vectors satisfy Eq. (13.35) too, save for a replacement of the factor 3 by a factor 2.

Now, the question is if the fourth order T has the same property. If $\mathbf{a}$, $\mathbf{b}$, $\mathbf{c}$ and $\mathbf{d}$ are arbitrary vectors, is $T = (\mathbf{a}, \mathbf{b}, \mathbf{c}, \mathbf{d}) = T_{\alpha\beta\gamma\delta}a_\alpha b_\beta c_\gamma d_\delta$ independent of the $\mathbf{c}_1$-orientation? A simple check, taking all vectors equal, answers the question for the square lattice, where c_i are simply plus or minus the orthogonal unit vectors e_x and e_y. It is then seen that

$$T(\mathbf{a}, \mathbf{a}, \mathbf{a}, \mathbf{a}) = \sum_i (\mathbf{c}_i \cdot \mathbf{a})^4 \qquad (13.51)$$

is 2 if $\mathbf{a}$ is aligned with $\mathbf{c}_1 = \mathbf{e}_x$, and 1 if it is oriented 45° to $\mathbf{c}_1$.

For the triangular lattice, the similar exercise, setting $\mathbf{a} \parallel \mathbf{c}_1$ and $\mathbf{a}$ at an angle of 30° to $\mathbf{c}_1$, shows that the two relative orientations give the same result, 9/4. This may be shown to hold for any orientation. Moreover, the result that T is independent of the orientation of the underlying lattice may be generalised to different vectors $\mathbf{a}$, $\mathbf{b}$, $\mathbf{c}$ and $\mathbf{d}$. Since $T = (\mathbf{a}, \mathbf{b}, \mathbf{c}, \mathbf{d})$ is clearly invariant under permutation of the vectors, it must have the form

$$T_{\alpha\beta\gamma\delta} = \frac{3}{4}(\delta_{\alpha\beta}\delta_{\gamma\delta} + \delta_{\alpha\gamma}\delta_{\beta\delta} + \delta_{\alpha\delta}\delta_{\beta\gamma}). \qquad (13.52)$$

This result ensures that Eq. (13.47) comes out without a dependence on the lattice orientation.[2] This property is quite remarkable. The triangular lattice is a *Bravais lattice*, that is, all its points may be reached by a combination of integer muliples of two basis vectors. It turns out that no isotropic Bravais lattice exists in three dimensions.

13.3.4 The Three-Dimensional Lattice Gas Model

The way to get around the lack of symmetries of three-dimensional Bravais lattices is to move on to four dimensions. In four dimensions, the face centered hyper cubic (FCHC) lattice given by the 24 basis vectors

$$\begin{aligned}
(\pm 1, \pm 1, 0, 0)\\
(\pm 1, 0, \pm 1, 0)\\
(\pm 1, 0, 0, \pm 1)\\
(0, \pm 1, 0, \pm 1)\\
(0, 0, \pm 1, \pm 1)\\
(0, \pm 1, \pm 1, 0)
\end{aligned} \qquad (13.53)$$

fulfills the necessary requirements. This lattice forms the basis for the lattice gas model introduced by d'Humières et al. (1986). Three-dimensional hydrodynamics is

---

[2] It turns out to have an interesting application in the simulation of elastic media, too: If a lattice structure is made out of Hookean springs, it is again the isotropy of T that determines the isotropy of the elastic properties of the network. So, the elastic properties of such a triangular lattice of springs is indistinguishable from that of a homogeneous, isotropic medium. A square lattice of springs, on the other hand, is well known to lack this property. If it is sheared, it does not offer any resistance to deformation whatsoever.

obtained by projecting the four-dimensional lattice back to three dimensions, giving a lattice that might be considered three-dimensional, but with double connections between sites. This added feature makes it something more than a simple Bravais lattice. The model is described by the same quantities (mass and momentum densities) as the two-dimensional one, but the number of velocity directions on every site requires a much longer collision table than in two dimensions. The boundary conditions are dealt with in the same manner as in the two-dimensional case.

13.3.5 Lattices Gases for more Complex Fluids

The main virtue of both the lattice gas and lattice Boltzmann models is the fact that it so easy to build in new physics at the particle level. Since the models are based on particle dynamics, new features may be added to these particles and the way they interact. The most obvious extension, perhaps, is to add a second particle species, distinguished from the first, say, by a different color. As long as the particle number in each species is conserved along with their combined momentum, such a model will simulate the simultaneous transport by hydrodynamic flow and molecular diffusion, that is, the flow of two miscible fluids, like coffee and cream.

Another, slightly more challenging extension is the introduction of surface tension, so that the two fluids actually represent imiscible phases, like oil and water. Such a model was introduced in the late 1980s by Rothman and Keller (1988). This model was designed with interaction rules that mimicked the molecular interactions in real liquids, where the molecular potentials cause attraction between like molecules and repulsion, or a weaker attraction, between unlike molecules. In the model, the molecules are red or blue, and the collision rules were generalized to include nearest neighbor effects. At a given lattice site, the number of red and blue particles at the nearest sites is calculated and the collisions carried out in such a way as to send blue toward blue and red toward red, while conserving mass and momentum. Such a rule gives rise to a surface tension that indeed fulfills Laplace's law for the pressure drop over curved interfaces. The model has been used to simulate thermally exited interface motion, as well as two-phase flow in porous media and droplet dynamics (Flekkøy and Rothman, 1995; Rothman and Zaleski, 1994).

Other extensions of the lattice gas model include models for chemical reactions, suspensions and surfactants. In the case of chemical reactions, two species may react to form a third species and vice versa. To simulate suspension, larger particles are introduced that interact with the lattice gas fluid in a momentum-conserving way. Surfactants are molecules, like soap molecules, that reside at the interface between two imiscible fluids, typically reducing the surface tension. Such molecules may be added to the lattice gas for immiscible fluids.

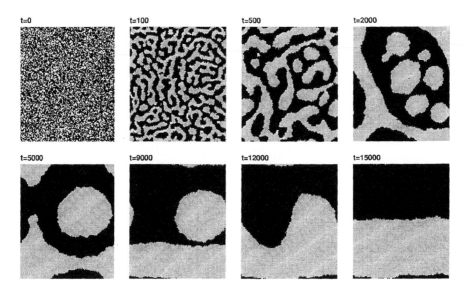

Figure 13.11 Phase separation simulated by means of the lattice gas for imiscible fluids introduced by Rothman and Keller (1988). The time t is in units of timesteps.

13.4 Lattice Boltzmann Models

The lattice Boltzmann models (LBM) work with the mean particle populations N_i instead of the discrete occupation numbers n_i. The idea of using the analytic description of the original lattice gas as an algorithm in its own right was first suggested by McNamara and Zanetti (1988).

The dynamics of the LBM is described by taking the average of Eq. (13.19) and applying the Boltzmann approximation: The Boltzmann approximation amounts to the assumption that the precollision occupation numbers n_i are uncorrelated. This is really an approximation, since such correlations do in fact exist. However, it implies that averages of products may be replaced by products of averages. Since the collision operator $\Omega_i(\{n_i\})$ consists of different terms that are products of the n_i's, this means that the average of the collision operator $\Omega_i(\{n_i\})$ takes the form $\Omega_i(\{N_i\})$, and Eq. (13.19) can be written

$$N_i(\mathbf{x} + \mathbf{c}_i, t + 1) = N_i(\mathbf{x}, t) + \Omega_i(\{N_j\}). \qquad (13.54)$$

In the simpler cases where there are not too many collision rules, the collision operator can be cast in a simple algebraic form according to these rules. Eq. (13.54) can thus be used to update the N_i in a "collision" step given by $\Omega_i(\{N_j\})$ and a propagation step that is exactly as before. The resulting dynamics will be free from the statistical noise that is present in the LGA. When $\Omega_i(\{N_j\})$ is chosen to correspond to a particular set of collision rules, the dynamics of the LBM will be the same as the averaged dynamics of the LGA, to the extent that the Boltzmann approximation is valid. Simulations using the above scheme have been carried out by McNamara and Zanetti (1988).

It is, however, not required that the collision operator corresponds to a particular set of LGA collisions. This is most fortunate for three-dimensional simulations in which the explicit expression for $\Omega_i(\{N_j\})$ would have to contain millions of terms corresponding to all possible collisions. It suffices to demand that $\Omega_i(\{N_j\})$ satisfies the basic conservation laws. A practical way to achieve this is by linearizing the collision operator. This was first done by Higuera et al., who presented their derivation in a series of three papers (Higuera and Succi, 1989; Higuera et al., 1989; Higuera and Jiménez, 1989).

13.4.1 The BGK Models

A further step away from the LGA is represented by the *lattice BGK* (Bhatnagar, Gross and Krook) models first introduced by Qian et al. (1992) and well analyzed by Chen and Doolen (1998). In these models, all connections with the LGA collisions rules are abandoned. The main observation is that the collision term only needs to conserve mass and momentum, and vanish in equilibrium when $N_i^{\text{neq}} = 0$. In practice, only BGK models are actually used, and nomally referred to as lattice Boltzmann models.

The defining equation for the BGK model, which is analogous to Eq. (13.54), has the form

$$\boxed{N_i(\mathbf{x} + \mathbf{c}_i, t + 1) = N_i(\mathbf{x}, t) - \omega N_i^{\text{neq}}(\mathbf{x}, t),} \tag{13.55}$$

where $N_i^{\text{neq}}(\mathbf{x}, t) = N_i(\mathbf{x}, t) - N_i^{\text{eq}}(\mathbf{x}, t)$, and the equilibrium distribution now has the independent definition

$$N_i^{\text{eq}} = \frac{\rho}{6} \left(1 + 2\mathbf{c}_i \cdot \mathbf{u} + 4Q_{i\alpha\beta}u_\alpha u_\beta\right). \tag{13.56}$$

Note that this is the same expression as that given for the equilibrium distributions of the LGA in Eq. (13.31) except for the g factor, which is now simply chosen to be one.

The key observation for the analyses of the BGK model is that the mass and momentum conservation, which are crucial for the macroscopic equations, are ensured by the choice of equilibrium distributions: It is easily shown that

$$\sum_i N_i^{\text{eq}} = \sum_i N_i \equiv \rho, \tag{13.57}$$

$$\sum_i \mathbf{c}_i N_i^{\text{eq}} = \sum_i \mathbf{c}_i N_i \equiv \rho. \tag{13.58}$$

Hence, the "collision" term on the right-hand side of Eq. (13.55) does not alter the mass or momentum contained in the N_i's. The derivation of the fluid dynamic equations of motion can thus be carried out just as for the LGA. In the present case, however, the incompressible Navier–Stokes equation without the g-factor will result.

In Eq. (13.55), the collision operator is replaced by the single relaxation parameter ω.

13.4.2 Chapman–Enskog Expansion

In the following, we will derive the momentum conservation equation in the same way as the mass conservation Eq. (13.41) was derived, that is, by a Chapman–Enskog expansion in the Knudsen number ϵ.

In order to do this we shall make use of the assumption that the fluid motion decays without soundwaves, so that $\partial_t \sim \epsilon^2$ in addition to the $\partial_\alpha \sim \epsilon$ scaling that follows from the definition of the Knudsen number. Also, we shall require a small Mach-number (the ratio of u to the speed of sound), so that $u \sim \epsilon$ as well. In the expansion of the momentum conservation equation,

$$\sum_i \mathbf{c}_i [N_i(\mathbf{x} + \mathbf{c}_i, t + 1) - N_i(\mathbf{x}, t)] = 0, \tag{13.59}$$

we shall keep terms up to order ϵ^3. Note that from Eq. (13.55) it follows immediateley that

$$N_i^{neq}(\mathbf{x}, t) \approx -\frac{\mathbf{c}_i \cdot \nabla N^{eq}}{\omega} \sim \epsilon \tag{13.60}$$

to leading order. Then, to order ϵ^3 we may write Eq. (13.59) as

$$\sum_i \mathbf{c}_i \left[(\partial_t + \mathbf{c}_i \cdot \nabla)(N_i^{eq} + N_i^{neq}) + \frac{1}{2}(\mathbf{c}_i \cdot \nabla)^2 N_i^{eq} \right] = 0, \tag{13.61}$$

which by Eq. (13.60) takes the form

$$0 = \sum_i \mathbf{c}_i \left[(\partial_t + \mathbf{c}_i \cdot \nabla)N_i^{eq} + \left(\frac{1}{2} - \frac{1}{\omega} \right)(\mathbf{c}_i \cdot \nabla)^2 N_i^{eq} \right]. \tag{13.62}$$

Now, in N_i^{eq}, terms are either even or odd in the $\mathbf{c}_i$'s, so we need only keep the parts that produce even order terms in Eq. (13.62). Doing this gives the result

$$0 = \sum_i \mathbf{c}_i \left[\mathbf{c}_i \cdot \nabla \left(\frac{\rho}{6} + \frac{2\rho}{3} Q_{i\alpha\beta} u_\alpha u_\beta \right) \right.$$
$$\left. + \left(\partial_t + \left(\frac{1}{2} - \frac{1}{\omega} \right)(\mathbf{c}_i \cdot \nabla)^2 \right) \frac{\rho \mathbf{u} \cdot \mathbf{c}_i}{3} \right]. \tag{13.63}$$

By our assumptions that $\partial_t \sim \epsilon^2$ and $u \sim \epsilon$, all terms in this equation, apart from the first one, are $\mathcal{O}(\epsilon^3)$ or smaller. This means that the $\nabla\rho$-term too must be $\mathcal{O}(\epsilon^3)$, and for this reason, the density deviations from the average $\delta\rho \sim \epsilon^2$. This in fact means that all other derivatives of ρ may be neglected in the following.

The different terms may now be evaluated as

$$\sum_i \mathbf{c}_i \mathbf{c}_i \cdot \nabla \frac{\rho}{6} = \frac{1}{2}\nabla\rho, \tag{13.64}$$

$$\sum_i \mathbf{c}_i c_{i\beta} \partial_\beta \left(\frac{2\rho}{3} Q_{i\gamma\delta} u_\gamma u_\delta \right) = \frac{1}{2}\nabla(\rho u^2) + \rho \mathbf{u} \cdot \nabla \mathbf{u}, \tag{13.65}$$

$$\sum_i \mathbf{c}_i \left[\partial_t + \left(\frac{1}{2} - \frac{1}{\omega} \right) (\mathbf{c}_i \cdot \nabla)^2 \right] \left(\frac{\rho \mathbf{u} \cdot \mathbf{c}_i}{3} \right) = \rho \partial_t \mathbf{u} - \rho \nu \nabla^2 \mathbf{u}, \tag{13.66}$$

where we have used Eq. (13.52) and introduced the kinematic viscosity

$$\nu = \frac{1}{4} \left(\frac{1}{\omega} - \frac{1}{2} \right). \tag{13.67}$$

By introducing also the pressure, through the equation of state

$$P = \frac{\rho}{2}(1 - u^2), \tag{13.68}$$

summing the above terms gives the Navier–Stokes equations

$$\frac{\partial \mathbf{u}}{\partial t} + \mathbf{u} \cdot \nabla \mathbf{u} = -\frac{1}{\rho}\nabla P + \nu \nabla^2 \mathbf{u}. \tag{13.69}$$

Also, to order ϵ^3, Eq. (13.41) gives

$$\nabla \cdot \mathbf{u} = 0 \tag{13.70}$$

as the mass conservation equation. The fact that the pressure depends on the velocity comes from the fact that a velocity $\mathbf{u}$ is set up not by small additions to the thermal velocities, but by moving mass into the direction of $\mathbf{u}$ from the transverse directions. However, it does not affect the solution for $\mathbf{u}$ since mathematically, the only role of the pressure is to ensure the $\nabla \cdot \mathbf{u} = 0$ condition.

13.4.3 The D2Q9 and D3Q19 Models

As a final step away from the lattice gases and towards an efficient computational tool for two-dimensional and three-dimensional hydrodynamics, we return to square and cubic lattices, thus leaving behind the somewhat impractical triangular and

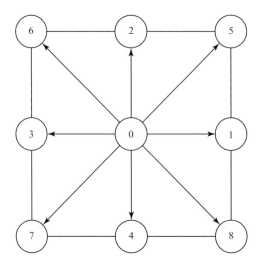

Figure 13.12 The D2Q19 lattice.

FCHC lattices. As a consequence of this, we also have to redefine the equilibrium distributions in order to keep isotropy as well as mass- and momentum conservation. So, the evolution equation is still Eq. (13.55), while the lattice and N_i^{eq} are different.

The D2Q9 Model

The *D2Q9 model* works on the square lattice but has connections to both nearest and next-nearest neighbors, the velocities being

$$\mathbf{c}_i = \begin{cases} 0, & \text{when } i = 0, \\ (\pm 1, 0), (0, \pm 1), & \text{when } i = 1, 2, 3, 4, \\ (\pm 1, \pm 1), & \text{when } i = 5, 6, 7, 8, \end{cases} \tag{13.71}$$

with the corresponding equilibrium distributions

$$N_i^{\mathrm{eq}}(\rho, \mathbf{u}) = W_i \rho \left(1 + 3\mathbf{c}_i \cdot \mathbf{u} + 3Q'_{i\alpha\beta} u_\alpha u_\beta\right), \tag{13.72}$$

where the lattice tensor

$$Q'_{i\alpha\beta} = \frac{3}{2} c_{i\alpha} c_{i\beta} - \frac{1}{2} \delta_{\alpha\beta} \tag{13.73}$$

takes a slightly different form than before due to the new choice of lattice. The weights,

$$W_i = \begin{cases} \frac{4}{9}, & \text{when } i = 0, \\ \frac{1}{9}, & \text{when } i = 1, 2, 3, 4, \\ \frac{1}{36}, & \text{when } i = 5, 6, 7, 8, \end{cases} \tag{13.74}$$

are chosen to fulfill the following requirements: First

$$\sum_{i=0}^{8} W_i Q'_{i\alpha\beta} = 0 \qquad (13.75)$$

and

$$\sum_{i=0}^{8} W_i = 1, \qquad (13.76)$$

so that

$$\sum_{i=0}^{8} N_i^{\text{eq}} = \rho; \qquad (13.77)$$

and, second, in order to make the hydrodynamic equations isotropic, we must have

$$\sum_{i=0}^{8} W_i c_{i\alpha} c_{i\beta} c_{i\gamma} c_{i\delta} = \frac{1}{9}(\delta_{\alpha\beta}\delta_{\gamma\delta} + \delta_{\alpha\gamma}\delta_{\beta\delta} + \delta_{\alpha\delta}\delta_{\beta\gamma}), \qquad (13.78)$$

where the pre-factor of $1/9$ is dictated by Eq. (13.76).

The D3Q19 Model

The *D3Q19 model* works on the cubic lattice, and like the two-dimensional model it has connections to both nearest and next-nearest neighbors. The velocities are

$$\mathbf{c}_i = \begin{cases} 0, & \text{when } i = 0, \\ (\pm 1, 0, 0), \ (0, \pm 1, 0), \ (0, 0, \pm 1), & \text{when } i = 1, \ldots, 6, \\ (\pm 1, \pm 1, 0), \ (\pm 1, 0, \pm 1), \ (0, \pm 1, \pm 1), & \text{when } i = 7, \ldots, 18, \end{cases} \qquad (13.79)$$

with the corresponding equilibrium distributions given by Eq. (13.72) again, but different weights,

$$W_i = \begin{cases} \frac{1}{3}, & \text{when } i = 0, \\ \frac{1}{18}, & \text{when } i = 1, \ldots, 6, \\ \frac{1}{36}, & \text{when } i = 7, \ldots, 18, \end{cases} \qquad (13.80)$$

are chosen to fulfill the same requirements as in Eqs (13.75), (13.76) and (13.77). When the Chapman–Enskog analysis is applied, the Navier–Stokes equations emerge with the kinematic viscosity

$$\boxed{\nu = \frac{1}{3}\left(\frac{1}{\omega} - \frac{1}{2}\right)} \qquad (13.81)$$

and an equation of state

$$P = \frac{\rho}{3}$$

(13.82)

for both the D2Q9 and D3Q19 models. Clearly, for the viscosity to be positive and finite, $0 < \omega < 2$.

Finally, we perform a simple stability analysis on the basis of Eq. (13.55). In the case in which there are no spatial variations in any of the variables, only N^{neq} can evolve with time. Since $N^{\text{eq}}(\rho, \mathbf{u})$ will be constant over the entire lattice, Eq. (13.55) takes the form

$$N_i^{\text{neq}}(\mathbf{x}, t + 1) = (1 - \omega)N_i^{\text{neq}}(\mathbf{x}, t).$$

(13.83)

It follows by iteration that

$$N_i^{\text{neq}}(\mathbf{x}, t) = (1 - \omega)^t N_i^{\text{neq}}(\mathbf{x}, 0).$$

(13.84)

In order for a small random perturbation of N_i^{neq} caused, say, by a round-off error, to be attenuated with time, we must have

$$|1 - \omega| < 1 \quad \text{or} \quad 0 < \omega < 2.$$

(13.85)

If this condition is not fulfilled, any perturbation will grow exponentially. Note that the stability condition is equivalent to the condition that the viscosity be positive and finite. The LGA is unconditionally stable and contains no round-off errors.

Lattice Boltzmann Algorithm

While the conceptual developments and analysis leading up to the lattice Boltzmann models may be comprehensive, the actual algorithm and its implementation are quite simple. In an actual simulation, the following steps must be carried out:

1. Compute $N_i^{\text{eq}}(\rho, \mathbf{u})$ at every site using Eqs (13.20) and (13.21).
2. Do the collisions step $N_i(\mathbf{x}, t) \rightarrow N_i(\mathbf{x}, t) - \omega \left(N_i(\mathbf{x}, t) - N_i^{\text{eq}}(\mathbf{x}, t) \right)$.
3. Do the propagation step $N_i(\mathbf{x}, t) \rightarrow N_i(\mathbf{x} + \mathbf{c}_i, t + 1)$.
4. Go to step 1.

13.4.4 Lattice BGK Model for Miscible Fluids, Thermal Gradients and Buoyancy

In the following we introduce a field C that is transported passively on a background velocity field $\mathbf{u}(\mathbf{x})$ by the combined action of advection and diffusion. As such it will be described by the advection–diffusion eq. (8.48). The algorithm was introduced on a triangular lattice by Flekkøy (1993) and is here modified to the D2Q9 and

D3Q19 lattices. Also, we will introduce a modification to the collision term for the N_i's that allows us to introduce a force per unit volume $\mathbf{F}$ into the flow equations.

This forcing is introduced quite simply as the last term in the following equation:

$$N_i(\mathbf{x} + \mathbf{c}_i, t + 1) = N_i(\mathbf{x}, t) - \omega\left(N_i(\mathbf{x}, t) - N_i^{\text{eq}}(\mathbf{x}, t)\right) + 3W_i\mathbf{c}_i \cdot \mathbf{F}. \qquad (13.86)$$

It should be noted that this term introduces the momentum

$$3\sum_i W_i\mathbf{c}_i\mathbf{c}_i \cdot \mathbf{F} = \mathbf{F} \qquad (13.87)$$

every timestep.

Now the diffusing field $C(\mathbf{x}, t)$ may represent both the concentration of a passive tracer, like salt, moving with the flow, or a temperature field doing the same thing. The algorithm must be conservative in the sense that it conserves the mass

$$C = \sum_i C_i, \qquad (13.88)$$

but there is no need to conserve a current $\sum_i \mathbf{c}_i C_i$ analogous to the momentum.

The rest of the algorithm is the standard hydrodynamic one given by Eq. (13.55), which produces the velocity field $\mathbf{u}(\mathbf{x})$ everywhere. This velocity field is then taken as input for the follwing updating scheme for C. First, we define the equilibrium distribution

$$C_i^{\text{eq}} = W_i C(1 + 3\mathbf{c}_i \cdot \mathbf{u}), \qquad (13.89)$$

which is used in the following Boltzmann equation:

$$\boxed{C_i(\mathbf{x} + \mathbf{c}_i, t + 1) = C_i(\mathbf{x}, t) - \omega_D\left(C_i(\mathbf{x}, t) - C_i^{\text{eq}}(\mathbf{x}, t)\right).} \qquad (13.90)$$

The fact that the collision part of this equation conserves $C(\mathbf{x}, t)$ (the propagation step is automatically conservative) may be used to show that the following macroscopic equations hold:

$$\nabla \cdot \mathbf{u} = 0, \qquad (13.91)$$

$$\partial_t \mathbf{u} + \mathbf{u} \cdot \nabla\mathbf{u} = -\frac{1}{\rho}\nabla P + \nu\nabla^2\mathbf{u} + \mathbf{F}, \qquad (13.92)$$

$$\partial_t C + \mathbf{u} \cdot \nabla C = D\nabla^2 C, \qquad (13.93)$$

with

$$\boxed{D = \frac{1}{3}\left(\frac{1}{\omega_D} - \frac{1}{2}\right)} \qquad (13.94)$$

and ν given by Eq. (13.81), as before.

The algorithm is summarized as follows:

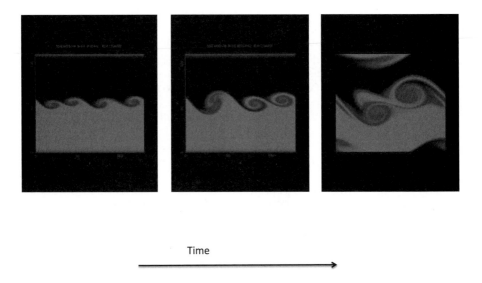

Time

Figure 13.13 Evolution of the Kelvin–Helholtz instability where the initial conditions are set up with velocity shear across the interface. The simulation is based on
Eqs (13.86) and (13.90). (A black and white version of this figure will appear in
some formats. For the colour version, refer to the plate section.)

1. Compute $N_i^{\text{eq}}(\rho, \mathbf{u})$ at every site using Eqs (13.20) and (13.21) and $C^{\text{eq}}(C, \mathbf{u})$
   using Eq. (13.89).
2. Do the collisions step $N_i(\mathbf{x}, t) \to N_i(\mathbf{x}, t) - \omega\left(N_i(\mathbf{x}, t) - N_i^{\text{eq}}(\mathbf{x}, t)\right) + 3W_i \mathbf{c}_i \cdot \mathbf{F}$.
3. Do the collisions step $C_i(\mathbf{x}, t) \to C_i(\mathbf{x}, t) - \omega\left(C_i(\mathbf{x}, t) - C_i^{\text{eq}}(\mathbf{x}, t)\right)$.
4. Do the propagation step $N_i(\mathbf{x}, t) \to N_i(\mathbf{x} + \mathbf{c}_i, t + 1)$.
5. Do the propagation step $C_i(\mathbf{x}, t) \to C_i(\mathbf{x} + \mathbf{c}_i, t + 1)$.
6. Go to step 1.

As an illustration, this model is applied to two particular instabilities: The first
one is the Kelvin–Helmholtz instability, which is responsible for wind-generated
waves, and sets in where there is a tangential velocity shear between two layers of
fluids. The fluids may be different phases or not. In the simulations of Figure 13.13,
the same fluid is used throughout the system, but the passive tracer is introduced
only in the bottom half, so as to illustrate the mixing created by the flow.

The other instability, the so-called Rayleigh–Benard instability, which is responsible for the formation of cumulus clouds, is caused by buoyancy as a fluid is heated
from below. This is easily introduced as top and bottom boundary conditions on the
temperature C along with a forcing term

$$\mathbf{F} = -\alpha \rho C \mathbf{g}. \tag{13.95}$$

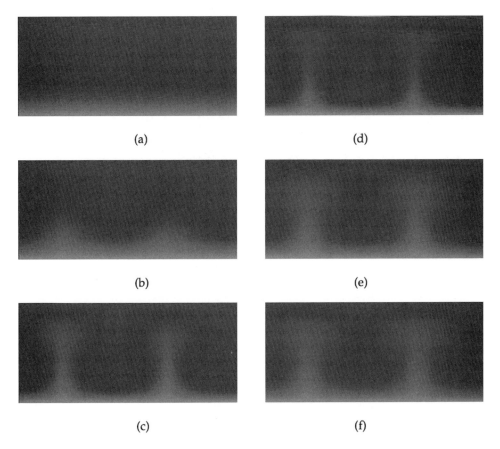

Figure 13.14 Evolution of the Rayleigh–Benard instability where the boundary condition is a larger temperature C at the bottom than at the top, and buoyancy is created by Eq. (13.95). The color represents temperature, red being hot and blue cold. Simulations are on a 128×64 lattice, $\alpha g = 0.0005$, $\nu = 0.25$ and $D = 0.25$. The timesteps are (a) $t = 6000$, (b) $t = 10\,000$, (c) $t = 12\,000$, (d) $t = 13\,000$, (e) $t = 70\,000$ and (f) $t = 154\,000$. (A black and white version of this figure will appear in some formats. For the colour version, refer to the plate section.)

Such a forcing will create an upwards force against gravity, $\mathbf{g}$, which is larger where C is larger. The constant α contains the thermal expansion coefficient and determines the magnitude of the buoyancy effect. Figure 13.14 shows a simulation where this model is used.

13.4.5 Lattice BGK Models with Surface Tension

The way to introduce surface tension follows the route first introduced by Gunstensen et al. (1991) for the model on a triangular lattice. We will, however, adapt it to the D2Q9 and D3Q19 models.

First, we need two phases, which will be represented by the masses of two colored populations (red and blue), R_i and B_i, which together make up the mass in each direction

$$N_i = R_i + B_i. \tag{13.96}$$

These colors are propogated just like the N_i's ($N_i(\mathbf{x}) \rightarrow N_i(\mathbf{x} + \mathbf{c}_i)$), and they are used to define the color gradient

$$\mathbf{f} = \sum_i \mathbf{c}_i \sum_j (R_j(\mathbf{x} + \mathbf{c}_i) - B_j(\mathbf{x} + \mathbf{c}_i)). \tag{13.97}$$

The collision rule for the mass populations N_i brings N_i to N_i' and has the form

$$\boxed{N_i' = N_i(\mathbf{x}, t) - \omega(N_i - N_i^{\text{eq}}) + A|\mathbf{f}| W_i Q_{i\alpha\beta}' \hat{f}_\alpha \hat{f}_\beta,} \tag{13.98}$$

where $\hat{\mathbf{f}} = \mathbf{f}/|\mathbf{f}|$. In this rule, ω sets the viscosity according to Eq. (13.81), and A is a free parameter that is used to set the surface tension.

The surface tension part works by moving mass into the directions $\pm\mathbf{f}$ and removing it from the directions perpendicular to $\mathbf{f}$. This is inspired by real molecular interactions that cause the stress in the direction along the interface (perpendicular to $\mathbf{f}$) to be smaller than the stress in the direction perpendicular to it. It is easy to check that it conserves mass and momentum since

$$\sum_i W_i Q_{i\alpha\beta}' = 0 \quad \text{and} \quad \sum_i W_i Q_{i\alpha\beta}' c_i = 0. \tag{13.99}$$

Then there is a recoloring step that conserves the amount of both colors, leaves the N_i's unchanged and sends blue toward blue neighbors and red toward red neighbors. This step is anti-diffusive in nature and is responsible for the phase separation. Once $\mathbf{f}$ is known, this procedure is quite simple and consists of the following step, first introduced by Latva-Kokko and Rothman (2005): First the total amount of color to be distributed is calculated, $R(\mathbf{x}) = \sum_i R_i(\mathbf{x})$ and $B(\mathbf{x}) = \sum_i B_i(\mathbf{x})$. Then, every population is modified by the replacemenent

$$R_i \rightarrow \frac{R}{R+B} N_i' + \beta \frac{RB}{(R+B)^2} N_i^{\text{eq}}(\rho, \mathbf{0}) \cos \varphi_i, \tag{13.100}$$

$$B_i \rightarrow \frac{B}{R+B} N_i' - \beta \frac{RB}{(R+B)^2} N_i^{\text{eq}}(\rho, \mathbf{0}) \cos \varphi_i, \tag{13.101}$$

where β is a new parameter that determines the interface width, the angle φ_i between $\mathbf{f}$ and $\mathbf{c}_i$ is taken to be 0 for $i = 0$, and $N_i^{\text{eq}}(\rho, \mathbf{0})$ is given in Eq. (13.72). Note that the first term on the right-hand side only distributes the color in proportion to its amount, so that when $\beta = 0$, there is no color separation. The two β-terms add to zero, so that the mass and momentum carried by $R_i + B_i$ is unchanged. Also, since $\sum \mathbf{c}_i = 0$, it does not change the total amounts of R and B. Normally, $0 \leq \beta \leq 1$, and the larger the β-value, the sharper the interface. It turns out that when $\beta \approx 1$, one starts to get some unwanted numerical artifacts like pinning of very small bubbles and history dependence of wetting properties of walls (Latva-Kokko and Rothman, 2005). For these reasons it may be useful to choose a smaller β-value in some contexts.

For too large A-values, the simulations will be unstable. Also, while the hydrodynamics itself is isotropic, the surface tension is not quite so. This gives rise to spurious currents in the vicinity of fluid interfaces, which may resemble the Marangoni effect and are caused by slight differences in surface tension with the orientation relative to the underlying lattice.

The algorithm may be summarized as follows:

1. Compute $N_i^{\text{eq}}(\rho, \mathbf{u})$ at every site using Eqs (13.20) and (13.21).
2. Do the collisions step $N_i(\mathbf{x}, t) \rightarrow N_i(\mathbf{x}, t) - \omega \left(N_i(\mathbf{x}, t) - N_i^{\text{eq}}(\mathbf{x}, t) \right) + A|\mathbf{f}| W_i Q'_{i\alpha\beta} \hat{f}_\alpha \hat{f}_\beta$.
3. Do the antidiffusive recoloring step.
4. Do the propagation step $R_i(\mathbf{x}, t) \rightarrow R_i(\mathbf{x}+\mathbf{c}_i, t+1)$ and $B_i(\mathbf{x}, t) \rightarrow B_i(\mathbf{x}+\mathbf{c}_i, t+1)$ and compute $N_i = R_i + B_i$.
5. Go to step 1.

Figure 13.15 shows a simulation using the above algorithm.

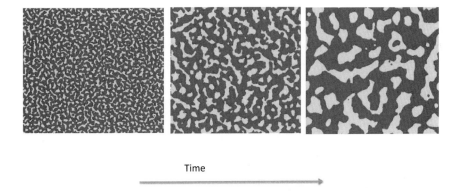

Time

Figure 13.15 Evolution of the phase separation process that results from an initial state with completely mixed phases. The simulations are done on a 128×128 lattice and are based on Eq. (13.98) and the subsequent steps of the imiscible lattice BGK model. (A black and white version of this figure will appear in some formats. For the colour version, refer to the plate section.)

Exercises

13.1 Kelvin–Helmholtz instability and viscosity measurements:

The Kelvin–Helmholtz instability is responsible for the way the wind forms water waves on an intially flat surface that and may sometimes be seen on top of clouds. Here we will start with a warm-up simulation to measure viscosity and to confirm that the code is working and end with one that can measure permeabilities of a porous medium. Set $L = 200$, the viscosity to $v = 0.01$ and let the initial velocity field be given by two counter-moving slabs of fluid:

$$u_x(y,0) = u_{0,x}\left[\tanh\left(\frac{y - 0.75L}{\delta}\right) - \tanh\left(\frac{y - 0.25L}{\delta}\right) - 1\right] \quad (13.102)$$

$$u_y(x,0) = u_{0,y}\sin(kx), \quad (13.103)$$

with $u_{0,x} = 0.1$, $u_{0,y} = 0.00001$ (small perturbation), and $\delta = 5$ (thin vorticity region). Allow the simulation to run for a time $T = 50\,000$.

13.2 Plot the vector field **u** at different times or produce a video of this field. Plot the vorticity field $\omega = \partial_x u_y - \partial_y u_x$ by using a central difference scheme to post-process your velocity fields. Explain the phenomena you observe qualitatively.

13.3 Permeability measurement:

Use the same code as above, but now with solid obstacles and a forcing of the flow as in Eq. (13.86), to measure the permeability of a porous medium. Use $L = 200$, and introduce a set of 9 randomly placed discs of radius as illustrated in Figure 13.16.

The discs are composed of solid sites where the bounce-back boundary condition is applied. You can allow overlap between the discs, but check that the obstacles do not entirely block the flow. The initial N_i distributions

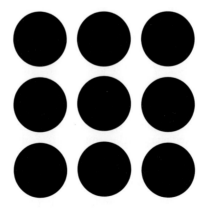

Figure 13.16 Solid obstacles.

can be set by choosing the density $\rho = 1.0$ and intial velocity $\mathbf{u} = 0$ and then $N_i = N_i^{eq}(\rho, \mathbf{u})$. Set $\nu = 0.1$ and the forcing in the x-direction to $F_x = 0.0001$, and calculate the average velocity U over the inlet boundary on the left side. Allow the flow to reach steady state where U appears to have reached its asymptotic value.

13.4 Measure the permeability $\kappa = \rho \nu U / F$ for $\nu = 0.1, 0.01$ and $F_x = 10^{-4}$, $10^{-5}, 10^{-6}$. Depending on your disc array, you may need to adjust F_x to stay inside the numerically stable regime. Comment on the result.

13.5 **Rayleigh–Benard convection:**
When a fluid is heated from below and expands, buoyancy forces (due to gravity) will eventually make the fluid move. Hot plumes start to rise, while the colder fluid above will sink. This is the mechanism that forms clouds and causes water to circulate in the sea and in lakes in spring and autumn. It is called Rayleigh–Benard convection. This phenomenon is described by the combination of the Navier–Stokes equation with a buoyancy term and the advection–diffusion equation for the temperature:

$$\rho \frac{d\mathbf{u}}{dt} = -\nabla P + \rho \nu \nabla^2 \mathbf{u} - \rho \alpha T \mathbf{g}, \qquad (13.104)$$

$$\frac{dT}{dt} = \kappa \nabla^2 T, \qquad (13.105)$$

where $\alpha = (1/V)(\partial V/\partial T)$ is the thermal expansion coefficient, and T is the difference between the local temperature and the temperature at the top of the fluid, or some other reference level where we take $T = 0$ (check that a change in the position of the reference level will only give an extra hydrostatic pressure like that of a constant gravity field). κ is the thermal diffusivity and plays the same role as the normal diffusion coefficient D does for a concentration field. Gravity $\mathbf{g}$ points downwards, so the force term acts upwards for positive T.

Use the lattice Boltzmann model for miscible fluids to solve the above equations for the case where the fluid is confined by two horizontal plates separated by a distance d, and where $\mathbf{u} = 0$, that is, no-slip boundary conditions. The system is heated from below so that $T = 1$ at the bottom plate and $T = 0$ at the top plate. In order to implement these boundary conditions, it is possible to impose $\Delta_i = \Delta_i^{eq}(T, \mathbf{u} = 0)$, where T is set to the boundary values. Sensible parameter values are $\alpha g = 0.0005$, $\nu = 0.25$, $\kappa = 0.025$ and lattice size 128×64. Initially, take $T = 0$ everywhere.

13.6 Explain why a sensible choice of the simulation time is $t = 2t_D$, where $t_D = d^2/(2\kappa)$; run such simulation and study the video showing the temperature field as a function of time. Describe qualitatively what happens.

13.7 Plot the temperature as a function of height, and plot $\log Q$, where $Q = \int dA \mathbf{u}^2$, as a function of t (Here $\int dA = \sum_{\mathbf{x}}$, which is the sum over all lattice sites). Interpret this graph.

13.8 Instabilities are characterized by exponential growth, where some amplitude like Q grows exponentially with time. For a certain time interval, $Q \sim \exp(-\gamma t)$. Identify this interval, and plot γ as a function of α and d. Let α and d vary at least by a factor of 5.

13.9 Theoretically it is known that convection only sets in for a certain minimum value of the temperature difference. This difference depends on the other properties of the system, and the quantity that needs to exceed a critical value in order to have convection is the Rayleigh number $\mathrm{Ra} = \alpha g d^3 / (\nu \kappa)$. Try and obtain the critical Ra value below which there is no flow but only diffusion. (Hint: you need to use rather small systems, say 64×32, to do this in a reasonable computation time.)

13.10 **Theory for a spring lattice:**
Elasticity and hydrodynamics are continuum theories that are closely related. We shall study the isotropy of the stress tensor on a square and triangular lattice of springs, all with the same spring constant k.

Define $\mathbf{q}(\mathbf{x})$ to be the displacement of the lattice site at $\mathbf{x}$ from an initial equilibrium position. Show that the tension in the link between the two sites at $\mathbf{x}$ and $\mathbf{x} + \mathbf{c}_i$ (these lattice vectors are those of Figure 13.8) is

$$t_i = k\mathbf{c}_i \cdot (\mathbf{q}(\mathbf{x} + \mathbf{c}_i) - \mathbf{q}(\mathbf{x})). \tag{13.106}$$

13.11 Interpret the stress tensor $\boldsymbol{\sigma}$, and express the force $\boldsymbol{\sigma} \cdot \mathbf{n}$ by the t_i's. (Hint: It may be useful to introduce the rate $f(\mathbf{c}_i) = (2/\sqrt{3})\mathbf{c}_i \cdot \mathbf{n}$ at which the red line in Figure 13.17 crosses the neighboring $\mathbf{c}_i$'s.)

13.12 Write $\mathbf{q}(\mathbf{x} + \mathbf{c}_i) - \mathbf{q}(\mathbf{x})$ to linear order in the Taylor expansion, and show how $\boldsymbol{\sigma} \cdot \mathbf{n}$ is given by the fourth order lattice tensor of Eq. (13.50).

13.13 Take $T_{\alpha\beta\gamma\delta}$ to describe a square lattice, and show that it has no resistance to displacements along a lattice direction. Show also that this is not true for the triangular lattice.

13.14 Introduce a unit vector $\mathbf{a}$ at an angle Θ to the horizontal, and prove that $T_{\alpha\beta\gamma\delta}a_\alpha a_\beta a_\delta a_\gamma$ is Θ-independent using standard geometric formulae.

13.15 **Simulation of the Saffman–Taylor instability:**
We will study the Saffman–Taylor instability using a two-dimensional lattice Boltzmann simulation. You may use the one described in Section 13.4.5 with the D2Q9 lattice, or a triangular lattice.

Argue that under the proper conditions, a three-dimensional Hele-Shaw cell of thickness b may be approximately described by the two-dimensional Stokes equation

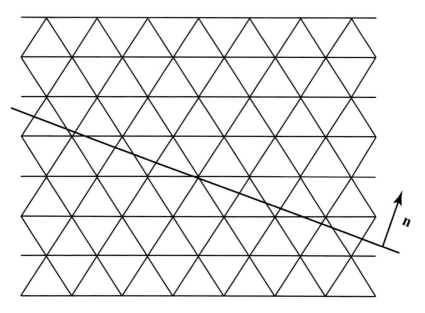

Figure 13.17 Triangular lattice of springs intersected by a line of normal **n**.

$$0 = -\nabla P + \mu \nabla^2 \bar{\mathbf{U}} - \frac{12\mu}{b^2} \bar{\mathbf{U}} + \mathbf{f}, \qquad (13.107)$$

where $\bar{\mathbf{U}}$ is the fluid velocity averaged in the b-direction. Explain under which conditions the left hand side of this equation is indeed zero and under which conditions the $\mu \nabla^2 \bar{\mathbf{U}}$-term may also be neglected, leaving a simple Darcy law.

13.16 Code a D2Q9 lattice Boltzmann model with the forcing term

$$\mathbf{F} = \mathbf{f} - \frac{12\mu}{b^2} \bar{\mathbf{U}}, \qquad (13.108)$$

where $\mathbf{f}$ is a constant vector in the x-direction. You may use periodic boundary conditions. A useful starting choice of parameters is $\rho = 3.0$, giving the more viscous (red) fluid a viscosity $\nu_r = 0.25$ and the less viscous (blue) fluid a viscosity $\nu_b = 0.07$, using a lattice of size $N = 100$ for development purposes and a larger one eventually. You may set $A = 0.0002$, $\beta = 1.0$ and the width $b = 1$.

First measure the surface tension using the equation of state Eq. (13.82) and the two-dimensional Laplace's law $\Delta P = \sigma / R$. Place a droplet in the middle of the system and measure the density difference between the inside and outside of the droplet using some different $R = 6, \ldots, 20$. Record the uncertainty.

13.17 Plot $R(x)$ across the interface for $\beta = 0.1, 0.25, 0.5, 0.74$ and 1.0.

13.18 Initialize a system of size $N \times N$ with blue fluid in the $x \leq N/2$ domain, and red in the $x > N/2$ domain, and apply the force $f = 0.0018$ in the x-direction. Start the system with the R_i or $B_i = N_i^{eq}(\rho, \mathbf{u})$, starting the instability with a small random perturbation on $\mathbf{u} \sim 0.01$.

13.19 Check by visual inspection or otherwise if the wavelength of maximum growth agrees with the prediction $\lambda_m = \sqrt{3}\lambda_c$, where λ_c is given by Eq. (10.26).

13.20 Vary σ and $\mathbf{f}$ to vary λ_m by a factor of 10, and compare the measured λ_m to Eq. (10.26).

13.21 What happens to the characteristic scale of the displacement pattern when it moves out of the linear regime? Discuss a possible physical reason for this.

13.22 Google *viscous fingering*, and compare qualitatively a pattern of yours with one taken from the literature that emerges in this way.

Appendix

Porosity Distributions

In the discussion of porous media, one often finds that the concept of *pore size distributions* is introduced. However, except for simple models, such distributions are not well defined. Pore volumes, pore necks, connectivity and other geometric details have to be specified with great care in order to make poresize distributions meaningful.

However, porosity, specific surface and tortuosity are only average measures of pore geometry. We need more detailed information if we are to calculate transport properties such as permeability, formation factor, capillary pressure, dispersivity and so on. In order to give a more complete description, one has to find a balance between excessive detail and oversimplification.

Hilfer (1991) introduced the notion of a *porosity distribution* $\mu(\phi)$ to characterize a porous medium in a practical way. He also used this distribution to derive an expression for *Archie's law* (Archie, 1942) for the formation factor. To illustrate the ideas, consider the two-dimensional "porous medium" illustrated in Figure A.1.

The porosity of a small piece of a homogeneous porous medium will in general be different from the porosity of the sample as a whole. If we measure the porosity of a small sample of volume $V_c = L \times L \times L$, we will find that the *local porosity* depends on the size of the measurement cell and the position $\mathbf{R}$ where the cell is located in the sample. We take $\mathbf{R} = n_1 \mathbf{a} + n_2 \mathbf{b} + n_3 \mathbf{c}$, so that with $n_i = 0, \pm 1, \pm 2, \ldots$, the set of points $\{\mathbf{R}\}$ define a regular lattice with the basis vectors $\mathbf{a}$, $\mathbf{b}$ and $\mathbf{c}$. This leads to the following definition of the local porosity:

$$\phi(\mathbf{R}, L) = L^{-d} \int d^d \mathbf{r}\, \chi_c(\mathbf{r}; \mathbf{R}, L) \chi(\mathbf{r}). \tag{A.1}$$

Here χ_c is the indicator function of the measurement cell, that is, $\chi_c = 1$ inside the cell and 0 outside the cell. Any shape of measurement cell is suitable, but a cubic cell (or square cell in two dimensions) is most convenient.

The *local porosity distribution* may now be introduced as the probability density $\mu(\phi; \mathbf{R}, L)$ to find the local porosity ϕ in the range from ϕ to $\phi + d\phi$ in a cell of linear dimension L at the point $\mathbf{R}$.

For a homogeneous porous medium, the porosity distribution must be independent of $\mathbf{R}$, so that $\mu(\phi; \mathbf{R}, L) = \mu(\phi; L)$. The function $\mu(\phi; L)$ will be called the *local porosity distribution* at scale L. This function gives a characterization of a porous medium at any

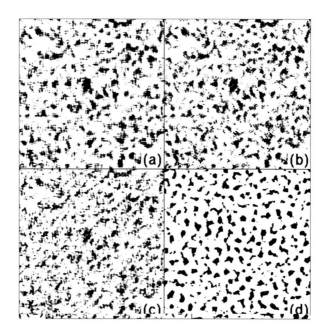

Figure A.1 Simulations of two-dimensional sections of porous media, Black represents the pore space and white the matrix. The bulk porosity is $\langle \phi \rangle = 0.2$ for each of the four images. The models were generated by starting with a set of random numbers on a 512×512 lattice. This random set was *filtered* using different filters that all fit inside a square of dimension ξ_F. Values inside the filter were averaged and assigned to the site at the center of the filter. The filtered set of numbers was discriminated at a level that gave a porosity of 0.2: (a) square filter; (b) circular filter; (c) L-shaped filter; (d) Fourier space filtering. (Boger et al. (1992).)

desired resolution L. In the limit of very large L, we simply recover the porosity of the complete sample:

$$\langle \phi \rangle = \phi(\mathbf{R}, L \rightarrow \infty) = \int_0^1 d\phi \, \phi \mu(\phi; L), \qquad (A.2)$$

independent of $\mathbf{R}$ and L.

The local porosity distributions $\mu(\phi; L)$ for the image in Figure A.1(a) are plotted as a function of ϕ for various values of L in Figure A.2. These results show clearly that the local $\mu(\phi; L)$ depends strongly on L. There are two competing effects. At small L, the local geometries are simple, and the measurement cell will in practice be either in the matrix, giving $\phi = 0$, or in the pore space with $\phi = 1$. Thus in the limit $L \rightarrow 1$, one simply recovers the characteristic function χ of the porous medium. For large L, the geometry inside the measurement cell is as complicated as the porous medium itself, and one finds $\phi = \langle \phi \rangle$ with negligible fluctuations. The two limiting behaviors are separated by some characteristic length scale ξ, which we call the *correlation length of the porous medium*. Thus we have the following limits:

$$\mu(\phi, L \gg \xi) = \delta(\phi - \langle \phi \rangle), \qquad (A.3)$$

$$\mu(\phi, L \ll \xi) = \langle \phi \rangle \delta(\phi - 1) + (1 - \langle \phi \rangle)\delta(\phi). \qquad (A.4)$$

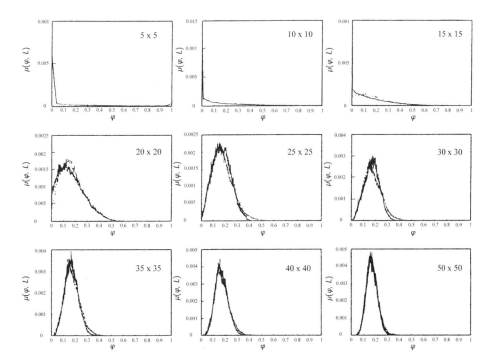

Figure A.2 The local porosity distributions $\mu(\phi, L)$ as a function of ϕ for increasing values of L for the image in Figure A.1(a). (Boger et al. (1992).)

There are many ways to choose the characteristic length scale ξ for a porous medium. One important length scale is related by the correlations between local porosities. Let ϕ_1 be the local porosity at position $\mathbf{R}_1$ and ϕ_2 the local porosity at $\mathbf{R}_2$. The correlations in local porosities are characterized by the two-cell distribution function $\mu_2(\phi_1, \mathbf{R}_1; \phi_2, \mathbf{R}_2; L)$, that is, the probability density for finding porosity ϕ_1 in the cell centered at $\mathbf{R}_1$ and simultaneously the porosity ϕ_2 in the cell centered at $\mathbf{R}_2$. Again the assumption of homogeneity gives the simplification that we may express μ_2 in terms of the difference $\mathbf{R} = \mathbf{R}_2 - \mathbf{R}_1$, so that $\mu_2(\phi_1, \mathbf{R}_1; \phi_2, \mathbf{R}_2; L) = \mu_2(\phi_1, \phi_2; \mathbf{R}, L)$. The porosity *autocorrelation function* at scale L is defined as

$$
\begin{aligned}
C(\mathbf{R}, L) &= \frac{\langle (\phi(\mathbf{R}_0) - \langle \phi \rangle)(\phi(\mathbf{R}_0 + \mathbf{R}) - \langle \phi \rangle) \rangle}{\langle (\phi(\mathbf{R}_0) - \langle \phi \rangle)^2 \rangle} \\
&= \frac{\int_0^1 d\phi_1 \int_0^1 d\phi_2 \, (\phi_1 - \langle \phi \rangle)(\phi_2 - \langle \phi \rangle) \mu_2(\phi_1, \phi_2; \mathbf{R}, L)}{\int_0^1 d\phi \, (\phi - \langle \phi \rangle)^2 \mu(\phi; R, L)}.
\end{aligned}
\tag{A.5}
$$

For isotropic homogeneous porous media, $C(\mathbf{R}, L)$ depends only on the distance R between the measurement cells, so that we may write $C(R, L)$ for the correlation function in this case. The correlation function is normalized so that $C(R \to 0, L) = 1$ independent of L. Also, as the separation of the measurement cells increases, their porosity becomes *uncorrelated*, and we have $C(R \to \infty, L) = 0$. The correlation function that is most convenient to evaluate is the pixel–pixel correlation function obtained by setting $L = 1$, that is, the box size L equals the pixel size.

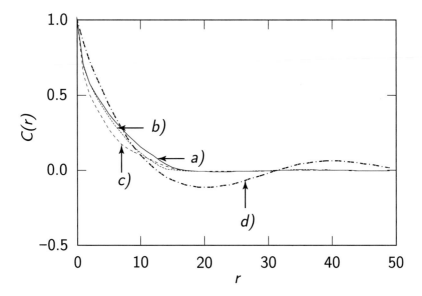

Figure A.3 The pixel–pixel correlation function $C(r, L = 0)$ as a function of r for the images in Figure A.1. (Boger et al. (1992).)

In Figure A.3, we show the correlation function $C(R, L = 0)$ calculated for the images shown in Figure A.1. Using the correlation function shown in Figure A.3, we use Eq. (A.6) to find the correlation length $\xi = 20$.

Hilfer (1991) chose to identify the correlation length ξ separating correlated from uncorrelated measurement cells by the expression

$$\xi^2 = \frac{\int d^3 R\, R^2 C(R, 1)}{\int d^3 R\, C(R, 1)}. \tag{A.6}$$

This definition is simple to evaluate in practice. A more robust definition of the characteristic length is related to the *information content* or *information entropy* S of the distribution (Boger et al., 1992)

$$S(L) = -\int_0^1 d\phi\, \mu(\phi; L)\, \ln[\mu(\phi; L)]. \tag{A.7}$$

Choosing the length scale to be ξ_S defined by $(dS/dL)|_{L=\xi_S} = 0$ that minimizes the information or maximizes the entropy consistent with the distribution μ being normalized leads to a unique length scale ξ_S with a distribution $\mu(\phi, \xi_S)$ that is as "wide" as possible and *unbiased* in the sense that it gives the least precise description of the actual pixel values in the given image being analyzed. This is illustrated in Figure A.4.

With the length scale ξ determined, we may define *the* local porosity distribution as the distribution

$$\mu(\phi) = \mu(\phi, L = \xi). \tag{A.8}$$

Simultaneously with this convention it will be assumed that the local geometries at scale $L < \xi$ are "simple." This is called the hypothesis of *local simplicity*.

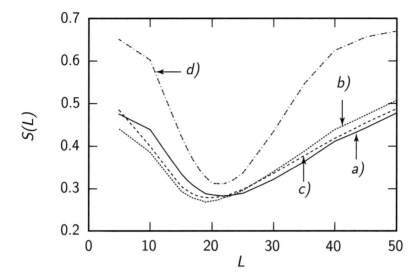

Figure A.4 The information entropy function $S(L)$ as a function of L for the images in Figure A.1. The image entropy function distinguishes very well the images in Figure A.1, which were constructed to differ only in small rather subtle details. This sensitivity to geometrical detail is a very useful aspect of $S(L)$ and permits the determination of a well defined length scale ξ_S at which $S(L)$ is a minimum. (Boger et al. (1992).)

The definition of the local porosity distribution as $\mu(\phi,\xi)$ is optimal in the sense that it contains a maximum amount of information based purely on the concept of porosity. The choices $L \gg \xi$ and $L \ll \xi$ yield the δ-function distributions given in Eq. (A.3), which contain little information. We expect that if we consider the porous medium to be composed of blocks of linear size $L = \xi$ then we may to a good first approximation consider these blocks to be statistically independent (uncorrelated), so that we may concentrate on the physical properties within each cell and then combine them using some effective medium or molecular field theory to calculate the global properties of the medium.

References

Adamson, A. W. 1982. *Physical Chemistry of Surfaces*. 4th ed. New York: John Wiley & Sons.

Ahmed, N., and Sunada, D. K. 1969. Nonlinear flow in porous media. *J. Hydraulics Div.*, **95**(6), 1847–1858.

Aker, E., Måløy, K. J., Hansen, A., and Batrouni, G. G. 1998. A two-dimensional network simulator for two-phase flow in porous media. *Transp. Porous Med.*, **32**, 163–186.

Ambegaokar, V., Halperin, B. I., and Langer, J. S. 1971. Hopping conductivity in disordered systems. *Phys. Rev. B*, **4**(8), 2612–2620.

Anderson, P. W. 1972. More is different. *Science*, **177**(4047), 393–396.

Archie, G. E. 1942. The electrical resistivity log as an aid in determining some reservoir characteristics. *Transactions of the AIME*, **146**(1), 54–62.

Armstrong, R. T., McClure, J. E., Robins, V., et al. 2018. Porous media characterization using Minkowski functionals: Theories, applications and future directions. *Transp. Porous Med.*, **130**(1), 305–335.

Aronovici, V. S., and Donnan, W. W. 1946. Soil-permeability as a criterion for drainage-design. *EOS, Trans. Amer. Geophys. Un.*, **27**(1), 95–101.

Avraam, D. G., and Payatakes, A. C. 1995. Flow regimes and relative permeabilities during steady-state two-phase flow in porous media. *J. Fluid. Mech.*, **293**, 207–236.

Bak, P., Tang, C., and Wiesenfeld, K. 1987. Self-organized criticality: An explanation of $1/f$ noise. *Phys. Rev. Lett.*, **59**(4), 381–384.

Bakke, J. Ø. H., and Hansen, A. 2007. Accuracy of roughness exponent measurement methods. *Phys. Rev. E*, **76**(3), 031136.

Barabási, A. L., and Stanley, H. E. 1995. *Fractal Concepts in Surface Growth*. Cambridge: Cambridge University Press.

Barabási, A. L., and Vicsek, T. 1991. Multifractality of self-affine fractals. *Phys. Rev. A*, **44**(4), 2730–2733.

Batchelor, G. K. 1967. *An Introduction to Fluid Dynamics*. Cambridge: Cambridge University Press.

Batrouni, G. G., Hansen, A., and Larson, B. 1996. Current distribution in the three-dimensional random resistor network at the percolation threshold. *Phys. Rev. E*, **53**, 2292–2297.

Bauer, D., Talon, L., Peysson, Y., et al. 2019. Experimental and numerical determination of Darcy's law for yield stress fluids in porous media. *Phys. Rev. Fluids*, **4**(6), 063301.

Bear, J. 1972. *Dynamics of Fluids in Porous Media*. Amsterdam: Elsevier Publishing Co. Republished by Dover Publications Inc., New York, 1988. ISBN 9780486656755.

Bear, J, and Bachmat, Y. 1990. *Introduction to Modeling of Transport Phenomena in Porous Media*. Dordrecht: Springer.

Bensimon, D., Kadanoff, L. P., Liang, S., Shraiman, B. I., and Tang, C. 1986. Viscous flows in two dimensions. *Rev. Mod. Phys.*, **58**(4), 977–999.

Berendsen, H. J. C. 2007. *Simulating the Physical World: Hierarchical Modeling from Quantum Mechanics to Fluid Dynamics*. Cambridge: Cambridge University Press.

Berlekamp, E. R., Conway, J. H., and Guy, R. K. 1982. *Winning Ways for Your Mathematical Plays*. New York: Academic Press.

Bernal, J. D. 1964. The Bakerian Lecture, 1962 The structure of liquids. *Proc. R. Soc. Lond. A*, **280**(1382), 299–333.

Bernal, J. D. 1965. The geometry of the structure of liquids. Pages 25–50 of: Hughel, T. J. (ed), *Liquids: Structure Properties, Solid Interactions*. Amsterdam: Elsevier.

Bernal, J. D., and Mason, J. 1960. Packing of spheres: Co-ordination of randomly packed spheres. *Nature*, **188**(4754), 910–911.

Berryman, J. G. 1985. Measurement of spatial correlation functions using image precessing techniques. *J. Appl. Phys.*, **57**(7), 2374–2384.

Berryman, J. G., and Blair, S. C. 1986. Use of digital image analysis to estimate fluid permeability of porous materials: Application of two-point correlation functions. *J. Appl. Phys.*, **60**(6), 1930–1938.

Bird, R. B., Armstrong, R. C., and Hassager, O. 1987. *Dynamics of Polymeric Liquids*. New York: John Wiley & Sons.

Birovljev, A., Furuberg, L., Feder, J., et al. 1991. Gravity invasion percolation in two dimensions: Experiments and simulations. *Phys. Rev. Lett.*, **67**(5), 584–587.

Blake, F. C. 1922. The resistance of packing to fluid flow. *Trans. Amer. Inst. Chem. Engrs.*, **14**, 415–421.

Blunt, M. J. 2017. *Multiphase Flow in Permeable Media*. Cambridge: Cambridge University Press.

Boger, F., Feder, J., Jøssang, T., and Hilfer, R. 1992. Microstructural sensitivity of local porosity distributions. *Physica A*, **187**(1–2), 55–70.

Bretherton, F. P. 1961. The motion of long bubbles in tubes. *J. Fluid Mech.*, **10**(2), 166–188.

Brinkman, H.C̃. 1949. A Calculation of the viscous force exerted by a flowing fluid on a dense swarm of particles. *Applied Scientific Research*, **1**, 27–34.

Broadbent, S. R., and Hammersley, J. M. 1957. Percolation processes I. Crystals and mazes. *Math. Proc. Camb. Philos. Soc.*, **53**(3), 629–641.

Brooks, R. H., and Corey, A. T. 1966. Properties of porous media affecting fluid flow. *J. Irrig. Grain. Div.*, **92**(2), 61–90.

Brown, R. L., and Bolt, R. H. 1942. The measurement of flow resistance of porous acoustic materials. *J. Acous. Soc. Amer.*, **13**(4), 337–344.

Buckley, S. E., and Leverett, M. C. 1942. Mechanism of fluid displacement in sands. *Trans. Am. Inst. Min. Eng.*, **146**(1), 107–116.

Buldyrev, S. V., Barabási, A. L., Caserta, F., et al. 1992. Anomalous interface roughening in porous media: Experiment and model. *Phys. Rev. A*, **45**, R8313–R8316.

Capuani, F., Frenkel, D., and Lowe, C. P. 2003. Velocity fluctuations and dispersion in a simple porous medium. *Phys. Rev. E*, **67**(5), 056306.

Carman, P. C. 1937. Fluid flow through granular beds. *Trans. Inst. Chem. Eng. Lond.*, **15**, 150–166.

Carman, P. C. 1938. Determination of the specific surface of powders I. Transactions. *J. Soc. Chem. Ind.*, **57**, 225–234.

Chen, L., De Luca, A., Rosso, A., and Talon, L. 2019. Darcy's law for yield stress fluids. *Phys. Rev. Lett.*, **122**, 245502.

Chen, S., and Doolen, G. D. 1998. Lattice Boltzmann method for fluid flows. *Ann. Rev. Fluid Mech.*, **30**(1), 329–364.

Chuoke, R. L., van Meurs, P., and van der Poel, C. 1959. The instability of slow, immiscible, viscous liquid-liquid displacements in permeable media. *Trans. Metall. Soc. AIME*, **216**(1), 188–194.

Darcy, H. 1856. *Les Fontaines Publiques de la Ville de Dijon*. Paris: Dalmont.

de Arcangelis, L., Redner, S., and Coniglio, A. 1985. Anomalous voltage distribution of random resistor networks and a new model for the backbone at the percolation threshold. *Phys. Rev. B*, **31**, R4725–R4727.

de Arcangelis, L., Redner, S., and Coniglio, A. 1997. Multiscaling approach in random resistor and random superconducting networks. *Phys. Rev. B*, **34**, 4656–4673.

de Gennes, P. G. 1983. Theory of slow biphasic flows in porous media. *Physico-Chem. Hydrodyn.*, **4**, 175–185.

de Gennes, P. G. 1985. Wetting: Statics and dynamics. *Rev. Mod. Phys.*, **57**(3), 827–863.

de Ligny, C. L. 1970. Coupling between diffusion and convection in radial dispersion of matter by fluid flow through packed beds. *Chem. Eng. Sci.*, **25**(7), 1177–1181.

Debye, P., and Bueche, A. M. 1949. Scattering by an inhomogeneous solid. *J. Appl. Phys.*, **20**(6), 518–525.

Debye, P., Anderson, H. R., and Brumberger, H. 1957. Scattering by an inhomogeneous solid. II. The correlation function and its application. *J. Appl. Phys.*, **28**(6), 679–683.

Devillard, P. 1993. Directed polymer modeling in hydrodynamic dispersion and fracture. *Phys. Rev. Lett.*, **70**(8), 1124–1127.

d'Humières, D., Lallemand, P., and Frisch, U. 1986. Lattice gas models for 3D hydrodynamics. *EPL*, **2**(4), 291–297.

Dowson, D. 1979. *History of Tribology*. London and New York: Longman.

Dullien, F. A. L. 1992. *Porous Media: Fluid Transport and Pore Structure*. 2nd ed. San Diego, CA 92101: Academic Press.

Dupuit, A. J. E. J. 1854. *Traité Théoretique et Practique de la conduite et de la Distribution des Eaux*. Paris: Dalmont.

Einstein, A. 1905. Über die von der molekularkinetischen Theorie der Wärme geforderte Bewegung von in ruhenden Flüssigkeiten suspendierten Teilchen. *Ann. Phys.*, **322**(8), 549–560.

Einstein, A. 1906. Eine neue Bestimmung der Moleküldimensionen. *Ann. Phys.*, **324**(2), 289–306.

Einstein, A. 1911. Berichtigung zu meiner Arbeit: "Eine neue Bestimmung der Moleküldimensionen." *Ann. Phys.*, **339**(3), 591–592.

Engelberts, W. F., and Klinkenberg, L. J. 1951. Laboratory experiments on the displacement of oil by water from packs of granular material. Pages 544–554 of: *3rd World Petroleum Congress, 28 May-6 June, The Hague, the Netherlands*. World Petroleum Congress.

Fairbrother, F., and Stubbs, A. E. 1935. 119. Studies in electro-endosmosis. Part VI. The "bubble-tube" method of measurement. *J. Chem. Soc.*, 527–529.

Feder, Jens. 1988. *Fractals*. New York: Springer.

Fleischer, R. L. 1995. Technological applications of ion tracks in insulators. *MRS Bull.*, **20**(12), 35–41.

Fleischer, R. L., and Price, P. B. 1963. Tracks of charged particles in high polymers. *Science*, **140**(3572), 1221–1222.

Fleischer, R. L., Price, P. B., and Symes, E. M. 1964. Novel filter for biological materials. *Science*, **143**(3603), 249–250.

Flekkøy, E. G. 1993. Lattice Bhatnagar-Gross-Krook models for miscible fluids. *Phys. Rev. E*, **47**(6), 4247–4257.

Flekkøy, E. G., and Rothman, D. H. 1995. Fluctuating fluid interfaces. *Phys. Rev. Lett.*, **75**(2), 260–263.

Flekkøy, E. G., Oxaal, U., Feder, J., and Jøssang, T. 1995. Hydrodynamic dispersion at stagnation points: Simulations and experiments. *Phys. Rev. E*, **52**(5), 4952–4962.

Flekkøy, E. G., Rage, T., Oxaal, U., and Feder, J. 1996. Hydrodynamic irreversibility in creeping flow. *Phys. Rev. Lett.*, **77**(20), 4170–4173.

Forchheimer, P. 1901. Wasserbewegung durch Boden. *Z. Ver. Deutsch. Ing.*, **45**, 1781–1787.

Frette, V., Christensen, K., Malthe-Sørenssen, A., et al. 1996. Avalanche dynamics in a pile of rice. *Nature*, **379**(6560), 49–52.

Frisch, U., Hasslacher, B., and Pomeau, Y. 1986. Lattice-gas automata for the Navier–Stokes equation. *Phys. Rev. Lett.*, **56**(14), 1505–1508.

Frisch, U., d'Humières, D., Hasslacher, B., et al. 1987. Lattice gas hydrodynamics in two and three dimensions. *Complex Syst.*, **1**(4), 649–707.

Furuberg, L., Hansen, A., Hinrichsen, E., Feder, J., and Jøssang, T. 1991. Scaling of overhang distribution of invasion percolation fronts. *Phys. Script.*, **T38**, 91–94.

Furuberg, L., Måløy, K. J., and Feder, J. 1996. Imtermittent behavior in slow drainage. *Phys. Rev. E*, **53**(1), 966–977.

Grassberger, P. 1999. Conductivity exponent and backbone dimension in 2-d percolation. *Physica A*, **262**, 251–263.

Gumbel, E. J. 2004. *Statistics of Extremes*. New York: Dover.

Gunstensen, A. K., and Rothman, D. H. 1993. Lattice-Boltzmann studies of immiscible two-phase flow through porous media. *J. Geophys. Res.*, **98**(B4), 6431–6441.

Gunstensen, A. K., Rothman, D. H., Zaleski, S., and Zanetti, G. 1991. Lattice-Boltzmann model of immiscible fluids. *Phys. Rev. A.*, **43**(8), 4320–4327.

Hagen, G. H. L. 1839. Ueber die Bewegung des Wassers in engen cylindrischen Röhren. *Ann. Phys.*, **122**(3), 423–442.

Hales, T. C. 1994. The status of the Kepler conjecture. *Math. Intell.*, **16**(3), 47–58.

Hales, T. C. 1998. *Sphere Packings I–V*. His computer based proof is now found at github. Version 12/26/15. His site has many interesting historical notes and pointers to other sites. `https://github.com/flyspeck/kepler98`.

Hansen, A. Hinrichsen, E.L̃. and Roux, S. 1991. Scale-invariant disorder in fracture and related breakdown phenomena. *Phys. Rev. B*, **43**, 665–678.

Hansen, A., Engøy, T., and Måløy, K. J. 1994. Measuring hurst exponents with the first return method. *Fractals*, **2**(4), 527–533.

Hansen, A., Schmittbuhl, J., and Batrouni, G. G. 2001. Distinguishing fractional and white noise in one and two dimensions. *Phys. Rev. E*, **63**(6), 062102.

Hansen, A., Batrouni, G. G., Ramstad, T., and Schmittbuhl, J. 2007. Self-affinity in the gradient percolation problem. *Phys. Rev. E*, **75**(3), 030102.

Hansen, A., Sinha, S., Bedeaux, D., et al. 2018. Relations between seepage velocities in immiscible, incompressible two-phase flow in porous media. *Transp. Porous Med.*, **125**(3), 565–587.

Hardy, J., de Pazzis, O., and Pomeau, Y. 1976. Molecular dynamics of a classical lattice gas: Transport properties and time correlation functions. *Phys. Rev. A*, **13**(5), 1949–1961.

Hele-Shaw, H. S. 1898. The flow of water. *Nature*, **58**(1489), 34–36.

Hiby, J. W. 1962. Longitudinal and transverse mixing during single-phase flow through granular beds. Pages 312–325 of: Rottenburg, P. A. & Sheperd, N. T. (eds), *Symposium on the Interaction between Fluids and Particles*. London: Inst. Chem. Engineers.

Higuera, F. J., and Jiménez, J. 1989. Boltzmann approach to lattice gas simulations. *EPL*, **9**(7), 663–668.

Higuera, F. J., and Succi, S. 1989. Simulating the flow around a circular cylinder with a lattice boltzmann equation. *EPL*, **8**(6), 517–521.

Higuera, F. J., Succi, S., and Benzi, R. 1989. Lattice gas dynamics with enhanced collisions. *EPL*, **9**(4), 345–349.

Hilfer, R. 1991. Geometric and dielectric characterization of porous media. *Phys. Rev. B*, **44**(1), 60–75.

Hinch, E. J. 1977. An averaged-equation approach to particle interactions in a fluid suspension. *J. Fluid Mech.*, **83**(4), 695–720.

Honarpour, M., Koederitz, L., and Harvey, A. H. 1986. *Relative Permeability of Petroleum Reservoirs*. Boca Raton, FL: CRC Press.

Hooke, R. 1661. Pamphlet. Hooke's first publication was a pamphlet on capillary action.

Hsiang, W.-Y. 1993. On the sphere packing problem and the proof of Kepler's conjecture. *Int. J. Math.*, **4**(5), 739–831.

Hsiang, W.-Y. 1995. A rejoinder to Hales's article. *Math. Intell.*, **17**(1), 35–42.

Israelachivili, J. N. 1992. *Intermolecular and Surface Forces*. 2nd ed. London: Academic Press.

Jaeger, H. M., and Nagel, S. R. 1992. Physics of the granular state. *Science*, **255**(5051), 1523–1531.

Jullien, R., Pavlovitch, A., and Meakin, P. 1992a. Random packings of spheres built with sequential models. *J. Phys. A*, **25**(15), 4103–4113.

Jullien, R., Meakin, P., and Pavlovitch, A. 1992b. Three-dimensional model for particle-size segregation by shaking. *Phys. Rev. Lett.*, **69**(4), 640–643.

Kalaydjian, F. 1990. Origin and quantification of coupling between relative permeabilities for two-phase flows in porous media. *Transp. Porous Med.*, **5**(3), 215–229.

Katz, A. J., and Thompson, A. H. 1986. Quantitative prediction of permeability in porous rocks. *Phys. Rev. B*, **34**(11), 8179–8181.

Kepler, J. 1941. *Gesammelte Werke*. Vol. 4. München: C. H. Beck.

Kepler, J. 1966. *The Six-Cornered Snowflake*. London: Oxford University Press. Edited by L. L. Whyte. The book contains Kepler's original (1611) text in latin with English translation by Colin Hardie. The book also contains comments by B. J. Mason: '*On the shape of snow crystals*', and by L. L. Whyte: '*Kepler's unsolved problem and the* FACULTAS FORMATRIX'.

Khanamiri, H. H., Berg, C. F., Slotte, P. A., Sclüter, S., and Torsæter, O. 2018. description of free energy for immiscible two-fluid flow in porous media by integral geometry and thermodynamics. *Water Resour. Res.*, **54**(11), 9045–9059.

Kinzel, W. 1983. Directed percolation. In: Deutscher, G., Zallen, R., and Adler, J. (eds), *Percolation Structures and Processes*. A. Hilger.

Kirkpatrick, S. 1971. Classical transport in disordered media: Scaling and effective-medium theories. *Phys. Rev. Lett.*, **27**, 1722–1725.

Kirkpatrick, S. 1973. Percolation and conduction. *Rev. Mod. Phys*, **45**(4), 574–588.

Kjelstrup, S., Bedeaux, D., Hansen, A., B., Hafskjold, and Galteland, O. 2019. Non-isothermal transport of multi-phase fluids in porous media. Constitutive equations. *Front. Phys. (Lausanne)*, **6**, 150.

Klinkenberg, L. J. 1941. The permeability of porous media to liquids and gases. Pages 200–213 of: *Drilling and Production Practice*. New York: American Petroleum Institute.

Koch, D. L., and Brady, J. F. 1985. Dispersion in fixed beds. *J. Fluid Mech.*, **154**, 399–427.

Kolmogorov, A. N. 1941a. Dissipation of energy in locally isotropic turbulence. *Dokl. Akad. Nauk SSSR*, **32**, 16–18. Republished: 1991. *Proc. R. Soc. Lond. A*, **434**(1890), 15–17.

Kolmogorov, A. N. 1941b. The local structure of turbulence in incompressible viscous fluid with very large Reynolds' numbers. *Dokl. Akad. Nauk SSSR*, **30**, 301–305. Republished: 1991. *Proc. R. Soc. Lond. A*, **434**(1890), 9–13.

Kolmogorov, A. N. 1962. A refinement of previous hypotheses concerning the local structure of turbulence in a viscous incompressible fluid at high Reynolds number. *J. Fluid Mech.*, **13**(1), 82–85.

Koocheki, A., Ghandi, A., Razavi, S. M. A., et al. 2009. The rheological properties of ketchup as a function of differenthydrocolloids and temperature. *Int. J. Food Sci. Tech.*, **44**, 569–602.

Kozeny, J. 1927. Über kapillare Leitung des Wassers im Boden (Aufstieg, Versickerung und Anwendung auf die Bewässerung). *Sitzungsber. Akad. Wiss. in Wien, Mathematisch-Naturwissenschaftliche Klasse, Abt. IIa*, **136**(2a), 271–306.

Landau, L. D., and Lifshitz, E. M. 1987a. *Fluid Mechanics*. Oxford, New York: Pergamon Press.

Landau, L. D., and Lifshitz, E. M. 1987b. *Statistical Mechanics*. Oxford, New York: Pergamon Press.

Latva-Kokko, M., and Rothman, D. H. 2005. Diffusion properties of gradient-based lattice Boltzmann models of immiscible fluids. *Phys. Rev. E*, **71**(5), 056702.

Lavallee, P., Boon, J. P., and Noullez, A. 1991. Boundaries in lattice gas flows. *Physica D*, **47**(1), 233–240.

Lenormand, R., and Zarcone, C. 1985. Invasion percolation in an etched network: measurement of a fractal dimension. *Phys. Rev. Lett.*, **54**(20), 2226–2229.

Leverett, M. C. 1941. Capillary behavior in porous solids. *Trans. AIME*, **142**(1), 152–169.

Liu, C.-h., Nagel, S. R., Schecter, D. A., et al. 1995. Force fluctuations in bead packs. *Science*, **269**(5223), 513–515.

Locke, L. C., and Bliss, J. E. 1950. Core analysis technique for limestone and dolomite. *World Oil*, **131**(4), 206–207.

Løvoll, G., Méhuest, Y., Toussaint, R., Schmittbuhl, J., and Måløy, K. J. 2004. Growth activity during fingering in a porous Hele-Shaw cell. *Phys. Rev. E*, **70**(2), 026301.

Maher, J. V. 1985. Development of viscous fingering patterns. *Phys. Rev. Lett.*, **54**(14), 1498–1501.

Måløy, K. J., Feder, J., and Jøssang, T. 1985. Viscous fingering fractals in porous media. *Phys. Rev. Lett.*, **55**(24), 2688–2691.

Mandelbrot, B. B. 1982. *The Fractal Geometry of Nature*. USA: W. H. Freeman and Co.

Maxwell, J. C. 1876. Capillary action. In: Baynes, T. S. (ed), *Encyclopædia Britannica*, 9th ed. Encyclopædia Britannica, Inc. Reprinted in *The Scientific Papers of James Clerk Maxwell*, **2**, 541–591 (Cambridge University Press, 2011).

Mazur, P., and Weisenborn, A. J. 1984. The Ossen drag on a sphere and the method of induced forces. *Physica A*, **123**(1), 209–226.

McLaughlin, J. F., and Goetz, W, H. 1955. Permeability, void content, and durability of bitumious concrete. *Highway Res. Board Proc.*, **34**, 274–286.

McLean, J. W., and Saffman, P. G. 1981. The effect of surface tension on the shape of fingers in a Hele-Shaw cell. *J. Fluid Mech.*, **102**, 455–469.

McNamara, G. R., and Zanetti, G. 1988. Use of the Boltzmann equation to simulate lattice-gas automata. *Phys. Rev. Lett.*, **61**(20), 2332–2335.

Mecke, K., and Arns, C. H. 2005. Fluids in porous media: A morphometric approach. *J. Phys. Condens. Matter*, **17**(9), S503–S534.

Mecke, K. R. 2000. Additivity, Convexity, and beyond: Applications of Minkowski functionals in statistical physics. Pages 111–184 of: Mecke, K. R. (ed), *Statistical Physics and Spatial Statistics*. Heidelberg: Springer.

Méheust, Y., Løvoll, G., Måløy, K. J., and Schmittbuhl, J. 2002. Interface scaling in a two-dimensional porous medium under combined viscous, gravity, and capillary effects. *Phys. Rev. A*, **66**(5), 051603.

Mehrabi, A. R., Rassamdana, H., and Sahimi, M. 1997. Characterization of long-range correlations in complex distributions and profiles. *Phys. Rev. E*, **56**, 712–722.

Misner, C. W., Thorne, K. S., and Wheeler, J. A. 2017. *Gravitation*. Princeton: Princeton University Press.

Mitchell, S. J. 2005. Discontinuities in self-affine functions lead to multiaffinity. *Phys. Rev. E*, **72**(6), 065103(R).

Mitton, R. G. 1945. The air pemeabilities of light leathers and their specific surfaces. *J. Int. Soc. Leather Trades' Chem.*, **29**, 255–263.

Moffatt, H. K. 1964. Viscous and resistive eddies near a sharp corner. *J. Fluid Mech.*, **18**, 1–18.

Mott, R. A. 1951. The laws of motion of particles in fluids and their application to the resistance of beds of solids to the passage of fluid. Pages 242–256 of: Lang, H. R. (ed), *Some Aspects of Fluid Flow*. London: Edward Arnold & Company, for The Institute of Physics.

Moura, M., Fiorentino, E. A., Maåløy, K. J., Schäfer, G., and Toussaint, R. 2015. Impact of sample geometry on the measurement of pressure-saturation curves: Experiments and simulations. *Water Resour. Res.*, **51**(11), 8900–8926.

Moura, M., Måløy, K. J., Flekkøy, E. G., and Toussaint, R. 2017. Verification of a dynamic scaling for the pair correlation function during the slow drainage of a porous medium. *Phys. Rev. Lett.*, **119**(15), 154503.

Muskat, M. 1937. *The flow of Homogeneous Fluids through Porous Media*. 1st ed. Ann Arbor: Edwards. 2nd printing, 1946.

Navier, C. L. M. H. 1822. Mémoire sur les Lois du Mouvements des Fluides. *Mémoires de l'Academie Royale des Sciences de l'Institut de France*, **VI**. Chez Firmin Didot, père et fils, libraires, Paris 1827.

Newton, I. 1867. *Philosophiae Naturalis Principia Mathematica*. Berkeley and Los Angeles: Imprimatur S. Pepys. Re. Soc. Praeses. See A. Motte's and 1729. Revised and supplied with historical and explanatory appendix by F. Cajori, Edited by R. T. Crawford (1934) and published by University of California Press, Berkeley and Los Angeles (1966).

Nosé, S. 1984. A unified formulation of the constant temperature molecular-dynamics methods. *J. Chem. Phys.*, **81**, 511–519.

Nutting, P. G. 1930. Physical analysis of oil sands. *Bull. Amer. Ass. Petr. Geol.*, **14**, 1337–1349.

Oak, M. J., Baker, L. E., and Thomas, D. C. 1990. Three-phase relative permeability of Berea sandstone. *J. Petr. Tech.*, **42**(8), 1054–1061.

Olafuyi, O. A., Cinar, Y., Knackstedt, M. A., and Pinczewski, W. V. 2008. Capillary pressure and relative permeability of small cores. In: *SPE Symposium on Improved Oil Recovery, 20–23 April 2008, Tulsa, Oklahoma, USA*. Society of Petroleum Engineers.

Onoda, G. Y., and Liniger, E. G. 1990. Random loose packings of uniform spheres and the dilatancy onset. *Phys. Rev. Lett.*, **64**(22), 2727–2730.

Oseen, C. W. 1910. Über die Stokes'sche Formel und über eine verwandte Aufgabe in der Hydrodynamik. *Arkiv för matematik, ast. och fysik*, **6**, 1–20.

Oseen, C. W. 1913. Über den Gültigkeitsbereich der Stokesschen Wiederstandsformel. *Arkiv för matematik, ast. och fysik*, **9**.

Oseen, C. W. 1945. Neuere Methoden und Ergebnisse der Hydordynamik. *Akademische Verlagsgesellschaft m.b.h.; Leipzig*, **1**, 325.

Oxaal, U., Flekkøy, E. G., and Feder, J. 1994. Irreversible dispersion at a stagnation point: Experiments and lattice Boltzmann simulations. *Phys. Rev. Lett.*, **72**(22), 3514–3517.

Pallmann, H., and Deuel, H. 1945. Über die Wasserdurchlässigkeit von Hydrogelen. *Experientia*, **1**(9), 325–326.

Pitts, E. 1980. Penetration of a fluid into a Hele-Shaw cell: The Saffman-Taylor experiment. *J. Fluid Mech.*, **97**, 53–64.

Poiseuille, J. L. M. 1840. Recherches expérimentales sur le mouvement des liquides dans les tubes de très petits diamètres; I: Influence de la pression sur la quantitté de liquide qui travese les tubes de très petits diamètres. *C. R. Acas. Sci.*, **11**, 1041–1048.

Porto, M., Bunde, A., Havlin, S., and Roman, H. E. 1997. Structural and dynamical properties of the percolation backbone in two and three dimensions. *Phys. Rev. E*, **56**, 1667–1675.

Press, W. H., Teukolsky, S. A., Vetterling, W. T., and Flannery, B. P. 2007. *Numerical Recipes: The Art of Scientific Computing*. 3rd ed. Cambridge: Cambridge University Press.

Proudman, I., and Pearson, J. R. A. 1957. Expansions at small Reynolds numbers for the flow past a sphere and a circular cylinder. *J. Fluid Mech.*, **2**(3), 237–262.

Qian, Y. H., d'Humières, D., and Lallemand, P. 1992. Lattice BGK models for Navier-Stokes equation. *EPL*, **17**(6), 479–484.

Reynolds, O. 1886. IV. On the theory of lubrication and its application to Mr. Beauchamp tower's experiments, including an experimental determination of the viscosity of olive oil. *Philosophical Transactions of the Royal Society of London*, **177**, 157–234.

Riguidel, F.-X., Hansen, A., and Bideau, D. 1994. Gravity driven motion of a particle on an inclined plane with controlled roughness. *EPL*, **28**(1), 13–18.

Rivet, J.-P., and Boon, J. P. 2001. *Lattice Gas Hydrodynamics*. Cambridge: Cambridge University Press.

Rothman, D. H., and Keller, J. M. 1988. Immiscible cellular-automaton fluids. *J. Stat. Phys.*, **52**(3–4), 1119–1127.

Rothman, D. H., and Zaleski, S. 1994. Lattice-gas models of phase separation: Interfaces, phase transitions, and multiphase flow. *Rev. Mod. Phys.*, **66**(4), 1417–1479.

Roux, S., and Hansen, A. 1987. Application of 'logical transport' to determine the directed and isotropic percolation thresholds. *J. Phys. A: Math. Gen.*, **20**, L873–L878.

Rowlinson, J. S., and Widom, B. 1982. *Molecular Theory of Capillarity*. Oxford: Oxford University Press.

Roy, S., Hansen, A., and Sinha, S. 2019a. Effective rheology of two-phase flow in capillary fiber bundles. *Front. Phys. (Lausanne)*, **7**, 92.

Roy, S., Pedersen, H., Sinha, S., and Hansen, A. 2021. The co-moving velocity in immiscible two-phase flow in porous media. *arXiv:2108.10187*.

Roy, S., Sinha, S., and A., Hansen. 2019b. Immiscible two-phase flow in porous media: Effective rheology in the continuum limit. *arXiv:1912.05248*.

Rumpf, H. C. H., and Gupte, A. R. 1971. Einflüsse der Porosität und Korngrössenverteilung im Wiederstandsgesetz der Porenströmung. *Chem. Ing. Tech.*, **43**(6), 367–375.

Saffman, P. G., and Taylor, G. I. 1958. The penetration of a fluid into a medium or Hele-Shaw cell containing a more viscous liquid. *Proc. R. Soc. Lond. A*, **245**(1242), 312–329.

Scheidegger, A. E. 1953. Theoretical models of porous matter. *Producers Monthly*, **17**, 17.

Scheidegger, A. E. 1958. The physics of flow through porous media. *Soil Sci.*, **86**, 355.

Scheidegger, A. E. 1974. *The Physics of Flow through Porous Media*. Toronto: University of Toronto Press.

Schick, M. 1990. Introduction to wetting phenomena. Pages 417–497 of: Charvolin, J., Joanny, J. F., and Zinn-Justin, J. (eds), *Liquids at Interfaces*. Amsterdam: Elsevier Science Publishers B.V. Les Houches, Session XLVIII, 1988.

Schmittbuhl, J., Vilotte, J.-P., and Roux, S. 1995. Reliability of self-affine measurements. *Phys. Rev. E*, **51**(1), 131–147.

Schriever, W. 1930. Law of flow for the passage of a gas-free liquid through a spherical-grain sand. *Trans. Amer. Inst. Min. Metall. Engrs., Pet. Div.*, **86**(1), 329–336.

Schweber, S. S. 1993. Physics, community and the crisis in physical theory. *Phys. Today*, **46**(11), 34–40.

Scott, G. D. 1960. Packing of spheres: Packing of equal spheres. *Nature*, **188**, 908–909.

Scott, G. D., and Kilgour, D. M. 1969. The density of random close packing of spheres. *J. Phys. D: Appl. Phys.*, **2**(6), 863–866.

Simonsen, I., and Hansen, A. 2002. Fast algorithm for generating long self-affine profiles. *Phys. Rev. E*, **65**(3), 037701.

Simonsen, I., Hansen, A., and Nes, O. M. 1998. Determination of the Hurst exponent by use of wavelet transforms. *Phys. Rev. E*, **58**(3), 2779–2787.

Sinha, S., and Hansen, A. 2012. Effective rheology of immiscible two-phase flow in porous media. *EPL*, **99**(4), 44004.

Sinha, S., Hansen, A., Bedeaux, D., and Kjelstrup, S. 2013. Effective rheology of bubbles moving in a capillary tube. *Phys. Rev. E*, **87**(2), 025001.

Sinha, S., Gjennestad, M. Aa., Vassvik, M., et al. 2019. Rheology of high-capillary number two-phase flow in porous media. *Front. Phys. (Lausanne)*, **7**, 65.

Sinha, S., Gjennestad, M. A., Vassvik, M., and Hansen, A. 2021. Fluid meniscus algorithms for dynamic pore-network modeling of immiscible two-phase flow in porous media. *Front. Phys.*, **8**, 548497.

Slichter, C. S. 1899. Theoretical investigations of the motions of ground waters. Pages 295–384 of: *Nineteenth Annual Report of the U. S. Geological Survey, Pt. 2*.

Stanton, T. E., and Pannell, J. R. 1914. V. Similarity of motion in relation to the surface friction of fluids. *Philos. Trans. R. Soc. A*, **214**(513), 199–224.

Stauffer, D., and Aharony, A. 1992. *Introduction To Percolation Theory*. 2nd ed. London: Taylor & Francis.

Stewart, I. 1991. How to succeed in stacking. *New Scientist*, **131**, 29–32.

Stokes, G. G. 1845. On the theories of the internal friction of fluids in motion, and of the equilibrium and motion of elastic solids. *Trans. Cambr. Phil. Soc.*, **8**, 287–319.

Strieder, W., and Aris, R. 1973. *Variational Methods Applied to Problems of Diffusion and Reaction*. Heidelberg: Springer-Verlag.

Strogatz, S. H. 1994. *Non-Linear Dynamics and Chaos*. Cambridge: Perseus Press.

Stull, R. T., and Johnson, P. V. 1940. Some properties of the pore system in bricks and their relation to frost action. *J. Res. Natl. Bur. Stand.*, **25**(6), 711–730.

Tallakstad, K. T., Løvoll, G., Knudsen, H. A., Flekkøy, E. G., and Måløy, K. J. 2009. Steady-state, simultaneous two-phase flow in porous media: An experimental study. *Phys. Rev. E*, **80**(3), 036308.

Talon, L., and Bauer, D. 2013. On the determination of a generalized Darcy equation for yield-stress fluid in porous media using a lattice-Boltzmann TRT scheme. *Eur. Phys. J. E*, **36**(12), 139.

Talon, L., Auradou, H., Pessel, M., and Hansen, A. 2013. Geometry of optimal path hierarchies. *EPL*, **103**(3), 30003.

Taneda, S. 1956. Experimental Investigation of the Wake behind a sphere at low Reynolds numbers. *J. Phys. Soc. Jpn.*, **11**(10), 1104–1108.

Taylor, G. I. 1953. Dispersion of soluble matter in solvent flowing slowly through a tube. *Proc. R. Soc. Lond. A*, **219**(1137), 186–203.

Taylor, T. D., and Acrivos, A. 1964. On the deformation and drag of a falling viscous drop at low Reynolds number. *J. Fluid Mech.*, **18**(3), 466–476.

Toussaint, R., Løvoll, G., Méheust, Y., Måløy, K. J., and Schmittbuhl, J. 2005. Influence of pore-scale disorder on viscous fingering during drainage. *EPL*, **71**(4), 583–589.

Tritton, D. J. 1977. *Physical Fluid Dynamics*. Wokingham: Van Nostrand Reinhold Co. Ltd.

Tryggvason, G., and Aref, H. 1983. Numerical experiments on Hele Shaw flow with a sharp interface. *J. Fluid Mech.*, **136**, 1–30.

van Genabeek, O., and Rothman, D. H. 1996. Macroscopic manifestations of microscopic flows through porous media: Phenomenology from simulation. *Annu. Rev. Earth Planet. Sci.*, **24**(1), 63–87.

van Genuchten, M. T. 1980. A closed-form equation for predicting the hydraulic conductivity of unsaturated soils. *Soil Sci. Soc. Am. J.*, **44**(5), 892–898.

van Kampen, N. G. 2007. *Stochastic Processes in Physics and Chemistry*. 3rd ed. Amsterdam: North Holland.

Vergeles, M., Keblinski, P., Koplik, J., and Banavar, J. R. 1995. Stokes drag at the molecular level. *Phys. Rev. Lett.*, **75**(2), 232–235.

Washburn, E. W. 1921. The dynamics of capillary flow. *Phys. Rev.*, **17**(3), 273–283.

Weissberg, H. L. 1963. Effective diffusion coefficient in porous media. *J. Appl. Phys.*, **34**, 2636.

Whitaker, S. 1986. Flow in porous media I: A theoretical derivation of Darcy's law. *Transp. Porous Media*, **1**(1), 3–25.

Wiggins, E. J., Campbell, W. B., and Maass, O. 1939. Determination of the specific surface of fibrous materials. *Can. J. Res.*, **17b**(10), 318–324.

Wilkinson, D., and Willemsen, J. F. 1983. Invasion percolation: A new form of percolation theory. *J. Phys. A*, **16**(14), 3365–3376.

Wold, I., and Hafskjold, B. 1999. Nonequilibrium molecular dynamics simulations of coupled heat and mass transport in binary fluid mixtures in pores. *Int. J. Thermophys.*, **20**(3), 847–856.

Wu, C., Xu, X., and Qian, T. 2013. Molecular dynamics simulations for the motion of evaporative droplets driven by thermal gradients along nanochannels. *J. Phys. Condens. Matter*, **25**(19), 195103.

Wu, X-l., Måløy, K. J., Hansen, A., Ammi, M., and Bideau, D. 1993. Why hour glasses tick. *Phys. Rev. Lett.*, **71**(9), 1363–1366.

Wyckoff, R. D., and Botset, H. G. 1936. The flow of gas-liquid mixtures through unconsolidated sands. *Physics*, **7**(9), 325–345.

Wyckoff, R. D., Botset, H. G., Muskat, M., and Reed, D. W. 1933. The measurement of the permeability of porous media for homogeneous fluids. *Rev. Sci. Instrum.*, **4**(7), 394–405.

Young, T. 1805. III. An essay on the cohesion of fluids. *Philos. Trans. Royal Soc.*, **95**, 65–87.

Zallen, R. 1983. *The Physics of Amorphous Solids*. New York: John Wiley & Sons.

Zhao, B., MacMinn, C. W., Primkulov, B., K., et al. 2019. Comprehensive comparison of pore-scale models for multiphase flow in porous media. *Proc. Nat. Acad. Sci.*, **116**, 13799–13806.

Zick, A. A., and Homsy, G. M. 1982. Stokes flow through periodic arrays of spheres. *J. Fluid. Mech.*, **115**, 13–26.

Index